Springer Series in
CHEMICAL PHYSICS 67

Springer-Verlag Berlin Heidelberg GmbH

Physics and Astronomy ONLINE LIBRARY

http://www.springer.de/phys/

Springer Series in
CHEMICAL PHYSICS

Series Editors: F. P. Schäfer   J. P. Toennies   W. Zinth

The purpose of this series is to provide comprehensive up-to-date monographs in both well established disciplines and emerging research areas within the broad fields of chemical physics and physical chemistry. The books deal with both fundamental science and applications, and may have either a theoretical or an experimental emphasis. They are aimed primarily at researchers and graduate students in chemical physics and related fields.

63 **Ultrafast Phenomena XI**
    Editors: T. Elsaesser, J.G. Fujimoto, D.A. Wiersma, and W. Zinth

64 **Asymptotic Methods in Quantum Mechanics**
    Application to Atoms, Molecules and Nuclei
    By S.H. Patil and K.T. Tang

65 **Fluorescence Correlation Spectroscopy**
    Theory and Applications
    Editors: R. Rigler and E.S. Elson

66 **Ultrafast Phenomena XII**
    Editors: T. Elsaesser, S. Mukamel, M.M. Murnane, and N.F. Scherer

67 **Single Molecule Spectroscopy**
    Nobel Conference Lectures
    Editors: R. Rigler, M. Orrit, T. Basché

Series homepage – http://www.springer.de/phys/books/chemical-physics/

Volumes 1–62 are listed at the end of the book

R. Rigler   M. Orrit   T. Basché

Single Molecule Spectroscopy

Nobel Conference Lectures

With 179 Figures

 Springer

Professor Dr. Rudolf Rigler

Karolinska Institutet, Department of Medical Biophysics,
S-17177 Stockholm, Sweden
e-mail: rudolf.rigler@mbb.ki.se

Dr. Michel Orrit

CNRS et Université Bordeaux I, C.P.M.O.H.
351 Cours de la Libération, 33405 Talence, France
e-mail: orrit@yak.cpmoh.u-bordeaux.fr

Professor Dr. Thomas Basché

Johannes Gutenberg-Universität, Institut für Physikalische Chemie
Jakob-Welder-Weg 11, 55099 Mainz, Germany
e-mail: thomas.basche@uni-mainz.de

Series Editors:

Professor F.P. Schäfer

Max-Planck-Institut für Biophysikalische Chemie
D-37077 Göttingen-Nikolausberg, Germany

Professor J.P. Toennies

Max-Planck-Institut für Strömungsforschung
Bunsenstrasse 10
D-37073 Göttingen, Germany

Professor W. Zinth

Universität München,
Institut für Medizinische Optik
Öttingerstr. 67
D-80538 München, Germany

ISSN 0172-6218

ISBN 978-3-642-62702-6

Library of Congress Cataloging-in-Publication Data

Single molecule spectroscopy : nobel conference lectures / [edited by] R. Rigler, M. Orrit, T. Basché.
p. cm. - - (Springer series in chemical physics, ISSN 0172-6218 ; 67)
ISBN 978-3-642-62702-6   ISBN 978-3-642-56544-1 (eBook)
DOI 10.1007/978-3-642-56544-1
1. Molecular spectroscopy. I. Rigler, Rudolf. II. Orrit, M. (Michel), 1956- III. Besché, T. IV. Springer series in chemical physics ; v. 67.
QC454.M6 S57 2002
535.8'4- -dc21   2001049365

http://www.springer.de

© Springer-Verlag Berlin Heidelberg 2001
Originally published by Springer-Verlag Berlin Heidelberg New York 2001
Softcover reprint of the hardcover 1st edition 2001

The use of general descriptive names, registered names, trademarks, etc. in this publication does not imply, even in the absence of a specific statement, that such names are exempt from the relevant protective laws and regulations and therefore free for general use.

Typesetting: PTP, Heidelberg Berlin
Cover concept: eStudio Calamar Steinen
Cover production: *design & production* GmbH, Heidelberg

Printed on acid-free paper   SPIN: 10847535   57/3141/yu - 5 4 3 2 1 0

Preface

One often hears that nanoscience or, in other words, the knowledge and control of matter at length scales of a few nanometers, will be the scientific frontier of the 21st century. Although it has become almost commonplace, this prediction deserves some justification. The technological and scientific stakes of nanoscience indeed encompass many fields of science: they include the ultimate miniaturization of electronic devices to acquire, store, and process information, and also such basic endeavors as understanding the microscopic processes and patterns responsible for the physical properties of materials, or the many unsolved questions raised by the astoundingly intricate workings of living matter. Although the dream of observing and controlling matter at molecular scales is nearly as old as the very concept of molecules, earlier attempts at practical realizations were hampered by a scarcity of suitable access to the nanoworld.

During the last two decades of the 20th century, owing to the several new tools which have been developed to address objects at nanometer scales, the nanoworld appears closer than ever, within our reach! A major class of methods in nanoscience are local probe microscopies such as scanning tunnelling or atomic force microscopies. They require scanning a sharp tip with molecular dimensions across the surface of the sample under study and, by direct action of the tip on the sample, they make nano-manipulations possible.

The present book is devoted to another class of methods, the selection and study of single, optically active nano-objects by purely optical means. The selected objects often are organic molecules, but the same techniques apply to nano-crystals, self-assembled quantum dots, metal nano-particles, etc. The central idea of these methods is to isolate the optical signal from a small area of a sample, where at most one active molecule can interact with the exciting laser light. The optical observable is most often fluorescence or luminescence, but other responses can also be used, for instance optical absorption, or surface-enhanced or resonance Raman scattering. The optical addressing of single objects offers specific capabilities that make these methods complementary, and sometimes even superior to scanning local microscopies, although their direct spatial resolution is inherently limited by the wavelength of light.

Optical illumination of a sample is usually non-intrusive, because it requires only very weak perturbations, which is essential for delicate biological structures. In contrast to scanning tips, photons can penetrate beyond surfaces and interrogate single molecules in their natural environments. In addition, a wide variety of spectroscopic optical techniques have been developed since the advent of lasers, for frequency-resolved and time-resolved spectroscopy. By applying laser radiation to single molecules, one can in principle transfer this whole toolbox of spectroscopies down to the nanoworld.

Single molecule spectroscopy and microscopy have spread at a quick pace in the past decade, and are poised for further expansion as major avenues in nanoscience, either by themselves or in combination with scanning probe microscopy and manipulation. Their specific advantages are those of optics combined with those of microscopy in general. Addressing single objects radically removes all averaging involved in conventional measurements on populations. Not only does this provide a much clearer picture of individual behaviors, and facilitate comparison with theoretical models, it also delivers novel observables such as statistical distributions, correlations, dynamical fluctuations, etc., which by definition vanish upon averaging over macroscopic populations.

The observation of fluctuations is particularly new and important, since all earlier kinetic studies of large populations in chemistry and biology invariably required one necessary step: the synchronization of all individuals. Single molecules for the first time provide direct insight into natural fluctuations, and the potential of this insight for biomolecules is tremendous. Similarly, following single molecules as a function of time can reveal rare but significant events that would disappear upon averaging. Rutherford's discovery of the atomic nucleus is a well-known illustration of the far-reaching insight provided by individual events. The same argument could soon apply to the fleeting intermediates involved in chemical reactions or in protein folding.

Much of the recent progress in optical science and applications is owed to lasers. More than four decades after its discovery, laser radiation still finds new uses, notably in studying and manipulating small objects in condensed matter. By trapping small particles optically (with so-called optical traps and optical tweezers), we can now measure the tiny forces causing bonding or conformational changes in single molecules, even when they are weak intermolecular interactions. A. Ashkin (Chap. 1) gives a historical perspective on the development of laser trapping, which has led to new insights in biophysics, as well as to the burgeoning physics of ultracold atoms. The general trend of addressing smaller and smaller particles has ultimately led to the manipulation and observation of single molecules, for which laser radiation is also central. The first single-molecule experiments were done with absorption, as W.E. Moerner recalls (Chap. 2), and later pursued by fluorescence excitation, which is the major method employed currently. The problem of how to measure the absorption signal of a single molecule under ambient conditions remains one of the major challenges in the field. The first single-molecule fluorescence experiments were done initially in 1990, either in a flowing solution at room temperature or in a solid matrix at cryogenic temperatures. In 1993, near-field optical microscopy led to the observation of single molecules immobilized on surfaces at room temperature. Soon thereafter, around 1994, confocal microscopy was successfully applied to the same end. Since confocal microscopy is a standard method, several groups started working on single molecules and observed the attending phenomena of blinking, photobleaching, bunching, etc. A major motivation in the work on single molecules under ambient conditions is the drive towards the exploration of biological processes. This has now become the most important field of application of single molecule spectroscopy.

At liquid helium temperatures, a particular object can often be observed for a long time, and be used as a test object for high-precision physics, as illustrated by several contributions. By combining cryogenic temperatures with single-frequency laser excitation, one achieves very high spectral resolutions, which can be used to investigate molecular physics very accurately. Some examples are provided in the review by Moerner (Chap. 2), in particular the magnetic resonance of single molecules with well resolved excitation and emission lines. Lower spectral resolutions (of the order of 1 cm^{-1}) can be extremely useful to elucidate the electronic structure of complex systems, as demonstrated in the case of single bacterial antenna complexes (Chap. 3). In the field of biophysics, low temperatures can help stabilize a structure and resolve complex electronic spectra. The work presented in Chap. 3 is a spectacular demonstration of the power of single-molecule spectroscopy: although the B800 and B850 absorption bands of a large population look similar, the spectra of single complexes and bandwidths differ profoundly, highlighting the differences in intermolecular interactions in these two coupled electronic systems.

A single molecule is also a sensitive probe for all dynamical processes taking place in its neighborhood. Using single molecules, it has been possible to investigate tunneling in crystals and in disordered solids, or to study the bistable conformations of a molecule and of its first solvent shells, as nicely illustrated with terrylene in p-terphenyl crystals. Several experiments and simulations converge to give a picture of the molecular motion involved, a flip of the central phenyl ring of a neighbor p-terphenyl molecule (Chap. 4). If the molecule and its solvent shell are stable enough, they can serve as a model optical two-level system, and be used to investigate light–matter interactions. This is illustrated with the description of a triggered source of single photons based on a single molecule (Chap. 5), where an adiabatic passage is used to bring the molecule into its excited state with high probability.

In physical chemistry, single molecules or small particles can probe inhomogeneous systems. The photophysical properties of conjugated polymers reflect the sample heterogeneities which are caused by molecular weight distributions and conformational defects on the polymer backbone. By studying single fluorescent conjugated polymer molecules (Chap. 6) a more detailed picture of intramolecular electronic energy relaxation is obtained, as well as a correlation between photophysical properties and specific polymer conformation. An artificial liposome is a minuscule test tube, in which single molecules can be trapped and preserved from external reactants. Wilson et al. (Chap. 7) show how such liposomes can be made and manipulated with micropipettes. Metal nanostructures can lead to strong enhancement of local optical fields, which enables surface-enhanced Raman scattering. As Kneipp et al. show (Chap. 8), SERS can reach single molecule sensitivity. It thus provides molecular fingerprints, with the added advantage that photobleaching is eliminated. In many fluorescence experiments, it is important to distinguish between different fluorophores. This is usually done by spectral filtering, or by lifetime measurements. Hübner et al. (Chap. 9) demonstrate that by combining spectral and time-resolved information, the number of photons required to discriminate two fluorescent labels can be reduced dramatically.

The developments which lead to the observation of single molecules in solution at room temperature have opened the field of single-molecule analysis in biosciences (Chap. 10). Fluorescence correlation spectroscopy (FCS) has proven to be a powerful technique which permits the observation of the dynamics of biomolecules down to the single-molecule level. Wohland et al. (Chap. 11) employ FCS to characterize receptor proteins. Extensions of the technique to dual-color cross-correlation schemes enhance the throughput rate for screening applications in biochemistry and evolutionary optimization processes (Chap. 12). The feasibility of analyzing the behavior of a single biomolecule in relation to an ensemble has given novel information on the function of biomolecules such as enzymes. Several contributions show that single enzyme molecules undergo a series of conformational transitions before catalysis can occur. Single molecule experiments show the existence of conformational fluctuations in the enzyme-cosubstrate complex (Chap. 13) and product formation (Chap. 10), as well in turnover rates (Chap. 14). These results are consistent with and lend strong support to the model of conformational substates proposed by Frauenfelder (Chap. 15). Closely related is the study of intramolecular dynamics in small enzyme populations (Chap. 16). Spectacular results have been achieved by the analysis of molecular motors at the single molecule level (Chap. 17), as well as from the analysis of the folding of single protein molecules (Chap. 18). The rapid development of single-molecule analysis now covers a broad variety of biological molecules, including nucleic acids (Chap. 19), and proteins of differing complexity, such as enzymes, contractile elements, and light-emitting proteins like GFP (Chap. 20). For biological systems, single-molecule analysis is of prime interest, since ensemble behavior can be deduced from the single molecule scenario but not vice versa. Fundamental biological processes such as self-replication, including evolution and selection, are basically single-molecule processes. In summary, the contributions in this field can be seen to be the starting point of very dynamic development in the near future.

This volume presents the current picture of activities in single-molecule spectroscopy at the dawn of the 21st century. This is an active and interdisciplinary field, which is likely to profoundly change in coming years, as single-molecule techniques propagate into different areas of application. Therefore, a prediction of future developments is particularly risky. The optical probing of complex systems with single molecules gives a detailed picture of molecular movements and structure at nanometer scale. This new tool can be applied to many problems in physical chemistry and material science, such as wetting, adhesion, and friction. A particularly promising application will be the combination of optical single-molecule microscopy on the one hand, and raster scanning techniques or electron microscopy on the other, to arrive at a correlation between the spectroscopic and structural properties of one and the same single nanoparticle.

Interestingly, while biochemical reactions have been studied at the single-molecule level, experiments at the very heart of chemistry, i.e. the study of classical chemical reaction mechanisms or product distributions, still appear mainly as a challenging endeavor. The main area of activity will probably remain molecular biology. The many unanswered mechanistic questions about the structure, dynamics, and functions of biomolecules in their natural environments could start to be answered in the next few years. Already, single enzymes have been investigated over long time scales, the docking of a substrate onto an enzymatic site and the subsequent conformational changes of the enzyme can be followed by fluorescence resonance energy transfer (FRET), in combination with force measurements, and the working of molecular motors has been considerably clarified, if not elucidated. The current challenges of single-molecule methods under ambient conditions are the blinking and photo-bleaching of the active particles, molecules, or nanocrystals. If these problems can be solved, the future of optical single-molecule methods will be quite bright indeed!

Mainz, Talence, Stockholm   *T. Basché*
August 2001   *M. Orrit*
                                                             R. Rigler

Contents

Preface   v

**1   History of Optical Trapping and Manipulation of Small
Neutral Particles, Atoms, and Molecules**
A. Ashkin   1

**2   Thirteen Years of Single-Molecule Spectroscopy
in Physical Chemistry and Biophysics**
W. E. Moerner   32

**3   The Electronic Structure of Single Photosynthetic
Pigment-Protein Complexes**
A. M. van Oijen, M. Ketelaars, J. Köhler, T. J. Aartsma, J. Schmidt . .   62

**4   Single-Molecule Optical Switching: A Mechanistic Study
of Nonphotochemical Hole-Burning**
F. Kulzer, T. Basché   82

5   Triggered Emission of Single Photons by a Single Molecule
C. Brunel, P. Tamarat, B. Lounis, M. Orrit   99

**6   Photophysics of Conjugated Polymers Unmasked by Single
Molecule Spectroscopy**
J. Yu, D.-H. Hu, P. F. Barbara 114

**7   Confining and Probing Single Molecules
in Synthetic Liposomes**
C. F. Wilson, D. T. Chiu, R. N. Zare, A. Strömberg, A. Karlsson,
O. Orwar 130

**8   Single Molecule Detection Using Near Infrared
Surface-Enhanced Raman Scattering**
K. Kneipp, H. Kneipp, I. Itzkan, R. R. Dasari, M. S. Feld 144

9   Single-Molecule Fluorescence – Each Photon Counts
C. G. Hübner, V. Krylov, A. Renn, P. Nyffeler, U. P. Wild 161

10   Fluorescence Correlation Spectroscopy
in Single-Molecule Analysis:
Enzymatic Catalysis at the Single Molecule Level
R. Rigler, L. Edman, Z. Földes-Papp, S. Wennmalm 177

11   The Characterization of a Transmembrane Receptor
Protein by Fluorescence Correlation Spectroscopy
T. Wohland, K. Friedrich-Bénet, H. Pick, A. Preuss, R. Hovius,
H. Vogel 195

12   Applications of Dual-Color Confocal Fluorescence
Spectroscopy in Biotechnology
A. Koltermann, U. Kettling, J. Stephan, M. Rarbach, T. Winkler,
M. Eigen 211

13   Single-Molecule Enzymology
X. S. Xie, H. P. Lu 227

14   Single-Molecule Enzymology
N. J. Dovichi, R. Polakowski, A. Skelley, D. B. Craig, J. Wong 241

15   The Energy Landscape
H. Frauenfelder, B. H. McMahon 257

16   Coherent Intramolecular Dynamics
in Small Enzyme Populations
A. S. Mikhailov, P. Stange, B. Hess 277

17   Single-Molecule Dynamics in Biosystems
T. Yanagida 293

18   Single-Molecule Dynamics Associated with Protein
Folding and Deformations of Light-Harvesting Complexes
D. S. Talaga, Y. Jia, M. A. Bopp, A. Sytnik, W. A. DeGrado,
R. J. Cogdell, R. M. Hochstrasser 313

19   The Study of Single Biomolecules
with Fluorescence Methods
T. Ha, X. Zhuang, H. Babcock, H. Kim, J. W. Orr, J. R. Williamson,
L. Bartley, R. Russell, D. Herschlag, S. Chu 326

20   Studying the Green Fluorescent Protein
with Single-Molecule Spectroscopy
A. Zumbusch, G. Jung, C. Bräuchle 338

Index 353

List of Contributors

Thijs J. Aartsma
Department of Biophysics
Huygens Laboratory
Leiden University
2300 RA Leiden, The Netherlands

A. Ashkin
Bell Laboratories
Lucent Technologies
Holmdel, NJ 07733, USA

Paul F. Barbara
Department of Chemistry
University of Texas
Austin, TX 78712, USA

Hazen Babcock
Department of Chemistry and Chemical
Biology
Harvard University
Cambridge, MA 02138, USA

Laura Bartley
Department of Biochemistry
B400 Beckman Center
Stanford University
Stanford, CA 94305-5307, USA

Thomas Basché
Institut für Physikalische Chemie der
Universität Mainz
Jakob-Welder-Weg 11
55099 Mainz, Germany

Martin A. Bopp
Chemistry Department and NIH Laser
Resource
University of Pennsylvania
Philadelphia, PA 19104, USA

Christoph Bräuchle
Institut für Physikalische Chemie
and Center for Nanoscience
Ludwig-Maximilians Universität
München
Butenandtstr. 11
81377 München, Germany

C. Brunel
C.P.M.O.H.
CNRS et Université Bordeaux I
351, Cours de la Libération
33405 Talence Cedex, France

Daniel T. Chiu
Department of Chemistry
Stanford University
Stanford, CA 94305-5080, USA
and
Department of Chemistry and Chemical
Biology
Harvard University
Cambridge, MA 02138, USA
Current address:
Department of Chemistry
University of Washington
Seattle, WA 98195-1700, USA

Steven Chu
Department of Applied Physics and
Physics
Stanford University
Stanford, CA 94305-4060

Richard J. Cogdell
Division of Biochemistry and Molecular
Biology
Institute of Biomedical and Life
Sciences
University of Glasgow
Glasgow, G12 8QQ, UK

Douglas B. Craig
Department of Chemistry
University of Winnipeg
Winnipeg, Manitoba R3B 2E9, Canada

Ramachandra R. Dasari
G.R. Harrison Spectroscopy Laboratory
Massachusetts Institute of Technology
Cambridge, MA 02139, USA

William A. DeGrado
Department of Biochemistry and
Biophysics
University of Pennsylvania
Philadelphia, PA 19104, USA

Norm J. Dovichi
Department of Chemistry
University of Washington
Seattle, WA 98195-1700, USA

Manfred Eigen
Max-Planck-Institut für Biophysikalische Kinetik
Am Faßberg 11
37077 Göttingen, Germany

Michael S. Feld
G.R. Harrison Spectroscopy Laboratory
Massachusetts Institute of Technology
Cambridge, MA 02139, USA

Hans Frauenfelder
Center for Nonlinear Studies
MS B-258
Los Alamos National Laboratory,
Los Alamos, NM 87545, USA

Kirstin Friedrich-Bénet
Department of Chemistry
LCPPM
Swiss Federal Institute of Technology
CH–1015 Lausanne, Switzerland

Taekjip Ha
Department of Physics
University of Illinois
Urbana, IL 61801-3080, USA

Daniel Herschlag
Department of Biochemistry
B400 Beckman Center
Stanford University
Stanford, CA 94305-5307, USA

B. Hess
Max-Planck-Institut für molekulare
Physiologie
Otto-Hahn-Str. 11
44227 Dortmund, Germany
Current address:
Max-Planck-Institut für medizinische
Forschung
Jahnstraße 29
69120 Heidelberg, Germany

Ruud Hovius
Department of Chemistry
LCPPM
Swiss Federal Institute of Technology
CH–1015 Lausanne, Switzerland

Dehong Hu
Department of Chemistry
University of Texas
Austin, TX 78712, USA

Christian G. Hübner
Physical Chemistry Laboratory
Swiss Federal Institute of Technology
ETH Zentrum
8092 Zurich, Switzerland

Irving Itzkan
G.R. Harrison Spectroscopy Laboratory
Massachusetts Institute of Technology
Cambridge MA 02139, USA
Physics Department
Technical University Berlin
10623 Berlin, Germany

Yiwei Jia
Chemistry Department and NIH Laser
Resource
University of Pennsylvania
Philadelphia, PA, 19104, USA

Gregor Jung
Institut für Physikalische Chemie
and Center for Nanoscience
Ludwig-Maximilians Universität
München
Butenandtstr. 11
81377 München, Germany

Anders Karlsson
Department of Chemistry
Göteborg University
Göteborg, SE-41296, Sweden

Martijn Ketelaars
Department of Biophysics
Huygens Laboratory
Leiden University
2300 RA Leiden, The Netherlands

Ulrich Kettling
DIREVO Biotech AG
Nattermannallee 1
50829 Köln, Germany .

Harold Kim
Department of Applied Physics and
Physics
Stanford University
Stanford, CA 94305-4060, USA

Harald Kneipp
Department of Physics
Technical University Berlin
10623 Berlin, Germany

Katrin Kneipp
Department of Physics
Technical University Berlin
10623 Berlin, Germany
and
G.R. Harrison Spectroscopy Laboratory
Massachusetts Institute of Technology

Cambridge, MA 02139, USA
and
Department of Electrical Engineering
and Computer Science
Massachusetts Institute of Technology
Cambridge MA 02139, USA

Jürgen Köhler
Centre for the Study of Excited States
of Molecules
Huygens Laboratory
Leiden University
2300 RA Leiden, The Netherlands
present address:
Universität Bayreuth
Lehrstuhl für Experimentalphysik IV
D–95440 Bayreuth, Germany

Andre Koltermann
DIREVO Biotech AG
Nattermannallee 1
50829 Köln, Germany

Vitali Krylov
Physical Chemistry Laboratory
Swiss Federal Institute of Technology
ETH Zentrum
8092 Zurich, Switzerland

Florian Kulzer
Institut für Physikalische Chemie der
Universität Mainz
Jakob-Welder-Weg 11
55099 Mainz, Germany

B. Lounis
C.P.M.O.H.
CNRS et Université Bordeaux I
351, Cours de la Libération
33405 Talence Cedex, France

H. Peter Lu
Pacific Northwest National Laboratory
Environmental Molecular Sciences
Laboratory
P.O. Box 999
Richland, WA 99352, USA

Benjamin H. McMahon
Theoretical Biology and Biophysics
Group
MS K-710
Los Alamos National Laboratory,
Los Alamos, NM 87545, USA

A. S. Mikhailov
Abteilung Physikalische Chemie
Fritz-Haber-Institut der Max-Planck-
Gesellschaft
Faradayweg 4-6
14195 Berlin, Germany

W. E. Moerner
Stanford University
Stanford, CA 94305, USA

Peter Nyffeler
Physical Chemistry Laboratory
Swiss Federal Institute of Technology
ETH Zentrum
8092 Zurich, Switzerland

Jeffrey W. Orr
Department of Molecular Biology
and The Skaggs Institute of Chemical
Biology
Scripps Research Institute
La Jolla, CA 92037, USA

Michel Orrit
Huygens Laboratory, Universiteit
Leiden
Postbus 9504
NL-2300 RA Leiden, The Netherlands

Owe Orwar
Department of Chemistry
Göteborg University
Göteborg, SE-41296, Sweden

Horst Pick
Department of Chemistry
LCPPM
Swiss Federal Institute of Technology
CH-1015 Lausanne, Switzerland

Robert Polakowski
Department of Chemistry
University of Alberta
Edmonton, Alberta T6G 2G2, Canada

Axel Preuss
Department of Chemistry
LCPPM
Swiss Federal Institute of Technology
CH-1015 Lausanne, Switzerland

Markus Rarbach
DIREVO Biotech AG
Nattermannallee 1
50829 Köln, Germany

Alois Renn
Physical Chemistry Laboratory
Swiss Federal Institute of Technology
ETH Zentrum
8092 Zurich, Switzerland

Rick Russell
Department of Biochemistry
B400 Beckman Center
Stanford University
Stanford, CA 94305-5307, USA

Jan Schmidt
Centre for the Study of Excited States
of Molecules
Huygens Laboratory
Leiden University
2300 RA Leiden, The Netherlands

Alison Skelley
Department of Chemistry
University of California
Berkeley, CA 94270-1460, USA

P. Stange
Abteilung Physikalische Chemie
Fritz-Haber-Institut der Max-Planck-
Gesellschaft
Faradayweg 4-6
14195 Berlin, Germany

Jens Stephan
DIREVO Biotech AG
Nattermannallee 1
50829 Köln, Germany

Anette Strömberg
Department of Chemistry
Göteborg University
Göteborg, SE-41296, Sweden

Alexander Sytnik
Chemistry Department and NIH Laser
Resource
University of Pennsylvania
Philadelphia, PA 19104, USA

Antoine M. van Oijen
Centre for the Study of Excited States
of Molecules
Huygens Laboratory
Leiden University
2300 RA Leiden, The Netherlands
present address:
department of Chemistry and Chemical
Biology
Harvard University
12 Oxford Street
Cambridge, MA 02138, USA

David S. Talaga
Chemistry Department and NIH Laser
Resource
University of Pennsylvania
Philadelphia, PA, 19104, USA

P. Tamarat
C.P.M.O.H.
CNRS et Université Bordeaux I
351, Cours de la Libération
33405 Talence Cedex, France

Horst Vogel
Department of Chemistry
LCPPM
Swiss Federal Institute of Technology
CH–1015 Lausanne, Switzerland

Urs P. Wild
Physical Chemistry Laboratory
Swiss Federal Institute of Technology
ETH Zentrum
8092 Zurich, Switzerland

James R. Williamson
Department of Molecular Biology
and The Skaggs Institute of Chemical
Biology
The Scripps Research Institute
La Jolla, CA 92037, USA

Clyde F. Wilson
Department of Chemistry
Stanford University
Stanford, CA 94305-5080, USA

Thorsten Winkler
Max-Planck-Institut für Biophysikalische Kinetik
Am Faßberg 11
37077 Göttingen, Germany

Thorsten Wohland
Department of Chemistry
LCPPM
Swiss Federal Institute of Technology
CH–1015 Lausanne, Switzerland
present address:
Department of Chemistry
Stanford University
Stanford, CA 94305-5080, USA
Thorsten.Wohland@epfl.ch

Jerome Wong
Department of Oncology
University of Alberta
Edmonton, Alberta T6G 2H7, Canada

X. Sunney Xie
Professor of Chemistry
Department of Chemistry and Chemical
Biology
Harvard University
12 Oxford Street
Cambridge, MA 02138, USA

Toshio Yanagida
Single Molecule Process Project
ICORP, JST
and
Department of Systems and Human
Science
Graduate School of Engineering Science
and
Department of Physiology and
Biosignaling
Graduate School of Medicine
Osaka University, Yamadaoka 2–2
Suita, Osaka, Japan

Ji Yu
Department of Chemistry
University of Texas
Austin, TX 78712, USA

Richard N. Zare
Department of Chemistry
Stanford University
Stanford, CA 94305-5080, USA

Xiaowei Zhuang
Department of Chemistry and Chemical
Biology
Harvard University
Cambridge, MA 02138, USA

Andreas Zumbusch
Institut für Physikalische Chemie
and Center for Nanoscience
Ludwig-Maximilians Universität
München
Butenandtstr. 11
81377 München, Germany

1 History of Optical Trapping and Manipulation of Small Neutral Particles, Atoms, and Molecules

A. Ashkin

I was asked by Prof. Rigler to help introduce this conference on single molecule spectroscopy by reviewing the history of optical trapping and manipulation of small neutral particles. Laser trapping techniques play a major role in single particle studies in physics, chemistry, and biology. The unique capabilities of these techniques have had a revolutionary impact in various subfields of these same sciences where single particles play a role. In the field of light scattering, it has led to the highest-resolution studies of Mie scattering, the first high-resolution observations of the resonant behavior of macroscopic spherical particles, and the use of these resonances in many applications in linear and nonlinear optics and lasers. The highest-Q optical resonances ever observed have been found in these so-called Mie resonance or "whispering gallery modes." In atomic physics, laser trapping and cooling techniques have led to the optical trapping of individual atoms, to atom cooling down to the lowest kinetic temperatures in the universe, to Bose–Einstein Condensation, and, more recently, to atom lasers. Practical advances in atomic clocks and measurement of gravitational forces have also been made. In the biological sciences and chemistry, use of laser techniques has led to the trapping and manipulation of single living cells, organelles within cells, single biological molecules, and measurement of the mechanical forces and elastic properties of cells and molecules.

Optical trapping has a long history, going back about thirty years to the early days of lasers. The forces involved are those of radiation pressure, which come from the momentum of light itself. With ordinary light sources these forces are quite small and usually play only a minor role in affecting the dynamics of particles. I will show how use of lasers in the first experiments in 1969 changed that picture. With lasers one can easily observe large changes in the dynamics of small particles. These effects give rise to the field of laser trapping and manipulation of particles as we know it today. Optical trapping has been observed over a particle size range from a few angstroms to 100's of micrometers, a range of $\sim 10^6$. In terms of temperature or energy, cooling of atoms was achieved from temperatures of $\sim 10^3$ K down to small fractions of a microkelvin, a change of $\sim 10^9$.

As we shall see, one needs to use only the simplest of concepts, such as momentum conservation, ray optics, and semiclassical rate equations, to understand the forces and optical traps. Indeed, it was only with the help of such simple concepts, simple experiments, and a little luck that trapping of particles was discovered in the first place.

As the various fields using optical traps advanced in time and complexity, many of the common problems that arose in different contexts tended to be obscured or even overlooked. Here I will do my best to emphasize the common features of laser trapping in the various disciplines. I feel qualified to do this since I discovered optical trapping of small particles, and proposed optical trapping of atoms. I invented optical tweezers and did pioneering work in all the major areas of optical manipulation. I will trace the development of the subject chronologically, starting in 1969 with the simple considerations leading to the first trap, and ending with some of the more complex work in Bose–Einstein condensation (BEC) and applications of tweezers to the study of single motor molecules, and DNA folding and sequencing. I will try to follow Prof. Norman Ramsey's advice on talks, "Everyone likes to understand a talk. Therefore, no talk can be too simple."

1.1   Basic Forces and the First Trap

My interest in the subject was aroused in 1969 by the following "back of the envelope" calculation of the magnitude of the radiation pressure force of light on a totally reflecting mirror. As we know, the momentum of a single photon is $h\nu/c$. If we have an incident power of P we have $P/h\nu = N$ photons striking the mirror per second. Since all these photons are assumed to reflect straight back, the total change in momentum of the light per second is (2N $P/h\nu)(h\nu/c) = 2$ P/c. By conservation of momentum this implies that the mirror acquires an equal momentum/sec, or force, in the direction of the light. That is, $F_{mirror} = 2P/c$. For P = 1 watt = 10^7 ergs/sec, one gets $F_{mirror} \cong 10^{-3}$ dynes = 10 nanonewtons, which is quite a small force in absolute terms. It represents the maximum force one can extract from the light momentum at a power of one watt. Suppose we have a laser and we focus our one watt to a small spot size of about a wavelength $\cong 1$ μm, and let it hit a particle of diameter also of ~1 μm. Treating the particle as a 100% reflecting mirror of density $\cong 1$ gm/cm^3, we get an acceleration of the small particle = A = F/m = 10^{-3} dynes/10^{-12} gm = 10^9 cm/sec^2. Thus, A $\cong 10^6$ g, where g $\cong 10^3$ cm/sec, the acceleration of gravity. This is quite large and should give readily observable effects, so I tried a simple experiment [1] to look for particle motion by laser radiation pressure. I used a sample of transparent latex spheres of density ~1, in water, to avoid any problems with heating or so-called radiometric effects. With just milliwatts of power, particle motion was observed in the direction of a mildly focused Gaussian beam. The particle velocity was in approximate agreement with our crude

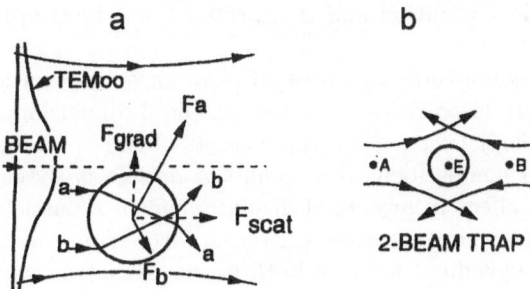

Fig. 1.1. a Origin of F_{scat} and F_{grad} for high index sphere displaced from TEM_{00} beam axis. **b** Geometry of 2-beam trap

force estimates, suggesting that this was indeed a radiation pressure effect. However, an additional unanticipated force component was soon discovered which strongly pulled particles located in the fringes of the beam into the high intensity region on the beam axis. Once on axis, particles stayed there and moved forward, even if the entire beam was slewed back and forth within the chamber. Particles were being guided by the light! They finally collected in a clump at the output face of the chamber. When the light was turned off, they wandered toward the fringes of the beam. If the light was turned on again, they were quickly pulled to the beam axis. Was this transverse force component light pressure, too?

Figure 1.1a shows that both these force components do indeed originate from radiation pressure. Imagine a high index of refraction sphere, many wavelengths in diameter, placed off axis in a mildly focused Gaussian beam. Consider a typical pair of rays "a" and "b" striking the sphere symmetrically about its center. Neglecting relatively minor surface reflections, most of the rays refract through the particle, giving rise to forces F_a and F_b in the direction of the momentum change. Since the intensity of ray "a" is higher than ray "b", the force F_a is greater than F_b. Adding all such symmetrical pairs of rays striking the sphere, one sees that the net force can be resolved into two components, F_{scat}, called the scattering force component pointing in the direction of the incident light, and F_{grad}, a gradient component arising from the gradient in light intensity and pointing transversely toward the high intensity region of the beam. For a particle on axis or in a plane wave, $F_a = F_b$ and there is no net gradient force component. A more detailed calculation of the sum of the forces of all the rays striking the sphere gave a net force in excellent agreement with the observed velocity. For a low index particle placed off-axis, the refraction through the particle reverses, F_a is less than F_b and such a particle should be pushed out of the beam. This behavior was seen using micron-sized air bubbles in glycerine. One also observes by mixing large and small diameter spheres in the same sample that the large spheres move faster and pass right by the smaller spheres as they proceed along the beam. This is a form of particle separation and is expected from the simple ray-optic calculations.

The understanding of the magnitude and properties of these two basic force components made it possible to devise the first stable 3-dimensional optical trap for single neutral particles [1]. The trap consists of two opposing moderately diverging Gaussian beams focused at points A and B as shown in Fig. 1.1b. The predominant effect in any axial displacement of a particle from the equilibrium point E is a net opposing scattering force. Any radial displacement is opposed by the gradient force of both beams. The trap was filled by capture of randomly diffusing small particles which wandered into the trap. The viscous damping of the liquid serves to dissipate all the kinetic energy gained from the trapping potential and particles come to rest at the trap center. If one blocks one beam, the particle is driven forward and guided by the second beam. If one restores the first beam the particle is pushed back to the equilibrium point E. It is surprising that this simple first experiment [1], intended only to show simple forward motion due to laser radiation pressure, ended up demonstrating not only this force but the existence of the transverse force component, particle guiding, particle separation, and stable three-dimensional particle trapping.

The success of these experiments on macroscopic particles prompted the hypothesis in reference [1] "that similar acceleration and trapping are possible with atoms and molecules using laser light tuned to specific optical transitions." It was shown that a scattering force should exist for atoms in the direction of the incident light due to the process of absorption and subsequent isotropic spontaneous emission of resonant photons. The low-intensity absorption cross-section of an atom is huge, about $\lambda^2/2$. This says that almost every photon of a strongly focused resonant beam hitting an atom is absorbed and subsequently scattered by the atom. Since the atomic mass is small, this implies a huge acceleration. For a sodium atom one gets a kick of ~3 cm/sec per photon scattered. This process, however, is strongly limited by saturation, even at very modest light intensities of hundreds of watts per square centimeter. The problem of saturation of the scattering force was treated phenomenologically using the so-called Einstein A and B coefficients to calculate the fraction of time f an atom spends in the excited state. The scattering force is given by the rate of scattering momentum $F_{\text{scat}} = hf/\lambda t$, where t is the spontaneous emission lifetime [2]. At high saturating intensities the population of a two-level atom equalizes and $f = \frac{1}{2}$. The magnitude of this saturated force is sufficient, however, to turn an atomic beam of sodium of average thermal velocity of $\sim 10^5$ cm/sec through a radius of curvature $\rho \cong 20$ cm, if applied continuously at right angles to the velocity to avoid any Doppler shifts. If one applies the saturated force in opposition to the atomic motion, one can stop atoms at the average velocity in a distance of $\rho/2 \cong 10$ cm, assuming one compensates for the large Doppler shift of the atomic resonance. It was suggested that one could use the scattering force to make an atomic beam velocity selector or an isotope separator [2]. A scheme for exerting significant optical pressure on a gas of atoms was also proposed [1].

No consideration was given in reference [2] to the gradient component of the force on atoms inasmuch as I did not understand how to treat the saturation of this force. The classical formula for the gradient force of an electromagnetic wave on a neutral atom, considered as a simple dipole, is the dipole force formula $\frac{1}{2}\alpha\nabla E^2$ where α is the optically induced polarizability of the atom or particle. α can be calculated using a simple harmonic oscillator model of an atom. For atoms the polarizability is dispersive and changes signs above and below resonance in analogy with the change in sign of the gradient force on high and low index particles. This gradient force formula, applied to atoms, electrons, and plasmas, was considered previously by Askar'yan [3], using lasers in a two-dimensional geometry in connection with self-focusing. Letokhov [4] also considered very weak off-resonant one-dimensional confinement of atoms in laser standing waves for spectroscopic purposes. Neither work discusses the possibility of stable three-dimensional trapping of atoms.

1.2 Optical Levitation and Applications

The next advance in optical trapping and manipulation was the demonstration of the optical levitation trap in air, under conditions where gravity plays a significant role [5]. In the levitation trap, as shown in Fig. 1.2, a single vertical beam confines a macroscopic particle at a point E where gravity and the upward scattering force balance. The equilibrium is stable because of the increase in axial scattering force with decreasing height near E and

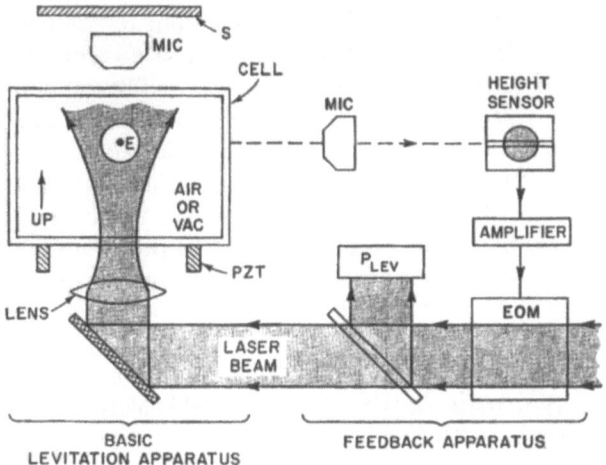

Fig. 1.2. Basic apparatus for optically levitating dielectric spheres and feedback stabilization apparatus for levitating in vacuum and measuring forces. PZT is a piezoelectric ceramic shaker; EOM is an electrooptic modulator

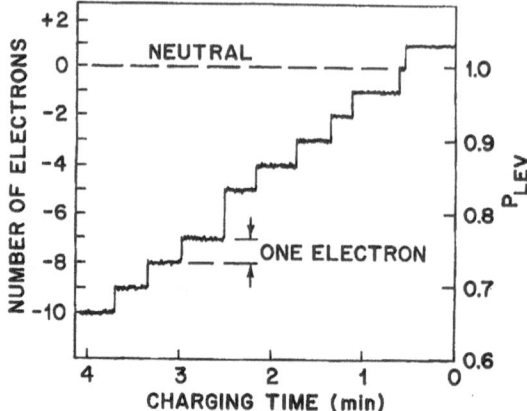

Fig. 1.3. Changes in the optical levitating power caused by the automatic feedback system as the charge on the sphere in an electric field increases by single-electron amounts

the transverse confinement of the gradient force. Once aloft, levitated particles can be freely manipulated by simply moving the beam. With a pair of movable beams one can combine the two beams and thus collide a pair of levitated particles and assemble compound particles. In this way one can fuse droplets and also form spheroids, teardrops, spherical doublets, triplets, etc. [6]. These complex particles align themselves in the beam and make ideal test particles for light scattering experiments. Levitation of hollow glass spheres, which are sometimes used as laser fusion targets, is also possible [7]. One uses TEM_{01}^* or doughnut mode beams having a hole in their center, since these hollow particles behave as air bubbles and are pushed out of the high intensity region of the light. Levitation in high vacuum is also possible [8]. However, feedback is needed to damp particle oscillations caused by random beam fluctuations [9]. See Fig. 1.2. The feedback scheme both locks the particle to a split photodiode height sensor and also varies the levitating power in proportion to the negative of the velocity to give strong optical damping. Importantly, feedback locking in either vacuum or air provides a means of automatically measuring forces on particles, since the change in power needed to hold the particle at a fixed height is a direct measure of an externally applied force [10].

The feedback force measuring technique was used to measure the electric force on oil drops as they accumulated single electron charges in a modern version of the Millikan oil drop experiment [10], where we could easily resolve the changes in levitating power associated with changes in the charge on a sphere by single electron amounts. See Fig. 1.3. We could also measure viscous drag forces on small particles as the velocity of a fluid varies, changes in radiometric forces with pressure, and changes in the optical scattering force

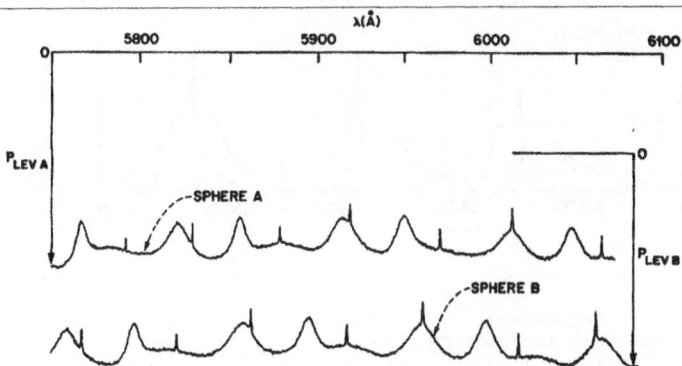

Fig. 1.4. Resonant behavior of light forces on dielectric spheres. The spectra show the variation with wavelength of P_{levA} and P_{levB}, the power needed to levitate oil drops A and B, which have index of refraction $n = 1.47$ and slightly different diameters ($\sim$10μm). The resonances of sphere A are shifted $\sim$50 Å higher in wavelength than the corresponding ones in sphere B

with axial position in the light beam [9]. We also used feedback to measure the wavelength dependence of the levitating force with a tunable dye laser [11]. This led to the experimental discovery of a complex series of high Q resonances in the force as shown in Fig. 4 of reference [10]. At first we wondered if they were even real. We suspected problems with the feedback circuit. According to the simple ray picture of the forces as I described it, one should not expect much of a change in force with wavelength. These resonances are in fact real and are predicted by Mie–Debye Theory [11]. They manifest themselves as dips in the radiation pressure force and peaks in the light scattering of a trapped spherical particle. Applications of these resonances, variously called surface-wave resonances, morphology-dependent resonances, or whispering gallery resonances, have had great impact on light scattering studies. Precision measurements of the resonant backscattered spectrum or "glory scattering" from silicone oil drops probably give the best experimental test of the Mie–Debye theory of electromagnetic scattering; see Fig. 5 of reference [12]. High-Q resonances yield 2–3 orders of magnitude improvement in absolute and relative size and index of refraction measurement of spheres [11,13]. More recently, drops have served as extremely high-Q dye laser and Raman laser resonators and as a medium for studying and enhancing a wide range of linear and nonlinear optical interactions [14]. Mirrorless optical resonators based on these modes have the highest Q's of any optical cavity ($> 2 \times 10^9$) [15] and have recently been used for low threshold semiconductor lasers [16,17] and for experiments in cavity electrodynamics [18].

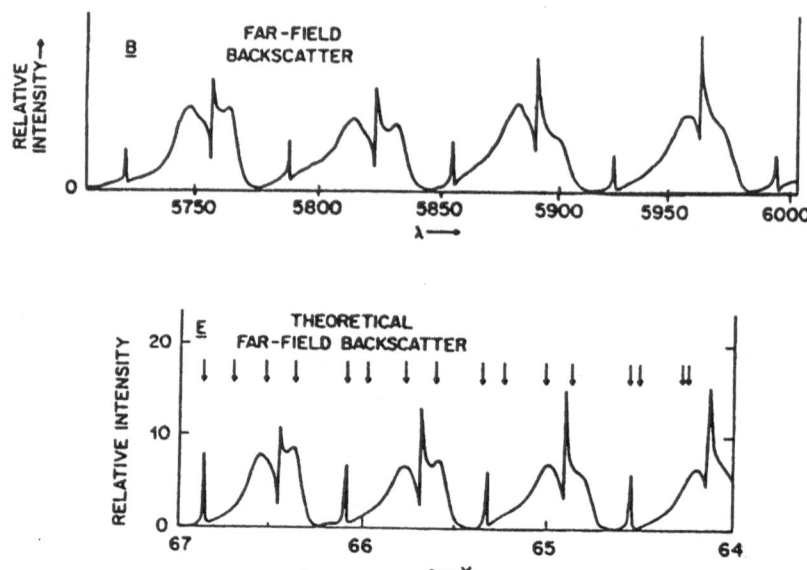

Fig. 1.5. Comparison of experimental Far-Field Backscatter vs. λ (*curve B*) with the theoretical Far-Field Backscatter (*curve E*) plotted vs. $x = 2\pi a/\lambda$

1.3   Origins of Optical Atom Trapping

Following the early work on light forces on atoms [1,2], experiments were performed demonstrating atomic beam deflection [19,20] and isotope separation [21] using the scattering force. In 1975 Hänsch and Schawlow made the important suggestion that it was possible to use the strong velocity dependence of the scattering force due to Doppler shift for the optical cooling or damping of atomic motions [22]. For example, in one dimension with a pair of identical opposing beams tuned below resonance, any atomic motion along the axis meets a net opposing force due to the strong Doppler shifts of the absorption. Three pairs of such opposing beams should damp all degrees of freedom. Does this imply the ability to cool atoms to absolute zero? Certainly not. Due to quantum fluctuations there is randomness in the directions of emission of successive photons and also in the times of emission, which corresponds to a constant heating process. The equilibrium temperature finally achieved is a balance of the optical cooling rate and the quantum heating rate. Letokhov and Minogin were the first to estimate the equilibrium temperature based on the fluctuations of the scattering force [23,24]. For a tuning $\gamma_n/2$ below resonance, which gives the optimum cooling rate, they estimate an equilibrium energy of $\sim h\gamma_n$. They also proposed at this time that one could use the same 6-beam cooling geometry for stably trapping atoms on the intensity maxima of the three-dimensional standing wave pattern by virtue of the

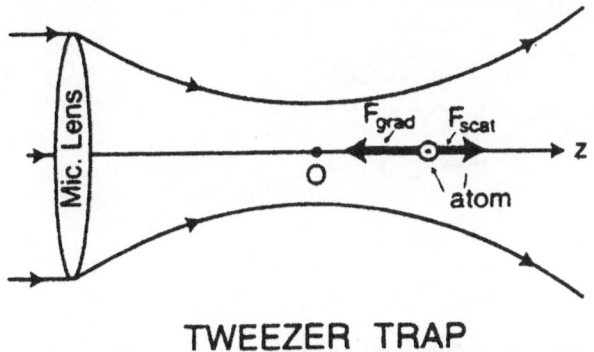

TWEEZER TRAP

Fig. 1.6. Tweezer trap for atoms. $F_{\text{grad}} > F_{\text{scat}}$ giving a net backward force toward E

gradient force. Unfortunately, they estimated that the trap depth was also $\sim h\gamma_n$ which implies a very leaky trap.

In order to see if one could succeed in getting a deep enough trap, I decided in 1978 to address the problem of saturation of the gradient force using the same semiclassical rate-equation approach used earlier for understanding saturation of the scattering force [2]. The key point was the realization that the classical value of the polarizability α in the formula $\frac{1}{2}\alpha\nabla E^2$ applies to an atom in its ground state, and that an atom in its excited state contributes polarizability of the opposite sign in proportion to the fraction of time f it spends in the excited state. With this approach one finds for the potential U of the gradient force for arbitrary tuning and arbitrary light intensity, $U = h/2(\nu-\nu_o) \ln (1 + p)$, where p is an intensity-dependent saturation parameter [25]. It is readily seen from the expression for U that one can greatly increase the potential depth U by factors of 10^2 or more for a given intensity by keeping the saturation modest ($p \cong 1$) and greatly increasing the detuning $(\nu-\nu_o)$ to values of about $10^2 \gamma_n$ or more. Use of detuning to reduce saturation made it possible, for the first time, to devise trapping geometries for atom traps which were stable in the Boltzmann sense, i.e. $U/kT \gg 1$. A 2-beam trap was proposed at this time in analogy with the first macroscopic particle trap. Also suggested was the simplest of all traps, the optical tweezers trap [25] consisting of a single strongly focused Gaussian beam; see Fig. 1.6. Although the tweezer trap at first sight seems counterintuitive, it is axially stable because of the dominance of the backward axial gradient force over the forward scattering force. With these new strongly detuned atom traps one cannot use the trapping beams to also provide cooling. This requires the use of optimally tuned auxiliary cooling beams to keep the atom temperature at $\sim h\gamma_n$ [26].

An experiment was performed to demonstrate the new concept of large gradient forces with detuned light [27]. An atomic beam was injected into the core of a Gaussian beam, as shown in Fig. 1.7. Depending on the tuning,

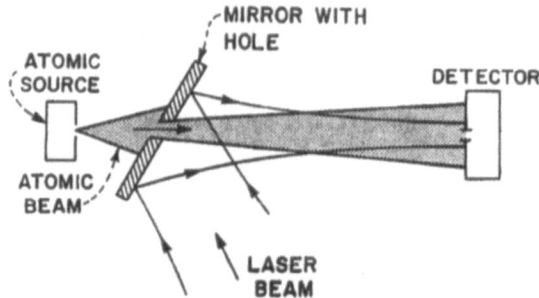

Fig. 1.7. Apparatus for observing focusing and defocusing of an atomic beam by the dipole force of a nearly resonant laser beam

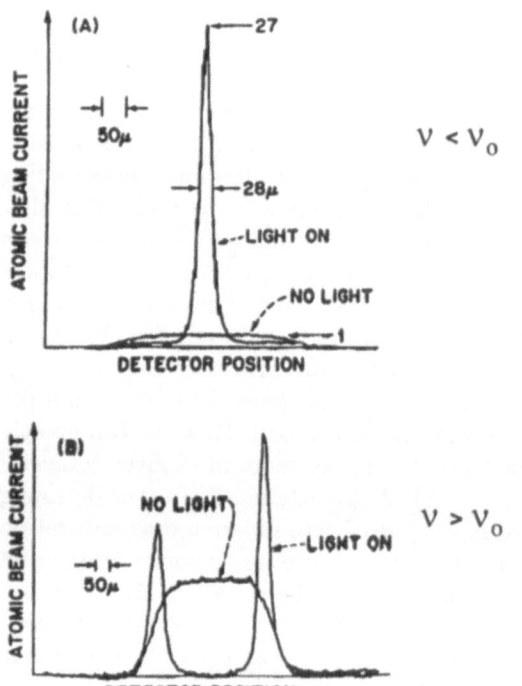

Fig. 1.8. a Focusing of an atomic beam by light tuned below resonance. **b** Defocusing of an atomic beam by light tuned above resonance

either below or above resonance, the atoms are strongly focused or defocused by the transverse gradient forces of the laser beam, as seen in [27,28], and in Fig. 1.8 from reference [10]. This work was the first experimental demonstration of the gradient force on atoms. It also represents a demonstration of two-dimensional trapping of atoms using light forces. In addition, it marks the beginning of the so-called field of atom optics in which atoms are guided by light. The Gaussian beam in effect acts as a gradient index or GRIN lens. Additional work studying the variation of the atomic beam focal spot size with light intensity gave the first evidence of quantum heating of atoms by light [29].

Prospects for optical atom trapping were bolstered by a theoretical analysis by Gordon et al. [30] entitled "Motion of atoms in a radiation trap." This work derived the basic optical forces on atoms, their saturation, and their fluctuations from first principles, using a fully quantal theory, and applied the results to traps. It confirmed the correctness of the scattering and gradient force components which I deduced from a combination of experiment, intuition, and semiclassical analysis. A new result of [10] was the derivation of the fluctuations of the gradient or dipole force. This is conceptually more difficult to understand than the scattering force fluctuations, but it contributes equally with the scattering force fluctuations to the quantum heating rate and the equilibrium temperature. This paper has become a basic reference on questions about optical forces on atoms.

A further big experimental step on the way to atom trapping was the gross slowing of atomic beams using the scattering force of an opposing laser beam by Phillips et al. [31,32]. See also Phillips' Nobel Lecture [33]. The major problem here was to compensate for the large Doppler shifts that occur as the atoms are slowed. This was done by magnetically tuning the resonant frequency of the atoms with a properly tapered magnetic field to keep the peak of the distribution of slowing atoms in resonance with the light. Chirping the light frequency was also suggested by Letokhov et al. [34] and subsequently demonstrated [35]. Although these one-dimensional techniques could slow the peak of the axial velocity distribution to zero, there was no transverse cooling and the lowest average temperature achieved was about 0.1°K. At this temperature relatively few atoms are available for filling small volume atom traps. One solution to the difficulty actively being pursued at that time at the National Bureau of Standards was a different type of trap [36] in which atoms were confined in a relatively large volume deep trap solely by the scattering force from mildly diverging beams. Unfortunately, this proposal was flawed. A theorem called the Optical Earnshaw Theorem was proven by Ashkin and Gordon [37] showing that any trap based solely on scattering forces, which are strictly proportional to the light intensity, is inherently unstable. This was proven in analogy with the Earnshaw Theorem in electrostatics.

In 1984 an experiment was started at Bell Laboratories in Holmdel on optical trapping of atoms. This was stimulated by the arrival of Steve Chu as a new department head with intentions of trapping atoms. See Chu's Nobel lecture [38]. The initial plan was to combine slowing, cooling, and trapping in a single experiment. Chu argued for a simpler first step, to first study the three-dimensional cooling scheme using the Doppler cooling technique [22] now referred to as "optical molasses." This was wise since molasses cooling succeeded so well it affected our subsequent choice of traps. The molasses experiment [39] produced a roughly 1 cm^3 volume of atomic vapor at a density of 10^9 atoms/cm^3, viscously confined at a temperature of about 250 μK, close to the Doppler limit [23,24,26,30] which persisted for times up to a second before diffusing away. Indeed, with this remarkable sample of cooled atoms it became possible to demonstrate the first three-dimensional stable atom trap [40] using the very simple tweezer trap consisting of just a single strongly focused Gaussian beam. Despite its small volume, tweezers placed anywhere within the sample of cold atoms proceeded to fill up to densities of about 10^{11} atoms/cm^2 by diffusion from the surrounding vapors, in analogy to the filling of the first particle trap by diffusion from the surrounding latex spheres [1]. Trapped atoms persisted in the trap and could be freely manipulated in space after the surrounding molasses atoms diffused away The success of these cooling and trapping experiments marked the beginning of a new era of experimentation which has revolutionized experimental atomic physics.

1.4   Recent Work on Atom Trapping and Manipulation

The achievement of optical cooling and trapping of a dense cloud of atoms in 1986 greatly stimulated interest in optical manipulation techniques. A new large volume magneto-optical trap (MOT) was developed using the scattering force [41]. A quadrupole Zeeman splitting field was used which made the resonance frequency, and α, the polarizability, position dependent. This results in a stable scattering force trap that doesn't violate Earnshaw's Theorem. This robust, large-volume, deep trap is widely used as a workhorse trap in spite of some poorly understood behavior [42].

Although the initial molasses temperature of 240 μK was close to the Doppler cooling limit for a two-level atom [39], disagreements soon arose. Unexpectedly, temperatures almost 10 times less than the Doppler limit were observed [43,33]. Explanations of this increased cooling are based on the multilevel nature of the cooling transition used [44,45]. Also refer to Phillips Nobel Lecture [33].

One might think the minimum possible temperature of cooled atoms would be T_r, which is the temperature due to the recoil of a single photon. For sodium the recoil velocity is about 3 cm/sec with a temperature T_r of about 2 μK. However, cooling below even T_r can be achieved. Chu et al. in 1986 [46] were the first to propose that one could reduce the temperature of atoms held in a tweezer trap at the molasses temperature to temperatures of ∼10^{-6} K by momentarily turning off the trap and then turning it on again. This allows fast atoms to escape and is the equivalent of evaporative cooling, as proposed by Hess for cooling hydrogen gas in magnetic traps [47]. Evaporative cooling techniques have the disadvantage that well over 90 % of the atoms originally in the trap are lost. Chu et al. [46] also proposed letting tightly confined atoms expand and cool into a larger volume harmonic trap with no atom loss. Sub-recoil temperatures were also calculated. This idea has never been implemented. Cooling below T_r also occurs using velocity selective coherent population trapping (VSCPT). In this technique, atoms randomly scatter photons until they fall into a superposition ground state with close to zero velocity where they are decoupled from the light [48,49]. See also Cohen-Tennoudji's Nobel Lecture [50]. Another practical scheme for cooling below T_r involves selective Raman cooling [51,52].

An early use of cold atoms was in the achievement of a practical "atomic fountain" [53]. Chu et al. showed that atoms optically launched vertically from a MOT trap could interact with a microwave cavity for long times to improve the accuracy of atomic clocks [53,38]. Another growing use of cold atoms is the study of ultracold atomic collisions. With ultracold collisions one can explore processes not seen at higher temperature [55–58].

A subfield of optical manipulation has developed called "atom optics." It loosely refers to the optical manipulation of atoms in ways similar to manipulation of light by conventional optical elements like lenses, mirrors, beam splitters, gratings, and interferometers. At times it makes use of the wave properties of atoms. The first experiments showed guiding and focusing of atoms using the distributed lens action of the gradient force in a long thin laser beam [27]. Other single optical lenses were demonstrated [59,60], but all suffer from chromatic aberration. Nevertheless, focusing to spot sizes of 20 Å has been seen [61]. Optical mirrors for atoms have been developed using reflection from the dipole force of evanescent waves of laser fields [62,63]. Atomic beam splitters have been extensively studied, based on a variety of interactions [64–67]. Atom optics techniques also have potential uses in technology for neutral atom lithography [68,69]. See a review of atom optics in reference [70]. Atom interferometers have been developed which are an important class of devices for making precision measurements. Interferometric measurements of the acceleration of gravity were made with an accuracy of 1 part in 10^6. Increases in accuracy to 10^{-10} g are anticipated [71]. Applications to geology, a search for the net charge on atoms, fifth force experiments, and a test of general relativity are suggested.

The most important recent development in the field has been the final achievement of Bose–Einstein condensation (BEC) of atomic vapors in atom traps. This was made possible by the realization of a combination of sufficiently cold and dense vapors of atomic bosons where the de Broglie wavelength of the atoms becomes large enough that individual atomic wave functions overlap and become coherent in a single ground state extending over the sample. Bose–Einstein condensation had previously been observed in superconducting solids and in superfluid liquids, but never in atomic vapor. In Bose–Einstein condensates in the vapor phase, the likelihood of three-body collisions, which might result in the formation of molecules or atomic clusters, is much reduced. In this sense the vapor condensate is metastable for times of seconds or minutes. However, once formed, the atomic vapor condensate is the purest macroscopic quantum system yet achieved. Almost all atoms are in the condensed state.

Bose–Einstein condensation was achieved by using evaporative cooling and specially designed magnetic traps [72,73]. Since its discovery, there has been an ever growing number of theoretical and experimental studies probing the novel properties of this new state of matter. The collective oscillation of the condensate has also been observed [74]. Mechanical measurements also have been made on the propagation of sound waves and phonons [75] in condensates. Ketterle et al. showed the coherence of the BEC by splitting it into two parts with a far off-resonance laser beam and observing macroscopic atom wave interference when they recombine [76]. More recently, measurements of the coherence length of a condensate were made by deducing the momentum uncertainty in the Heisenberg uncertainty principle [77]. The results showed the entire condensate was a coherent matter wave.

Techniques were devised for expelling atoms from condensates in traps to give an external beam of coherent atoms which is in fact a form of atom laser. For example, in 1997 Ketterle and colleagues flipped the spins of a condensate in a magnetic trap, causing condensed atoms to fall downward in gravity [78]. Hänsch and collaborators used a continuous radio-frequency signal to make a continuous output coupler. Phillips and his group used optical Raman transitions in which the photon recoil ejected atoms from the trap to give a directional output coupler [79]. These atom lasers make possible a new form of atom nonlinear optics in which the atoms of the condensate itself act as a nonlinear medium. This has resulted in the observation by Phillips et al. [80] of four-wave mixing with matter waves. In recent experiments Ketterle and also a combined NIST and Japanese group have observed real gain in the number of atoms in a coherent atom laser beam by using a pump laser to convert condensate atoms into the coherent incident signal beam [81,82]. Coherent atom laser beams may some day play an important role in atom lithography. The principal difficulty with use of present atom lasers for lithography is the lack of sufficient beam intensity. In time we expect that all the atom optics components that have been developed for incoherent atomic beams will have their coherent atom laser beam analogs.

One of the interesting recent developments in the BEC field is the realization of the importance of the simple single-beam dipole trap, or tweezer trap [83,84]. This is essentially the same trap used in the first atom trap in 1986 and subsequently used in biology and elsewhere for trapping and manipulating macroscopic particles and Rayleigh particles. These single-beam tweezer type traps, which have been largely neglected by the Bose–Einstein community, overcome many of the drawbacks of magnetic traps. Magnetic traps are large, fixed in space and limited to particular hyperfine levels. One can load more than 5×10^6 condensed atoms at densities as high as 3×10^{15} atoms/cm^3 into single-beam dipole traps from magnetic traps [83] and trap several hyperfine states using adjustable external fields. Experiments have been done with dipole traps on Feshbach resonances [84–86]. Experiments also have been done to measure scattering lengths [86], observe spin domains [87], and measure quantum tunneling across spin domains [88] Reversible formation of condensates has been studied using optical traps [84]. The field of BEC is growing rapidly. The number of papers per year on the subject is fast approaching a thousand [89].

1.5   Origins of Optical Trapping in Biology

Although the optical tweezers trap was originally designed as an atom trap and was used in the first optical trapping experiment [39], that experiment did not represent the first use of tweezers. During the atom trapping experiment, at a time of temporary difficulty, it was decided to try the tweezer trap on simpler Rayleigh dipole particles, such as submicron silica spheres. Using the known polarizability of a submicron Rayleigh sphere one can show that it is possible to satisfy the criterion for a tweezer trap that the backward gradient force exceeds the forward scattering force at the point of maximum axial gradient of a strongly focused Gaussian beam [90]. Submicron colloidal silica particles were placed in a simple water-filled chamber and irradiated with a strongly focused 5145 Å argon laser beam, as shown in Fig. 1.9. Scattered light at 90° was viewed with a low power microscope, either visually or with a photodetector. Individual particles were easily seen in the cone of the focused laser light and displayed beautiful Brownian motion. Whenever a particle wandered close to the beam focus, it was immediately pulled into the trap. It brightened up and all Brownian motion ceased. Particle after particle entered and we showed that they didn't just fuse into a single lump, but formed a fixed array in the focus, held apart by their colloidal charge. Particles as

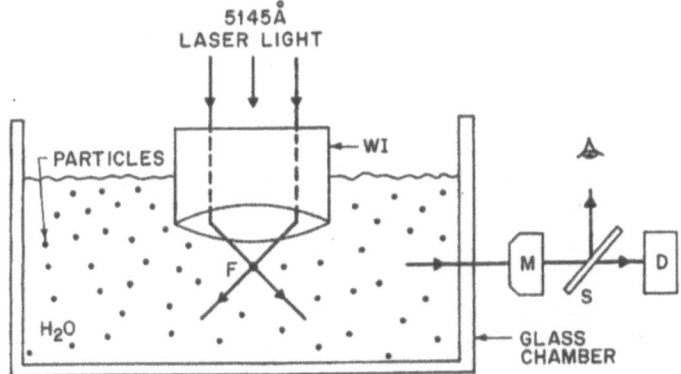

Fig. 1.9. Apparatus for trapping and observing submicron colloidal particles in water

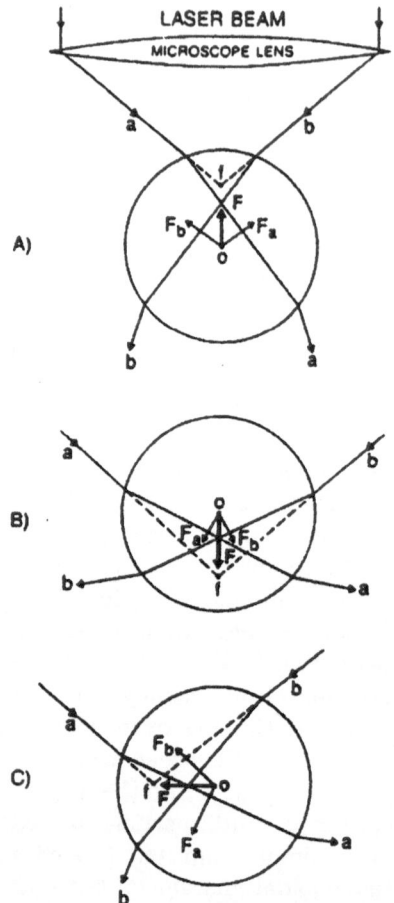

Fig. 1.10. Simple ray-optics picture of the stability of tweezer trap. Any displacement of a macroscopic sphere away from the focus, f, either axially as in **a** or **b**, or transversely as in **c**, results in a net restoring force

small as ~250 Å were trapped. We also showed that tweezers could trap micron sized spheres large compared with the wavelength. This extended the notion of a backward gradient force to large particles as well. The origin of the backward light force for tweezers in the ray-optic regime [91] is shown in Fig. 1.10. For many applications with macroscopic particles, tweezers are superior to the levitation trap. Levitation traps depend on gravity and have forces of ~mg, where m is the mass and g is the acceleration of gravity. Tweezers, however, are an all-optical trap and can have forces of thousands of mg, limited only by the optical power. This is useful for confining submicron particles in situations where gravity plays a minor role and Brownian motion dominates. The compact tweezer trap is also more tolerant of particle shape irregularities than the levitation trap. As we shall see, tweezer trapping of macroscopic particles has turned out to be widely useful in biology and the physical sciences.

Our next experiments involved trapping of colloidal tobacco mosaic virus [92]. Tobacco mosaic virus (TMV) is a rugged rod-like protein ~200 Å in diameter and ~3000 Å long. It is easily trapped by ~100 mW and it too can form arrays of aligned viruses in the beam focus. This may well be the first rod-like molecule to be fully aligned by an optical electric field. While we were doing these experiments a very serendipitous event occurred. We tended to keep these TMV samples a long time in a container open to the ambient atmosphere. With time we noticed the appearance of increasing numbers of strange, relatively large, apparently self-propelled particles. Some occasionally were trapped when they neared the focus, where they gave rise to a wild display of scattering. We called these new particles "bugs" and it turned out they were bugs, bacteria that is, which presumably contaminated the chamber from the air. At low powers of ~5 mW they continued scattering for many minutes. If we raised the power to ~100 mW, there was a huge burst of scattering and then nothing but a weak steady scattering. We interpreted this as "opticution," death by light. The weak scattering was from the empty carcasses of the bugs. To check this we introduced the trap into a high resolution microscope as shown in Fig. 1.11. We could then trap, observe, and manipulate bacteria which we grew from bits of Joe Dziedzic's ham sandwich. We readily confirmed our hypothesis. Our paper in Science [92] on laser trapping of viruses and bacteria was the first report of optical manipulation of living cells, although optical damage to bacteria cells was apparent.

We decided to try other laser wavelengths that might be less damaging to living cells [93]. Since the absorption of molecules like chlorophyll and hemoglobin falls rapidly in the near infrared, we decided to try trapping with a Nd:YAG laser at 1.06 μm. Also, the absorption of water at 1.06 μm is quite small. Using 1.06 μm YAG laser we immediately found a large decrease in damage to bacteria. We could collect large numbers of motile bacteria within a single trap for many minutes and see them swim away, apparently undamaged, with powers of 50 mW or more. As a test, we obtained E. coli bacteria and trapped one in the middle of the microscope chamber for a long time, far from other untrapped cells resting on the bottom surface. To our pleasant surprise we saw it grow in size and finally divide into two cells. We retrapped these two cells and after another hour or so we had four cells. E. coli was reproducing right in the high intensity trap focus! Clearly there was no serious optical damage occurring. We then showed that single yeast cells could also reproduce by budding right in the trap. We grew large clumps of cells this way, which we could freely manipulate within the chamber. Joe Dziedzic and I proceeded to trap all sorts of cells: pigmented red blood cells, green algae, diatoms, amoebas and other protozoans. In general, there was

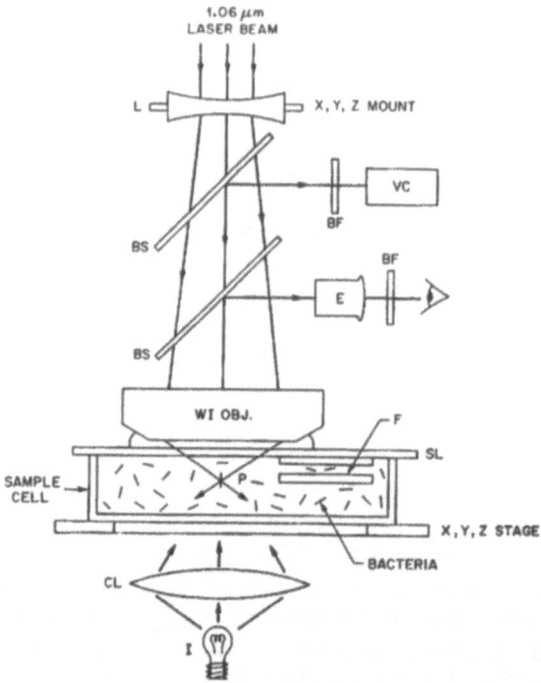

Fig. 1.11. Combined high-resolution optical microscope and 1.06 µm infra-red laser trap for observing, manipulating, and separating bacteria and other organisms

greatly reduced damage using the YAG laser light. In contrast, red blood cells and green cells with chlorophyll simply exploded with argon laser light at 5145 Å. With larger cells, such as scallion cells, we were able to manipulate small particles within the living cell. We could collect small micron-sized vesicles and other particles and probe the geometry of the chloroplasts and central vacuole by dragging the particles about, moving them deep in the cell from the bottom to the top, under and between cell structures. The tweezer trap is robust enough to tolerate partial shading of the beam by intervening structures.

We demonstrated the ability of tweezer traps to separate a single selected bacterium from a "gemisch" or collection of bacteria. In Fig. 1.11 we show an approximately 15-µm inner diameter fiber F attached to the top of the sample chamber. We were able to trap one or more bacteria and manipulated them into the core of the hollow fiber. Despite the rather severe optical distortions at the input of the fiber, one could still maneuver the bacteria into the fiber core without losing them. To complete the separation we removed the chamber lid, washed it, rinsed it, and dried it to remove any possible bacteria clinging to the lid, and then with a gentle air stream blew the liquid contents of the fiber into another water filled vessel. The fiber acted as a very

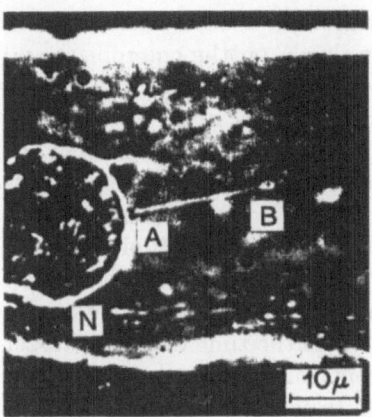

Fig. 1.12. Artificial cytoplasmic filaments in a scallion cell. The laser trap originally located on the surface of the nucleus (N) at A is moved to B, pulling out the viscoelastic filament AB into the central vacuole

convenient storage vessel for the separated bacteria. The ability to separate bacteria and other biological cells is a very simple but important capability of tweezer techniques. It was also shown that one could introduce two traps into the microscope. One can then grab a bacterium or cell at its ends and orient it in space or rotate it at will by moving one trap relative to the other. One can stretch it and observe its mechanical properties. A red blood cell is so pliable that simply running a tweezer trap over its surface makes a very noticeable distortion. As big as they are, one can squeeze many blood cells into a single trap. The elastic behavior of the cytoplasm is evident in almost all cells.

We made more careful observations on the elastic properties of the cytoplasm in subsequent work entitled "Internal cell manipulation using infrared traps" [94]. Using scallion cells we could generate what we called "artificial cytoplasmic filaments" pulled by the trap from the surfaces of most internal cell organelles. Fig. 1.12 shows such a filament pulled from the surface of the nucleus of the cell into the central vacuole of the cell. The filament stretches from the original location of the trap at A to its final location at B. If one quickly turns off the trap at B, the filament snaps right back to A. If one waits a minute or so at B, the filament snaps back, but more slowly. The longer one waits at B, the more slowly the filament returns to A. At longer times the filament only partially retracts and we are left with a sagging remnant of cytoplasm. This is classic viscoelastic behavior. Cytoplasm is like the toy "silly putty." For quick distortions it is highly elastic. For slow distortions or long times, the material flows and sags and is relatively weak. Using tweezers, evidence for viscoelasticity is seen everywhere within cells. We used tweezers to perform a new kind of internal cell surgery. If one tries to pull quickly with tweezers on some fairly large structure within a cell, such as the spiral chloroplast of a spirogyra cell, it looks quite rigid and it barely moves before it slips out of the trap. If, however, one pulls slowly on the chloroplast and continues to apply force after it moves somewhat, one finds that the tension slowly relaxes and that further motion is possible. This can be repeated and in time one can pull the entire chloroplast right off the cell wall and into the fluid of the central vacuole, where it is quite free. One can thus make gross changes in cell structure. If the chloroplast is placed against the side wall of the vacuole, it doesn't adhere to the inner membrane initially. If one uses the trap to continue to hold it against the wall, its membrane fuses with the wall membrane, completing the operation. The nucleus of the cell and its cytoplasmic supports can be similarly manipulated. At the time of these tweezer experiments there had been very few experiments showing the viscoelasticity of live cells. In 1950, the physicist Crick, of Watson and Crick fame, tried to study viscoelasticity by getting cells to ingest small magnetic particles, which he proceeded to move with a large electromagnet [95]. After much work, he barely observed any effect, probably due to the fairly rapid relaxation of the tension on the particles. Later a similar experiment with magnetic particles showed a somewhat larger effect [96].

With tweezers one can observe more viscoelastic behavior of the cytoplasm within a cell in a few minutes than has been seen from the time of the Crick experiments to the present. With tweezers we are able to probe the viscoelasticity of the different parts of the living cell. Tweezers have also been used recently to explore the viscoelastic behavior of single strands of polymeric molecules, such as DNA in many contexts [97,98]. With tweezers one can interact strongly with the cytoplasmic streaming that one sees so readily in scallion and other plant cells. If the trap is placed in some streaming channels, one can capture the moving particles of the stream and then release them and see them move on. Depending on the power, we can stop the entire stream, particles and all. Subsequent particles are stalled or can even turn around and move backwards. One can probe the relative viscosities of the moving streams and of the other more liquid regions of the cell. By rupturing the cell and manipulating the same particles in pure water, one can estimate the absolute viscosities of the cell contents. This type of information about cells was generally not known [99]. There is clearly much to be learned about the streaming process with trapping techniques.

The three papers discussed here on damage-free trapping [92–94] in a real sense mark the beginning of the new field of optical trapping in biology.

1.6   Recent Work on Optical Trapping and Manipulation in Biology

An early application of this new manipulative technique of tweezers in biology involved the measurement of the torsional compliance of bacterial flagella by twisting a bacterium about a tethered flagellum [100]. It was shown that this compliance was located within the bacterial motor itself [101]. Tweezers helped show that the flagella of spirochete bacteria also work by the rotary action of their motors [102]. Greulich and Berns were the first to use the tweezers technique in combination with the so-called "microbeam" technique of pulsed laser cutting (sometimes called "laser scissors" or "scalpel") for cutting and moving cells and organelles. Greulich's early work involved UV cutting and tweezer manipulation of pieces of chromosomes for gene isolation [103]. Tweezers were also used to bring cells into contact with one another in order to effect cell fusion by cutting the common wall [104]. Recently the spatial and temporal contact needed between antigen-presenting cells and T lymphocyte cells to initiate the activation process was studied with a new optical tweezer-based assay involving intracellular calcium signaling [105]. Berns and his group used tweezers, often combined with optical scissors, to manipulate chromosomes during cell division [106] as a new way to study the complexities of mitosis.

Experiments were performed with tweezers to manipulate live sperm cells in three dimensions [107,108] and to measure their swimming forces [109]. Applications of tweezers and scissors to all-optical *in vitro* fertilization are being considered [110]. UV drilling of channels in the zona pellucida of oocytes was performed and selected sperm were inserted into channels to effect fertilization [111,112]. Experiments by Berns' group measured the effects of wavelength on possible optical damage processes in sperm and in other contexts using tunable Ti sapphire lasers [113].

One of the most important biological applications of tweezers is in the study of molecular motors. These mechanoenzymes interact with the microtubules or actin filaments of the cell to generate the forces responsible for cell motility, muscle action, cell locomotion, and organelle movement within cells. In early work using the "handles" technique, Block et al. [114,115] attached single kinesin motor molecules to spheres and placed them directly onto microtubles where they could be activated by ATP. This new technique greatly improved on earlier *in vitro* motility assays which used many motors and relied on random diffusion for attachment to filaments. Ashkin and colleagues [116], using a related *in vivo* technique, estimated the force generated by a few dynein motors attached to mitochondria as they moved along microtubules in the giant amoeba reticulomyxa. Kuo and Sheetz [117] working *in vitro* with tweezers and handles attached to a microtubule filament estimated the force generated by a single kinesin molecule.

A major advance in the field was the resolution by Svoboda et al. [118] of the detailed motion of a single kinesin molecule into a sequence of 8 nm steps as it advanced along a microtubule. The first observation of this previously postulated stepping motion used an optical trapping interferometric position monitor with subnanometer resolution [118]. Proper damping of the Brownian motion of the sphere by the trap was also needed to see the steps [119]. Later, Svoboda and Block [120] measured the complete force–velocity relationship of single kinesin motors as a function of ATP concentrations. A maximum force of ~5–6 pN was observed. Finer et al. [121,122] shortly thereafter introduced a new feedback-enhanced tweezer trap with a detection capability of subnanometers in position, piconewtons in force, and ms in time response. They studied the interaction of actin with myosin in a dual trap scheme which suspended the actin filament over a single myosin molecule. They observed stepwise motion of about 11 nm and forces of about 3–4 pN. Malloy et al. [123] also used feedback to study the interaction of myosin with mutant drosophila actins. The unbinding force of a single myosin molecule and actin filaments in the absence of ATP was measured with tweezers by Nishizaka et al. [124]. Another form of myosin, called myosin-V, was recently studied with a pair of tweezer traps. Myosin-V was shown to be a processive actin-based motor that can move in large steps approximating the 36 nm pseudo-repeat distance of actin filaments [125,126]. Single-motor molecule experiments have triggered work on detailed models of motion, the ATP hydrolysis cycle, and single enzyme kinetics [119,127–130]. Recently Visscher, Schnitzer, and Block have developed a novel feedback technique called a molecular force clamp which maintains constant force on a single kinesin molecule as it moves along microtubules [131–133]. New data show that one molecule of ATP is hydrolyzed per 8 nm step of kinesin over wide ranges of the force. The new data on tight coupling between ATP hydrolysis and mechanical stepping seems to rule out many theoretical models for force generation by kinesin, including the so-called thermal ratchet model [134,135].

A recent exciting advance in the field was the extension of tweezer force measuring techniques to a new class of motors, nucleic acid motor enzymes. Using a handles technique the force generated by a single RNA polymerase enzyme was measured as it pulled itself along a DNA molecule while synthesizing an RNA transcript [136]. The motion is slow, but the motor is surprisingly powerful. It was observed to stall reversibly at 14 pN. Force versus velocity measurements were recently obtained for single RNA polymerase molecules using an optical feedback technique in which the position of the molecule was held fixed as the laser power was varied. Novel behavior was seen which can be modeled as a double potential well [137,138]. Use of tweezer techniques opens a new way of studying the transcription process of RNA polymerase [139]. Increasingly, tweezers is becoming the technique of choice for the study by many groups of the mechanics of the many types of motor molecules [135,140].

Work has also been done with tweezers to examine the mechanical properties of microtubules, actin filaments, and DNA biopolymers. Kurachi et al. [141] measured the flexural rigidity of microtubules by attaching polystyrene beads and bending them with tweezers. Feigner et al. [142] studied the rigidity directly by manipulating free-floating single microtubules. The torsional rigidity of actin was deduced from a measurement of the rotational Brownian motion of a single actin filament suspended from a freely rotating sphere held in a tweezers trap [143]. Chu et al. [144] made the first direct observation of the tube-like motion of a single extended fluorescently labeled DNA polymer strand as it relaxed through a dense entangled polymer solution. The behavior supports the reptation model of deGennes. The model explains the observed viscoelastic behavior of many biological materials [93,94]. The stretching of double stranded DNA was studied with optical forces. At 70 pN of force a reversible transition to a single-stranded unraveled form of DNA was seen [145]. Libchaber et al [146] have studied the role of ATP hydrolysis in RecA polymerization on double-stranded DNA using single-molecule manipulation with optical tweezers. RecA protein plays a complex role in DNA growth and DNA repair. Block and collaborators stretched DNA molecules with optical tweezers using feedback control of position. They achieved great improvements in accuracy for sample lengths as short as ~1 µm with forces ranging from ~0.1 pN to ~50 pN. An accurate value of the DNA persistence length of ~40 nm was obtained [147]. There have been recent reports in Science that tweezers can possibly play an important role in DNA sequencing [148]. Fürst and Gast have used a pair of tweezer traps with tethers as a powerful quantitative technique to study the micromechanics of dipole chains made from superparamagnetic particles in liquids. Rupture forces and yield stresses were measured [149]. A beautiful experimental *in vivo* study of vesicle transport was performed in drosophila embryos. Knowledge of the regulation of forces and kinetics within a cell is a key to understanding cell and embryo development [150].

The ability of tweezers to separate single bacteria from a mixed sample in a chamber was recently used for separating selected archaea bacteria under high temperature anaerobic conditions for cloning purposes [151]. This new technique has already yielded a new species of hyperthermophilic archaeum from the hot springs of Yellowstone Park. A high-temperature marine bacterium *Thermotoga maritima* has recently been isolated by tweezers and sequenced [152]. This has evolutionary significance. It is also a potential energy source, since it metabolizes carbohydrates to produce hydrogen gas and carbon. The hope is to find new high-temperature enzymes, possibly as valuable as Tac polymerase used in PCR [153,154]. There are vast numbers of unidentified water and soil bacteria which could be separated by similar tweezer techniques [155].

Burkhart et al. [156] used tweezers in a study to identify the mechanisms within killer cells and T lymphocytes by which so-called lytic particles move to attack target cells. They developed an *in vitro* assay that showed kinesin-dependent motility of these particles on microtubules. A study of the cell-substrate adhesive process was made by Sheetz et al. [157] using the ability of tweezer-manipulated coated microspheres to stick to the surface of moving fibroblast cells. They identified increased integrin–cytoskeleton adhesive interactions at the front of moving cells and increased deformability of the cell membrane at the rear of such cells. Measurement of changes in plasma membrane lipid structure and viscoelasticity during hypoxia were made by Kuo et al. [158] with tweezers. Results showed a transition to a more rigid state and the loss of membrane viscoelasticity during hypoxia. Sheetz et al. [159] made the first study of the mechanical properties of membranes on the leading edges of migrating neuronal growth cones by pulling out membrane tethers with tweezers. The force to extend the membrane and the membrane surface viscosity were determined.

The early work of Ashkin et al. [93] showed the ability of tweezers to distort the shape of red blood cells. Svoboda and Block [160] measured the elastic properties of isolated red blood cell membrane skeletons. Recently, using three tweezer traps, Brakenhoff et al. [161] developed a new assay to sensitively measure the shape recovery time of single red blood cells, using physiologically relevant shapes and conditions. Significant differences in relaxation times were found for old and young cells. Measurements were made in blood plasma and gave markedly different results from previous assays using pipettes in buffer solution. With automation, this may be a powerful technique for study of sub-populations of pathological cells. The three computer-controlled tweezer traps used a multiple scanning trap system developed by Visscher et al. [162].

Greulich et al. have used tweezers to simulate the effect of gravity on the growing tips of algal cells [163]. Dragging the statoliths or gravity sensors of the cell to one side can induce the cell to reorient its growth in that direction.

An assay to study the collision of two particles or cells under controlled biologically relevant conditions, called "OPTCOL," was developed with two tweezer traps [164]. The adhesion of influenza virus covered spheres to erythrocytes during collision with controlled velocities and controlled geometry was studied in the presence of various attachment inhibitors. The new extremely sensitive technique has identified the most potent known inhibitor of this process. The authors foresee wide usage of OPTCOL for studies of collisions of biological particles such as bacteria, viruses, T cells, ribosomes, liposomes, and even nonbiological objects.

1.7   Other Recent Work on Optical Trapping and Manipulation in Physics and Chemistry

Interesting applications of optical manipulation techniques exist in other diverse areas of physics and chemistry. In the field of statistical physics and nonlinear dynamics, Simon and Libchaber [165] used stochastic resonance to synchronize the escape of a Brownian particle from a pair of coupled tweezer traps. Ackerson et al. [166] studied phase transitions and crystallization of a random two-dimensional colloidal suspension to a colloidal crystal using the optical forces of a standing wave beam. Higurashi et al. [167] observed optically induced torques and rotations of micromachined micron-sized anisotropic particles held in a tweezers trap. Svoboda and Block [168] showed that small metallic Rayleigh particles have polarizabilities larger than dielectric particles and can be trapped by tweezers. Ghislain and Webb [169] have built a novel scanning force microscope based on a tweezer-trapped stylus particle having a much lower spring constant than a mechanical cantilever. Applications to imaging soft samples in water is anticipated. Tweezers were used to help measure the entropic forces of about 40 femtonewtons which control motion of colloidal particles at passive surface microstructures [170]. Entropic attraction and repulsive forces were directly observed with tweezers in binary colloids depending on the concentration of the small spheres [171].

Extensive use of optical trapping techniques has been made in the field of microchemistry, which studies the spectroscopy and chemistry of small micron-sized domains. Experiments combining trapping with fluorescnce, absorption spectroscopy, photochemistry and electrochemistry were performed. Polymerization, ablation, and other microfabrication techniques were demonstrated with micron samples. Beam scanning techniques were developed for trapping of micron-sized metal particles, low index particles, and moving of particle arrays in complex patterns. These experiments are by Masuhara, et al. [172] summarizing the results of a five-year ERATO project.

Bar-Zvi et al. [173,174] have used tweezers to study the physical properties of membranes and vesicles. The local unbinding of pinched membranes [173] and pressurization and entropic expulsion of inner vesicles from large vesicles [174] was studied. Direct measurements using tweezers showed that an attractive force can exist between like-charged particles in a colloidal suspension near a surface, contrary to theory [171]. Metastable colloidal crystals were made based on this attractive potential. This has importance on theoretical and possibly practical grounds [176]

1.8 Conclusion

The study of single molecules in chemistry, biology and physics is growing rapidly. A prime example is the special section in Frontiers in Chemistry on single molecules in Science [177]. Tweezers and other optical methods figure prominently in this work. See also the Nature News and Views article by Magdalena Helmer, "Singular take on molecules," which is a report on the conference on "Single Molecule Biophysics" in Tours, France, 8–15 July, 1999 [178].

As we saw here, the precise degree of control made possible by optical trapping and manipulation of small neutral particles has caught the imagination of experimentalists in diverse areas of science, especially atomic physics and biology. Many ingenious and previously impossible experiments have been devised, some having revolutionary impact. The scope of applications is still growing and the future looks bright.

References

1. A. Ashkin, Phys. Rev. Lett. **24**, 156 (1970)
2. A. Ashkin, Phys. Rev. Lett. **25**, 1321 (1970)
3. V. S. Askar'yan, Zh. Eksp. Teor. Fiz. **42**, 1567 (1962) [Sov. Phys. JETP Lett. **15**, 1088 (1962)]
4. V. L. Letokhov, Pis'ma Zh. Eksp. Teor. Fiz. **7**, 348 (1968) [JETP Lett. **7**, 272 (1968)]
5. A. Ashkin and J. M. Dziedzic, Appl. Phys. Lett. **19**, 283 (1971)
6. A. Ashkin and J. M. Dziedzic, Appl. Optics **19**, 660 (1980)
7. A. Ashkin and J. M. Dziedzic, Appl. Phys. Lett. **24**, 586 (1974)
8. A. Ashkin and J. M. Dziedzic, Appl. Phys. Lett. **28**, 333 (1976)
9. A. Ashkin and J. M. Dziedzic, Appl. Phys. Lett. **30**, 202 (1977)
10. A. Ashkin, Science **210**, 1081 (1980)
11. A. Ashkin and J. M. Dziedzic, Phys. Rev. Lett. **38**, 1351 (1977)
12. A. Ashkin and J. M. Dziedzic, Appl. Optics **20**, 1803 (1981)
13. P. Chylek, V. Ramaswamy, A. Ashkin and J. M. Dziedzic, Appl. Optics **22**, 2303 (1983)
14. S. C. Hill and R. E. Benner, "Morphology-Dependent Resonances," in: Optical Effects Associated with Small Particles, ed. by P.W. Barber and R.K. Chang (World Scientific, Singapore, 1988) pp. 3–61.
15. V. Sandoghder, F. Treussart, J. Hare, V. Lefever-Seguin, J.-M. Raimond, and S. Haroche, Phys. Rev. A **54**, 1777 (1996)
16. Y. Yamamoto and R. E. Slusher, Physics Today **46**, 66 (1993)
17. C. Gmachl, F. Capasso, E. E. Narimanov, J. U. Nöckel, A. D. Stone, J. Faist, D. L. Sivco, and A. Y. Cho, Science **280**, 1556 (1998)
18. S. Haroche, in Cavity Electrodynamics, Proceedings of the Les Houches Summer School of Theoretical Physics, Session LIII, 1990, ed. by J. Dalibard, J.-M.Raimond, and J. Zinn-Justin (North-Holland, Amsterdam, 1992).
19. R. Schieder, H. Walther, and L. Woste, Opt. Comm. **5**, 337 (1972)
20. P. Jacquinot, D. Liberman, J. L. Pigne, and J. Pinard, Opt. Comm. **8**, 163 (1973)
21. A. F. Bernhardt, Appl. Phys. **9**, 19 (1976)
22. T. W. Hänsch and A.L. Schawlow, Opt. Comm. **13**, 68 (1975)
23. V. L. Letokhov, V. G. Minogin, and B. D. Pavlik, Zh. Eksp. Teor. Fiz. **72**, 1328 (1977) [Sov. Phys. JETP **45**, 698 (1977)]
24. V. L. Letokhov and V. G Minogin, Appl. Phys. **17**, 99 (1978)
25. A. Ashkin, Phys. Rev. Lett. **40**, 729 (1978)
26. A. Ashkin and J.P. Gordon, Opt. Lett. **4**, 161 (1979)
27. J. E. Bjorkholm, R. R. Freeman, A. Ashkin, and D. B. Pearson, Phys. Rev. Lett. **41**, 1361 (1978)
28. D. B. Pearson, R. R. Freeman, J. E. Bjorkholm and A. Ashkin, Appl. Phys. Lett. **36**, 99 (1980)
29. J. E. Bjorkholm, R. R. Freeman, A. Ashkin, and D. B. Pearson, Opt. Lett. **5**, 111 (1980)
30. J. P. Gordon and A. Ashkin, Phys. Rev. A **21**, 606 (1980)
31. W. D. Phillips and H. Metcalf, Phys. Rev. Lett. **48**, 596 (1982)
32. P. Prodan, A. Migdall, W. D. Phillips, I. So, H. Metcalf, and J. Dalibard, Phys. Rev. Lett. **54**, 992 (1985)

33. W. D. Phillips, Rev. Mod. Phys. **70**, 21 (1998)
34. V. S. Letokhov and V. G. Minogin, Phys. Rep. **73**, 1 (1981)
35. W. Ertmer, R. Blatt, J. L. Hall, and M. Zhu, Phys. Rev. Lett. **54**, 996 (1985)
36. V. G. Minogin and J. Javanainen, Opt. Commun. **43**, 119 (1982) .
37. A. Ashkin and J. P. Gordon, Opt. Lett. **8**, 511 (1983)
38. S. Chu, Rev. Mod. Phys. **70**, 685 (1998)
39. S. Chu, L. Holberg, J. E. Bjorkholm, A. Cable, and A. Ashkin, Phys. Rev. Lett. **55**, 48 (1985)
40. S. Chu, J. E. Bjorkholm, A. Cable, and A. Ashkin, Phys. Rev. Lett. **57**, 314 (1986)
41. E. L. Raab, M. Prentiss, A. Cable, S. Chu, and D. E. Pritchard, Phys. Rev. Lett. **59**, 2631 (1987)
42. T. Walker, D. Sasko, and C. Wieman, Phys. Rev. Lett. **64**, 408 (1990)
43. P. D. Lett, R. N. Watts, C. I. Westbrook, W. D. Phillips, P. L. Gould, and H. J. Metcalf, Phys. Rev. Lett. **61**, 169 (1988)
44. P. J. Ungar, D. S. Weiss, E. Riis, and S. Chu, Opt. Soc. Am. B **6**, 2058 (1989)
45. J. Dalibard, and C. Cohen-Tannoudji, Opt. Soc. Am. B **6**, 2023 (1989)
46. S. Chu, J. E. Bjorkholm, A. Ashkin, J. P. Gordon, and L. Holberg, Opt. Lett. **11**, 73 (1986)
47. H. Hess, G. P. Kochenski, D. Kleppner, and T. J. Greytak, Phys. Rev. Lett. **59**, 672 (1987)
48. A. Aspect, E. Arimondo, R. Kaisor, N. Vansteenkiste, and C. Cohen-Tannoudji, Phys. Rev. Lett. **61**, 826 (1988)
49. J. Lawall, S. Kulin, B. Saubamea, N. Bigelow, M. Leduc, and C. Cohen-Tannoudji, Phys. Rev. Lett. **75**, 4194 (1995)
50. C. N. Cohen-Tannoudji, Rev. Mod. Phys. **70**, 707 (1998)
51. M. Kasevich and S. Chu, Phys. Rev. Lett. **69**, 1741 (1992)
52. H. J. Lee, C.S. Adams, M. Kasevich and S. Chu, Phys. Rev. Lett. **76**, 2658 (1996)
53. M. A. Kasevich, E Riis, S. Chu, and R. G. DeVoe, Phys. Rev. Lett. **63**, 612 (1989)
54. R. Drullinger, APS News Vol. **5**, No. 6, p. 4. (1996)
55. M. Prentiss, A. Cable, J. E. Bjorkholm, S. Chu, E. Raab, and D. E. Pritchard, Opt. Lett. **13**, 452 (1988)
56. D. Sesko, T. Walker, C. Monroe, A. Gallagher, and C. Wieman, Phys. Rev. Lett. **63**, 961 (1989)
57. H. R. Thorsheim, J. Weiner, and P. S. Julienne, Phys. Rev. Lett. **58**, 2420 (1987)
58. P. S. Julienne and F. H. Mies, J. Opt. Soc. Am. B **6**, 2257 (1989)
59. T. Sleator, T. Pfau, V. Balykin, and J. Mlynek, Appl. Phys. B **54**, 375 (1992)
60. V. Balykin et al., J. Mod. Opt. **35**, 17 (1988)
61. G. M. Gallatin and P. L. Gould, J. Opt. Soc. Amer. B **8**, 502 (1991)
62. R. J. Cook and R. K. Hill, Opt. Comm. **43**, 258 (1982)
63. R. Kaiser, et al., Opt. Comm. **104**, 234 (1993)
64. P. J. Martin, B. G. Oldaker, A. H. Miklich, and D. E. Pritchard, Phys. Rev. Lett. **60**, 515 (1988)
65. K. S. Johnson, A. Chu, T. W. Lynn, K. K. Berggren, M. S. Shahriar, and M. Prentiss, Opt. Lett. **20**, 1310 (1995)
66. J. Lawall and M. G. Prentiss, Phys. Rev. Lett. **72**, 993 (1994)

28   A. Ashkin

67. M. Weitz, B. C. Young, and S. Chu, Phys. Rev. Lett. **72**, 2563 (1994)
68. G. Timp, R. E. Behringer, D. M. Tennant, J. E. Cunningham, M. Prentiss, and K. Berggren, Phys. Rev. Lett. **69**, 1636 (1992)
69. J .J. McClelland, R. E. Scholten, E. C. Palm, and R. J. Celotta, Science **262**, 877 (1993)
70. V. I. Balykin and V. S. Letokhov, Physics Today **42**, 23 (1989)
71. M. Kasevich and S. Chu, Phys. Rev. Lett. **67**, 181 (1991)
72. M. H. Anderson, J. R. Ensher, M. R. Mathews, C. E. Wieman, and E. A. Cornell, Science **269**, 198 (1995)
73. K. B. Davis, M.-O. Mewes, M. R. Andrews, N. J. van Druten, D. S. Durfee, D. M. Kurn, and W. Ketterle, Phys. Rev. Lett. **75**, 3969 (1996)
74. D. S. Jin, M. R. Mathews, J. R. Ensher, C. E. Wieman, and E. A. Cornell, Phys. Rev. Lett. **78**, 764 (1997)
75. D. M. Stamper-Kurn et al., Phys. Rev. Lett. **83**, 2876 (1999)
76. M. R. Andrew, C. G. Townsend, H.-J. Miesner, D. S. Durfee, D. M. Kurn, and W. Ketterle, Science **275**, 637 (1997)
77. J. Stenger et al., Phys. Rev. Lett. **82**, 4569 (1999)
78. M.-O. Mews et al., Phys. Rev. Lett. **78**, 582 (1997)
79. E. W. Hagley, L. Deng, M. Kozuma, J. Wen, K. Helmerson, S. L. Rolston, and W. D. Phillips, Science **283**, 1706 (1999)
80. L. Deng, E. W. Hagley, J. Wen, M. Trippenback, Y. Band, P. S. Julienne, J. E. Simsarian, K. Helmerson, S. L. Rolston, and W. D. Phillips, Nature **398**, 218 (1990)
81. S. Inouye, T. Pfau, S. Gupta, A. P. Chikkatur, A. Görlitz, D. E. Pritchard, and W. Ketterle, Nature **402**, 641 (1999)
82. M. Kozumi, Y. Suzuki, Y. Torii, T. Sugiwa, T. Kuga, E. W. Hagley and L. Deng, Science **286**, 2309 (1999)
83. D. M. Stamper-Kurn, M. R. Andrews, A. P. Chikkatur, S. Inouye, H.-J. Miesner, J. Stenger, and W. Ketterle, Phys. Rev. Lett. **80**, 2027 (1998)
84. J. Stenger, D.M. Stamper-Kurn, M.R. Andrews, A.P. Chikkatur, S. Inouye, H.-J. Miesner and W. Ketterle, J. Low Temp. Phys. **113**, 167 (1998)
85. S. Inouye, M. R. Andrews, J. Stenger, H.-J. Miesner, D. M. Stamper-Kurn, A. P. Chikkatur, and W. Ketterle, Nature **392**, 151 (1998)
86. J. Stenger, S. Inouye, M. R. Andrews, H.-J. Miesner, D. M. Stamper-Kurn, and W. Ketterle, Phys. Rev. Lett. **82**, 2422 (1999)
87. J. Stenger, S. Inouye, D. M. Stamper-Kurn, H.-J. Miesner, A. P. Chikkatur, and W. Ketterle, Nature **396**, 345 (1998)
88. D. M. Stamper-Kurn, H.-J. Miesner, A. P. Chikkatur, S. Inouye, J. Stenger, and W. Ketterle, Phys. Rev. Lett. **83**, 661 (1999)
89. W. Ketterle, Phys. Today **52**, 30 (1999)
90. A. Ashkin, J. M. Dziedzic, J. E. Bjorkholm, and S. Chu, Opt. Lett. **11**, 288 (1986)
91. A. Ashkin, Biophys. J. **61**, 569 (1992)
92. A. Ashkin and J. M. Dziedzic, Science **235**, 1517 (1987)
93. A. Ashkin, J. M. Dziedzic and T. Yamane, Nature **330**, 769 (1987)
94. A. Ashkin and J. M. Dziedzic, Proc. Natl. Acad. Sci. USA **86**, 7914 (1989)
95. F. H. C. Crick and A. F. W. Hughes, Exp. Cell Res. **1**, 37 (1950)
96. M. Sato, T. Z. Wong, and R. D. Allen, J. Cell Biol. **97**, 1089 (1983)
97. T. T. Perkins, D. E. Smith, and S. Chu, Science **264**, 822 (1994)

98. S. B. Smith, Y. Cui, and C. Bustamante, Science **271**, 795 (1996)
99. M. Sato, T. Z. Wong, and R. D. Allen, "A Preliminary Investigation of Living Physarum Endoplasm," in: The Application of Light Scattering to the Study of Biological Motion, J. C. Earnshaw and M. W. Stear (eds.) (Plenum Press, New York, 1983)
100. S. M. Block, D. F. Blair, and H. C. Berg, Nature **338**, 514 (1989)
101. S. M. Block, D. F. Blair, and H. C. Berg, Cytometry **12**, 492 (1991)
102. N. W. Charon, S. F. Goldstein, S. M. Block, K. Curci, and J. D. Ruby, J. Bacteriol. **174**, 832 (1992)
103. S. Seeger, S. Manojembaski, K. J. Hutter, G. Futterman, J. Wolfrum, and K. O. Greulich, Cytometry **12**, 497 (1991)
104. R. W. Steubing, S. Cheng, W. H. Wright, Y. Namajiri, and M. W. Berns, Cytometry **12**, 505 (1991)
105. X. Wei, B. J. Tromberg, and M. D. Cahalan, Proc. Natl. Acad. Sci. USA **96**, 18471 (1999)
106. H. Liang, W. H. Wright, W. He, and M. W. Berns, Exp. Cell Res. **204**, 110 (1993)
107. Y. Tadir, W. H. Wright, O. Vafa, T. Ord, R. H. Asch, and M. W. Berns, Fertility & Sterility **52**, 870 (1989)
108. J. M. Colon, P. Sarosi, P. G. McGovern, A. Ashkin, J. M. Dziedzic, J. Skurnick, G. Weiss, and E. M. Bonder, Fertility & Sterility **57**, 695 (1992)
109. E. M. Bonder, J. M. Colon, J. M. Dziedzic, and A. Ashkin, J. Cell Biol. **111**, 421a (1990)
110. Y. Tadir, W. H. Wright, O. Vafa, L. H. Liaw, R. Asch, and M. W. Berns, Human Reproduct. **6**, 1011 (1991)
111. K. Schütze, A. Clement-Sengewald, and A. Ashkin, Fertility & Sterility **61**, 783 (1994)
112. A. Clement-Sengewald, K. Schütze, A. Ashkin, G. A. Palma, G. Kerlen, and B. Brem, J. Assisted Reprod. & Genetics **13**, 259 (1996)
113. H. Liang, K. T. Vu, P. Krishnan, T. C. Trang, D. Shin, S. Kimel, and M. W. Berns, Biophys. J. **70**, 1529 (1996)
114. S. M. Block, L. S. B. Goldstein, and B. J. Schnapp, Nature **348**, 348 (1990)
115. K. Svoboda and S. M. Block, Ann. Rev. Biophys. Biomol. Struct. **23**, 247 (1994)
116. A. Ashkin, K. Schütze, J. M. Dziedzic, U. Eutenauer, and M. Schliwa, Nature **348**, 346 (1990)
117. S. C. Kuo and M. P. Sheetz, Science **260**, 232 (1993)
118. K. Svoboda, C. F. Schmidt, B. J. Schnapp, and S. M. Block, Nature **365**, 721 (1993)
119. S. M. Block, Trends in Cell Biology **5**, 169 (1995)
120. K. Svoboda and S. M. Block, Cell **77**, 773 (1994)
121. J. T. Finer, R. M. Simmons, and J. A. Spudich, Nature **368**, 113 (1994)
122. R. M. Simmons, J. T. Finer, S. Chu, and A. Spudich, Biophys. J. **70**, 1813 (1996)
123. J. E. Molloy, J. E. Burns, J. C. Sparrow, R. T. Tregear, J. Kendrick-Jones, and D. C. White, Biophys. J. **68**, 2985 (1995)
124. T. Nishizaka, H. Miyata, H. Yoshikawa, S. Ishiwata, and K. Kinosita Jr., Nature **377**, 251 (1995)
125. A. D. Mehta, R. S. Rock, M. Rief, J. A. Spudick, M. S. Moosekor, and R. E. Cheney, Nature **400**, 590 (1999)

126. H. Sakakibara, H. Kojima, Y. Sakai, E. Katayana, and K. Oiwa, Nature **400**, 586 (1999)

127. H. Kojima, E. Muto, H. Higuichi, and T. Yanagida, Biophys. J. **73**, 2012 (1997)

128. B. G. Levi, Phys. Today **48 No. 4**, 17 (1995)

129. K. Svoboda, P. P. Mitra, and S. M. Block, Proc. Natl. Acad. Sci. USA **91**, 11782 (1994)

130. M. J. Schnitzer and S. M. Block, Nature, **388**, 386 (1997)

131. A. E. Knight and J. E. Malloy, Nature Cell Biol. **1**, E87 (1999)

132. K. Visscher, M. J. Schnitzer, and S. M. Block, Nature **400**, 184 (1999)

133. K. Visscher and S. M. Block, Meth. Enzymol. **298**, 460 (1998)

134. R. D. Asturmian and M. Bier, Phys. Rev. Lett. **72**, 1766 (1994)

135. L. P. Faucheux, L. S. Bourdieu, P. D. Kaplan, and A. D. Libchaber, Phys. Rev. Lett. **74**, 1504 (1995)

136. H. Yin, M. D. Wang, K. Svoboda, R. Landick, S. M. Block, and J. Gelles, Science **270**, 1653 (1995)

137. M. D. Wang, M. J. Schnitzer, H. Yin, R. Landrick, J. Gelles, and S. M. Block, Science **282**, 902 (1998)

138. R. F. Service, Science **283**, 1668 (1999)

139. C. O'Brien, Science **270**, 1668 (1995)

140. A. D. Mehta, M. Rief, J. A. Spudick, D. A. Smith, and R. H. Simmons, Science **283**, 1689 (1999)

141. M. Kurachi, M. Hoshi, and H. Tashiro, Cell Motil. Cytoskel. **30**, 221 (1995)

142. H. Feigner, R. Frank, and M. Schliwa, J. Cell Sci. **109**, 509 (1996)

143. Y. Tsuda, H. Yasutake, A. Ishijima, and T. Yanagida, Proc. Natl. Acad. Sci. USA **93**, 12937 (1996)

144. T. T. Perkins, D. E. Smith, and S. Chu, Science **264**, 822 (1994)

145. S. B. Smith, Y. Cui, and C. Bustamante, Science **271**, 795 (1996)

146. G. V. Shivashankar, M. Feingold, O. Krichevsky, and A. Libchaber, Proc. Natl. Acad. Sci. USA **96**, 7916 (1999)

147. M. D. Wang, H. Yin, R. Landrick, J. Gelles, and S. M. Block, Biophys. J. **72**, 1335 (1997)

148. R. F. Service, Science **283**, 1669 (1999)

149. E. M. Furst and A. P. Gast, Phys. Rev. Lett. **83**, 4130 (1999)

150. M. A. Welte, S. P. Gross, M. Postner, S. M. Block, and E. F. Wieschaus, Cell, **92**, 547 (1998)

151. R. Huber, S. Burggraf, T. Mayer, S. M. Barns, P. Rossnagel, and K. O. Stetter, Nature **376**, 57 (1995)

152. K. E. Nelson et al., Nature **399**, 323 (1999)

153. M. Milstein, Science **270**, 226 (1995)

154. S. Barker, Nature **381**, 455 (1996)

155. J. G. Mitchell, R. Weller, M. Beconi, J. Sell, and J. Holland, Microbial Ecology **25**, 113 (1993)

156. J. K. Burkhardt, J. M. McIlvain Jr., M. P. Sheetz, and Y. Argon, J. Cell Sciences **104**, 151 (1993)

157. C. E. Schmidt, A. F. Horwitz, D. A. Lauffenburger, and M. P. Sheetz, J. Cell Biol. **123**, 977 (1993)

158. X. F. Wang, J. J. Lemaster, B. Herman, and S. Kuo, Opt. Engin. **32**, 284 (1993)

159. J. Dai and M. P. Sheetz, Biophys. J. **68**, 988 (1995)
160. K. Svoboda, C. F. Schmidt, D. Branton, and S. M. Block, Biophys. J. **63**, 784 (1992)
161. P. J. H. Bronkhorst, G. J. Streekstra, J. Grimbergen, E. J. Nijhof, J. J. Sixma, and G. J. Brakenhoff, Biophys. J. **69**, 1666 (1995)
162. K. Visscher, G. J. Brakenhoff, and J. J. Krol, Cytometry **14**, 105 (1993)
163. G. Leitz, E. Schnepf, and K. O. Greulich, Planta **197**, 278 (1995)
164. M. Mammer, K. Helmerson, R. Kishore, C. Seok-Ki, W. D. Phillips, and G. M. Whitesides, Chem. & Biology **3**, 757 (1996)
165. A. Simon and A. Libchaber, Phys. Rev. Lett. **68**, 3375 (1992)
166. A. Choudhury, B. J. Ackerson, and N. A. Clark, Phys. Rev. Lett. **55**, 833 (1985)
167. E. Higurashi, H. Ukita, H. Tanaka, and O. Ohguchi, Appl. Phys. Lett. **64**, 2209 (1994)
168. K. Svoboda and S. M. Block, Opt. Lett. **19**, 930 (1994)
169. L. P. Ghislain and W. W. Webb, Opt. Lett. **18**, 1678 (1993)
170. A. D. Dinsmore, A. G. Yodh, and D. J. Pine, Nature **383**, 239 (1996)
171. J. C. Crocker, J. A. Mattes, A. D. Dinsmore, and A. G. Yodh, Phys. Rev. Lett. **82**, 4352 (1999)
172. H. Masuhara, F. C. deSchryver, N. Kitamura, and N. Tamai, Microchemistry – Spectroscopy and Chemistry in Small Domains (North-Holland Delta Series) (1994)
173. R. Bar-Ziv, R. Menes, E. Moses, and S. A. Safran, Phys. Rev. Lett.**75**, 3356 (1995)
174. R. Bar-Ziv, T. Frisch, and E. Moses, Phys. Rev. Lett. **75**, 3481 (1995)
175. A. E. Larsen and D. G. Grier, Nature **385**, 230 (1997)
176. C. A. Murray, Nature **385**, 203 (1997)
177. "Frontiers in Chemistry: Single Molecules," Science **283**, 1667 (1999)
178. M. Helmer, "Single take on molecules," Nature **401**, 225 (1999)

2 Thirteen Years of Single-Molecule Spectroscopy in Physical Chemistry and Biophysics

W. E. Moerner

2.1 Why Study Single Molecules in Condensed Phases?

The Nobel Conference on *Single Molecule Spectroscopy in Physics, Chemistry and Biology* was held at the Södergarn Mansion in Stockholm, Sweden on June 5–9, 1999. This meeting gathered researchers from all over the globe who utilize the optical properties of individual molecules to explore a wide range of problems spanning numerous fields of science. The breadth of interest and large number of workers in this new field provided a particularly stimulating intellectual environment for all. Thanks are due to the Nobel Foundation for making this conference possible, and in particular to the organizer, Professor Rudolf Rigler, for preparing a most exciting and enjoyable event.

In this paper, selected accomplishments in the area of optical single-molecule detection and spectroscopy are reviewed, with examples taken from the work of the Moerner group at IBM Research (1987–1995), at the University of California San Diego (1995–1998), and at Stanford University (1998–present). After a brief summary of the introductory concepts, the regime of liquid-helium-temperature spectroscopy of single molecules in solids is described, followed by a summary of more recent applications of single-molecule optical techniques to chemical, biophysical, and quantum optical problems at room temperature.

Single-molecule spectroscopy (SMS) allows *exactly one* molecule hidden deep within a condensed phase sample to be observed by tunable optical radiation (Fig. 2.1a). This represents detection and spectroscopy at the ultimate sensitivity level of $\sim 1.66 \times 10^{-24}$ moles of the molecule of interest (1.66 yoctomole), or a quantity of moles equal to the inverse of Avogadro's number. Detection of the single molecule of interest must be done in the presence of billions to trillions of solvent or host molecules. To achieve this, a light beam (typically a laser) is used to pump an electronic transition of the one molecule resonant with the optical wavelength (Fig. 2.1b), and either the resulting optical absorption [1] or fluorescence [2] is detected. Several comprehensive reviews of this area may be consulted for details beyond the scope of this article [3-13].

How can SMS provide new information? Clearly, standard ensemble measurements which yield the average value of a parameter for a large number of (presumably identical) copies of the molecule of interest still have great

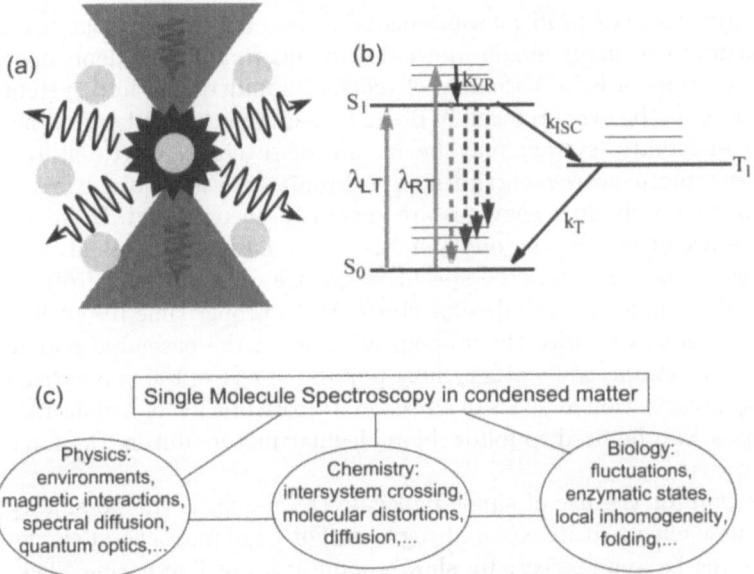

Fig. 2.1. a Schematic of an optical beam pumping a single resonant molecule, which subsequently emits fluorescence. **b** Typical energy level scheme for single-molecule spectroscopy. S_0, ground singlet state; S_1, first excited singlet; T_1, lowest triplet state or other intermediate state. For each electronic state, several levels in the vibrational progression are shown. Typical low-temperature studies use wavelength λ_{LT} to pump the (0–0) transition, while at room temperature shorter wavelengths λ_{RT} are more common. The intersystem crossing or intermediate production rate is k_{ISC}, and the triplet decay rate is k_T. Fluorescence emission shown as dotted lines originates from S_1 and terminates on various vibrationally excited levels of S_0 or S_0 itself. **c** Some connections associated with single-molecule spectroscopy

value. By contrast, SMS completely removes the ensemble averaging, which allows construction of a frequency histogram of the actual distribution of values for an experimental parameter. All would agree that the distribution contains more information than the average value alone. The shape of the full distribution can be examined to see if it has multiple peaks, or whether it has a strongly skewed shape. Such details of the underlying distribution become crucially important when the system under study is heterogeneous. This would be expected to be the case for many complex condensed matter environments such as real crystals, polymers, or glasses. Fortunately, a single molecule can be a local reporter of its "nanoenvironment", that is, the exact constellation of functional groups, atoms, ions, electrostatic charges and/or other sources of local fields in its immediate vicinity. For biomolecules, heterogeneity easily arises, for example, if the various individual copies of a protein or oligonucleotide are in different folded states, different configurations, or different states of an enzymatic cycle.

Another advantage of SMS measurements is that they remove the need for synchronization of many single molecules undergoing a time-dependent process. For example, a large ensemble of molecules undergoing intersystem crossing events must be synchronized in order to measure the triplet lifetime. Similarly, an enzymatic system may be in one of several catalytic states, and in an ensemble measurement initial synchronization is required but is quickly lost as the individual enzymes are generally uncorrelated. However, if single copies are observed, any one member of the ensemble is in only one state at a given time, and thus the specific sequence of binding, hydrolysis, and other catalytic steps is available for study. With proper time resolution, rare intermediates can be directly probed, whereas in the ensemble regime these forms can be swamped by other, more populated forms. Using polarized excitation and polarization analysis of emission, the orientation of a molecular transition dipole can be used to follow biomolecular motors during the force generation process.

A final reason for the use of single-molecule techniques is the possibility of observing new effects in unexplored regimes. For example, several single-molecule systems have unexpectedly shown some form of fluctuating, flickering, or stochastic behavior [14]. The absorption frequency of the single molecule can change as a result a change in its photophysical parameters or a change in local environment; this behavior has been termed "spectral diffusion," and it can produce spectral shifts or fluctuations. Such fluctuations are now becoming important diagnostics of the single-molecule regime, and they provide unprecedented insight into behavior which is generally obscured by ensemble averaging.

Stepping back for a moment, it is clear that the last few decades have witnessed a dramatic increase in interest in the "nanoworld" of single atoms, ions, and molecules, for both scientific and technological reasons. Indeed, SMS as defined is related to, but distinct from, several recently successful lines of research on individual species: (i) the spectroscopy of single electrons or ions confined in electromagnetic traps [15–17], (ii) scanning tunneling microscopy (STM) [18] and atomic force microscopy (AFM) [19], (iii) the study of ion currents in membrane-embedded single channels [20], (iv) single polymer studies with high concentrations of fluorophores [21], and (v) force measurements on single molecular motors using optical traps [22]. Perhaps the closest relative to SMS is the area of fluorescence correlation spectroscopy (FCS) [23–25], in which dynamical properties are sensed during the diffusion-driven passage of a low concentration of the molecules of interest through the focus of a laser beam. FCS provides useful dynamical information in the ns–ms range, but to obtain a high contrast autocorrelation, the dynamics of many molecules must be summed.

At present, the impact of SMS spans several fields, from physics, to chemistry, to biology (Fig. 2.1c), and a selection of examples will be described in this paper and in other papers in this volume.

2.2   Basic Principles of Single-Molecule Spectroscopy

To put it simply, to achieve SMS at any temperature, one must: (a) guarantee that only one molecule is in resonance in the volume probed by the laser, and (b) provide a signal-to-noise ratio (SNR) for the single-molecule signal that is greater than unity for a reasonable averaging time.

Guaranteeing only one molecule in resonance is generally achieved by dilution. For example, at room temperature one need only work with roughly 10^{-10} mole/liter concentration in a probed volume of 10 μm^3. At liquid helium temperatures, the phenomenon of inhomogeneous broadening ([26,27], described in the next section) can be used to achieve dilution factors from $\sim 10^4$–10^5 simply by tuning the laser frequency to a spectral region where only one molecule is in resonance.

Achieving the required SNR can be done by several methods. To obtain as large a signal as possible, one needs a combination of small focal volume, large absorption cross section, high photostability, weak bottlenecks into dark states such as triplet states, operation below saturation of the molecular absorption, and high fluorescence quantum yield if fluorescence is detected. For absorption methods, achieving a low noise level from background effects follows from careful reduction of residual signals and operation at a power level sufficient to reduce the relative contribution from laser shot noise. For fluorescence methods, one must rigorously exclude fluorescent impurities, minimize the volume probed to avoid Raman scattering, and scrupulously reject any scattered radiation at the pumping wavelength.

Several optical configurations have been demonstrated to satisfy these requirements. At low temperatures, high resolution optical methods can be applied such as laser frequency-modulation spectroscopy [1] or fluorescence excitation spectroscopy [2]. At room temperature, one can detect the burst of light as a molecule passes through the focus of a laser beam [28], or one can use optical microscopy to observe the same single molecule for an extended period, measuring signal strength, lifetime, polarization, fluctuations, and so on, all as a function of time. Successful microscopic techniques include scanning methods such as near-field scanning optical microscopy (NSOM) [29] and confocal microscopy [30], as well as the wide-field methods of epifluorescence [31] and total internal reflection microscopy [32].

Table 2.1 lists specific SMS milestones of the Moerner group for the low-temperature regime, and a selection of these will now be described.

2.3   Early Steps toward Single-Molecule Spectroscopy

Our first approach to the single-molecule regime arose from work at IBM in the early 1980's on persistent spectral hole-burning effects in the optical transition of impurities in solids (for a review, see [33]). Briefly, if a molecule with a strong zero-phonon transition and minimal Franck–Condon distortion

Table 2.1. Low-temperature single-molecule spectroscopy milestones: Moerner group

Experiment	Refs.
Observation of statistical fine structure scaling as $\sqrt{N}$	[34]
Optical detection and spectroscopy of a single molecule in a solid	[1][37]
Optical dephasing, nonlinear optical saturation	[43]
Spectral diffusion and measurement of spectral trajectories	[37][43]
Photon antibunching for a single molecule	[61]
Single molecule in a polymer, photoinduced kinetics	[46][55]
Vibrational spectroscopy (resonance Raman)	[67][68]
Magnetic resonance of a single molecular spin	[64]
Near-field optical spectroscopy of a single molecule in a solid	[70]
Pumping single molecules with morphology-dependent resonances of a microsphere	[71]

is doped into a solid and cooled to liquid helium temperatures, the optical absorption becomes inhomogeneously broadened (Fig. 2.2a). The width of the lowest electronic transition for any one molecule becomes very small because few phonons are present, while at the same time different copies of the impurity molecule acquire slightly different absorption energies due to local strains and other defects in the solid. Spectral hole-burning occurs when light-driven physical or chemical changes are produced only in those molecules resonant with the light, which yields a dip or "spectral hole" in the overall absorption profile that may be used for optical recording of information in the optical frequency domain.

One goal of the research in the Moerner group at IBM was the exploration of ultimate limits to the spectral hole-burning optical storage process. A particularly interesting limit on the signal-to-noise ratio of a spectral hole results from the finite number of molecules that contribute to the absorption line near the hole. Due to unavoidable number fluctuations in the density of molecules in any spectral interval, there should exist a "spectral noise" on an inhomogeneous absorption profile scaling as the square root of the number of molecules in resonance. We named this effect "statistical fine structure" (SFS), and Fig. 2.2a shows how the relative size of SFS scales as $1/\sqrt{N}$, while the absolute root-mean-square (rms) size of the fine structure scales as $\sqrt{N}$. Surprisingly, prior to the late 1980's, SFS had not been detected.

In 1987, SFS was observed for the first time [34], using a powerful zero-background absorption technique, laser frequency-modulation (FM) spectroscopy [35] and the system composed of pentacene dopant molecules in a p-terphenyl crystal. FM spectroscopy probes the sample with a phase-modulated laser beam; when a narrow spectral feature is present, the imbalance in the laser sidebands leads to amplitude modulation in the detected photocurrent at the modulation frequency. A key feature of the method is

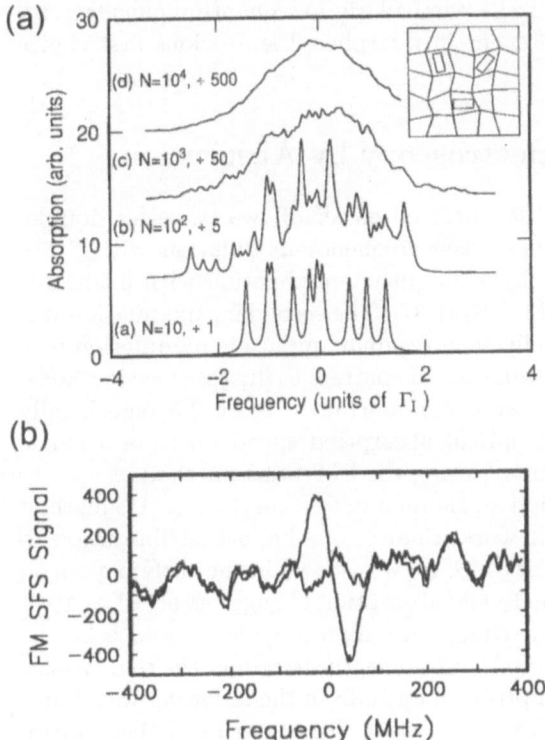

Fig. 2.2. (**a**, upper) Illustration of the source of SFS using simulated absorption spectra with different total numbers of absorbers N, where a Gaussian random variable provides center frequencies for the inhomogeneous distribution. Subtraces (a) through (d) correspond to N values of 10, 100, 1000, and 10,000, respectively, and the traces have been divided by the factors shown. The traces have been vertically offset for clarity, $\gamma_H = \Gamma_I/10$. Inset: several guest molecules are sketched as rectangles with different local environments produced by strains, local electric fields, and other imperfections in the host matrix. (**b**, lower) SFS detected by FM spectroscopy for pentacene in p-terphenyl at 1.4 K, with a spectral hole at zero relative frequency for one of the two scans

that it senses only the deviations of the absorption from the average value, so that detection of SFS could be easily accomplished. In Fig. 2.2b, SFS is the repeatable spectral noise over the entire range of the two scans, one of which includes a spectral hole burned at the center, showing directly that the size of a spectral hole must be larger than the SFS in order to be detected [36]. SFS is clearly an unusual spectral feature, in that its *size* depends not upon the total number of resonant molecules, but rather upon the square root of the number. Due to three factors, (a) $\sqrt{N}$ scaling, (b) insensitivity to any scattering background and (c) quantum-limited detection sensitivity, FM spectroscopy seemed an appropriate method to push to the single-molecule limit. It must also be stated that the particularly low quantum efficiency for spectral hole-burning made pentacene in p-terphenyl an obvious first choice for single-molecule detection.

2.4 Single-Molecule Spectroscopy by Absorption

The first SMS experiments in 1989 utilized either of two powerful double-modulation absorption techniques, laser frequency-modulation with Stark secondary modulation (FM–Stark) or frequency-modulation with ultrasonic strain secondary modulation (FM–US) [1,37]. The secondary modulation was required in order to remove the effects of residual amplitude modulation produced by the imperfect phase modulator. In contrast to fluorescence methods, Rayleigh and Raman scattering were unimportant. Figure 2.3 (specifically trace d) shows examples of the optical absorption spectrum from a single molecule of pentacene in p-terphenyl using the FM–Stark method.

Although this early observation and similar data from the FM–US method served to stimulate much further work, there is one important limitation to the general use of FM methods for SMS. As was shown in the early papers on FM spectroscopy [35,38], extremely low absorption changes as small as 10^{-7} can be detected in a 1 s averaging time, but only if large laser powers on the order of several mW can be delivered to the detector to reduce the relative size of the shot noise. This presents a problem for SMS in the following way. Since the laser beam must be focused to a small spot, the power in the laser beam must be maintained below the value which would cause saturation broadening of the single-molecule lineshape. As a result, it is quite difficult to utilize laser powers in the mW range for SMS of allowed transitions at low temperatures – in fact powers below 100 nW are generally required. This is one reason why the SNR of the original data on single molecules of pentacene in p-terphenyl in Fig. 2.3 was only on the order of 5. (The other reason was the use of relatively thick cleaved samples, which produced a larger number of weak out-of-focus molecules in the probed volume. This problem has been overcome with much thinner samples in modern experiments.) In recent experiments [39], frequency modulation of the absorption line itself (rather than the laser) was produced by an oscillating (Stark) electric field alone, and this method has also been used to detect the absorption from a single molecule at liquid helium temperatures. While successful, these methods are limited by the quantum shot noise of the laser beam, which is relatively large at the low laser intensity required to prevent saturation of the single-molecule absorption.

2.5 Single-Molecule Spectroscopy
by Fluorescence Excitation

The optical absorption experiments on pentacene in p-terphenyl showed that this system has sufficiently weak hole-burning that it should be a useful model

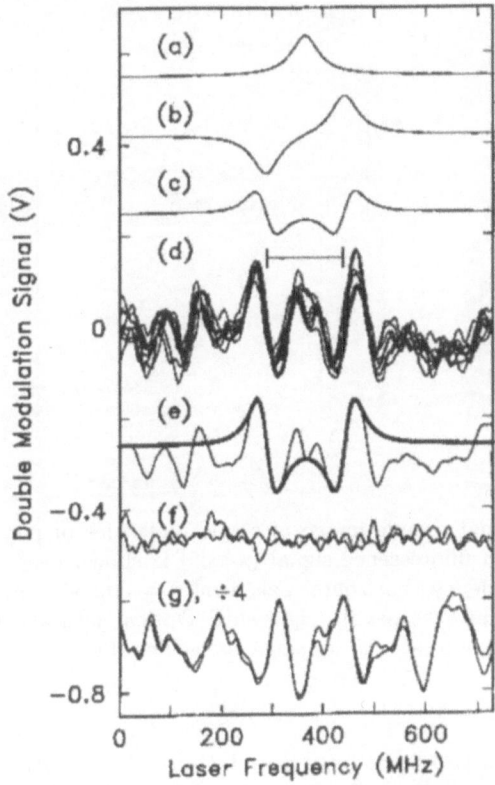

Fig. 2.3. Single molecules of pentacene in p-terphenyl detected by FM-Stark optical absorption spectroscopy. The "W"-shaped structure in the center of trace (d) is the absorption from a single pentacene molecule. For details, see [1]

system for single-molecule spectroscopy. In 1990, Orrit et al. demonstrated that fluorescence excitation produces superior signal-to-noise if the emission is collected efficiently and the scattering sources are minimized [2]. Subsequent experiments have almost exclusively used this method. In fluorescence excitation, a tunable narrowband single-frequency laser is scanned over the absorption profile of the single molecule, and the presence of absorption is detected by measuring the fluorescence emitted. A long-pass filter is used to block the pumping laser light, and the fluorescence shifted to long wavelengths from the tunable pump is detected with a photon-counting system. The detection is background-limited and the shot noise of the probing laser is only important for the signal-to-noise of the spectral feature, not the signal to background. For this reason, it is critical to efficiently collect photons (as with a paraboloid or other high numerical aperture collection system), and to reject the pumping laser radiation. To illustrate, suppose a single molecule of pentacene in p-terphenyl is probed with 1 mW/cm^2, near the onset

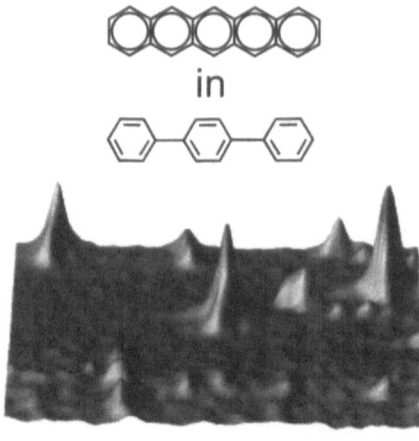

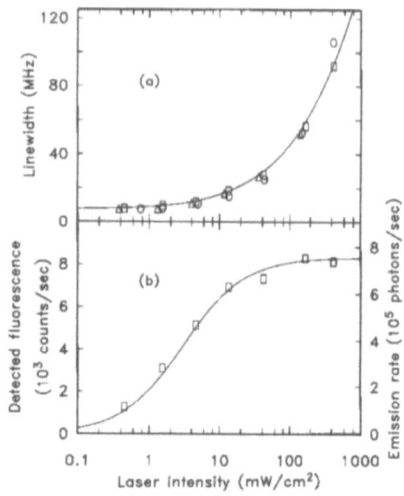

Fig. 2.4. (left side) Three-dimensional pseudo-image of single molecules of pentacene in p-terphenyl. The measured fluorescence signal (z-axis) is shown over a range of 300 MHz in excitation frequency (horizontal axis, center = 592.544 nm) and 40 µm in spatial position (axis into the page). (right side) Optical saturation behavior of the linewidth (**a**) and emission rate (**b**) for single molecules of pentacene in p-terphenyl. For details, see [44]

of saturation of the absorption due to triplet level population. The resulting incident photon flux of 3×10^{15} photons/s-cm^2 will produce about 3×10^4 excitations per second. With a fluorescence quantum yield of 0.8 for pentacene, about 2.4×10^4 emitted photons/s can be expected. At the same time, 3×10^8 photons/s illuminate the focal spot 3 µm in diameter. Considering that the resonant 0–0 fluorescence from the molecule must be thrown away along with the pumping light, rejection of the pumping radiation by a factor greater than 10^5 to 10^6 is generally required, with minimal attenuation of the fluorescence. This is often accomplished by low-fluorescence glass filters or by holographic notch attenuation filters.

In early experiments on the model system of pentacene in p-terphenyl at low pumping intensity, the lifetime-limited homogeneous linewidth of 7.8 ± 0.2 MHz was observed [40]. This linewidth is the minimum value allowed by the lifetime of the S_1 excited state of 24 ns in agreement with photon echo measurements on large ensembles [41,42]. Such narrow single-molecule absorption lines are wonderful for the spectroscopist: many detailed studies of the local environment can be performed, because such narrow lines are much more sensitive to local perturbations than are broad spectral features.

A hybrid image of a single molecule can be obtained by acquiring spectra as a function of the position of the laser focal spot in the sample. Figure 2.4 (left side) shows such a three-dimensional "pseudo-image" of single molecules of pentacene in p-terphenyl [43]. The z-axis of the image is the (Stokes-shifted) emission signal, the horizontal axis is the laser frequency detuning (300 MHz range), and the axis going into the page is one transverse spatial dimension produced by scanning the laser focal spot (40 μm range). For sections of this image along the spatial dimension the single molecule is actually serving as a highly localized nanoprobe of the laser beam diameter itself (here ~5 μm). However, in the frequency dimension the features are fully resolved as the laser linewidth is negligible (~3 MHz).

Single-molecule spectra as a function of laser intensity provide details of the incoherent saturation behavior and the influence of the dark triplet-state dynamics. Figure 2.4 (right side) shows both the linewidth and the detected emission rate for three single molecules in the wings of the inhomogeneous line, yielding a (free-space) value for the saturation intensity of $I_s = 2.5$ mW/cm^2. This value is smaller than expected from the ensemble-measured intersystem crossing parameters [44], indicating that the individual molecules experience modifications due to differences in local environments. In the high intensity limit, the peak fluorescence emission rate saturates at $7.2 \pm 0.7 \times 10^5$ photons/second, and the homogeneous linewidth broadens according to the expected form. It was also possible to measure the linewidths of single pentacene molecules as a function of temperature in order to probe dephasing effects produced by coupling to a local phonon mode [44].

2.6   Observation of Spectral Diffusion of a Single Molecule

In the course of the early SMS studies on pentacene in p-terphenyl in the Moerner group, an unexpected phenomenon appeared: resonance frequency shifts of individual pentacene molecules in a crystal at 1.5 K [40,43], called "spectral diffusion" in deference to similar shifting behavior long postulated for amorphous systems [45] (although this behavior is not diffusion in the strict sense, i.e., it is not governed by a diffusion equation). Here, spectral diffusion means changes in the center (resonance) frequency of a the probe molecule due to configurational changes in the nearby host which affect the frequency of the electronic transition via guest–host coupling. For example, Fig. 2.5a shows a sequence of fluorescence excitation spectra of a single pentacene molecule in p-terphenyl taken as fast as allowed by the available SNR. The spectral shifting or hopping of this molecule from one resonance frequency to another from scan to scan is clearly evident.

Spectral shifting can be studied by (a) recording the observed lineshapes for many single molecules [46], (b) autocorrelation [47], and (c) measurement of the *spectral trajectory* $\omega_o(t)$ [43]. To record $\omega_o(t)$, one analyzes many sequentially acquired fluorescence excitation spectra like those in Fig. 2.5a and plots a trajectory or trend of the center frequency versus time as shown in Fig. 2.5b. The spectral trajectory shows that, for this molecule, the opti-

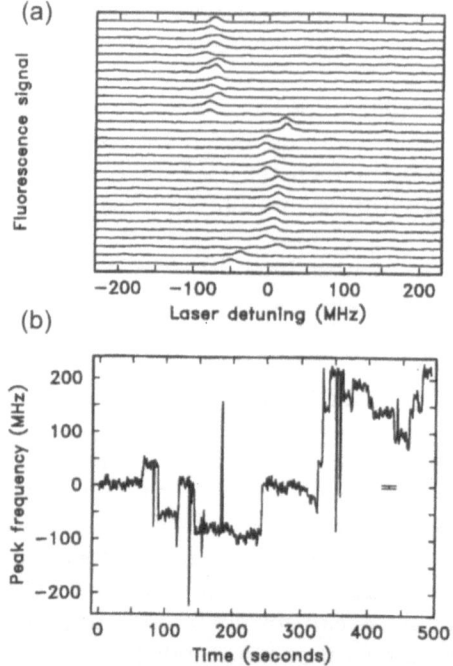

(a)

Fluorescence signal

Laser detuning (MHz)

(b)

Peak frequency (MHz)

Time (seconds)

Fig. 2.5. Examples of single-molecule spectral diffusion for pentacene in p-terphenyl at 1.5 K. **a** A series of fluorescence excitation spectra each 2.5 s long spaced by 0.25 s showing discontinuous shifts in resonance frequency, with zero detuning = 592.546 nm. **b** Trend or trajectory of the resonance frequency over a long time scale for the molecule in **a**. For details, see [43]

cal transition energy appears to have a preferred set of values and performs spectral jumps between these values that are discontinuous on the 2.5 s time resolution of the measurement. The behavior of other single molecules was qualitatively and quantitatively different, ranging from a wandering to creeping [44]. This is a direct demonstration of hidden heterogeneity from molecule to molecule that can only be observed by SMS – such spectral trajectories cannot be obtained when a large ensemble of molecules is in resonance. The individual jumps are generally uncorrelated, thus the behavior of an ensemble-averaged quantity such as a spectral hole in this system would show only a broadening and smearing of the line.

The first question which should be asked when such behavior is observed is this: is the effect spontaneous, occurring even in the absence of the probing laser radiation, or is it light-driven, i.e., produced by the optical excitation itself? To distinguish spontaneous vs. light-driven behavior, the spectral diffusion was explored as a function of laser power, and for pentacene in p-terphenyl crystals, the effect was found to be predominantly spontaneous. Since the single-molecule absorption is extremely sensitive to the local strain field, it is reasonable to expect that the spectral jumps are due to internal dynamics of some configurational degrees of freedom in the surrounding lattice, driven by the phonons present at the experimental temperature. The situation is analogous to that for amorphous systems, where the local dynamics result from the two-level systems near the guest. The dynamics results from phonon-assisted tunneling or thermally activated barrier crossing. One possible source for the tunneling states in this crystalline system could be discrete torsional librations of the central phenyl ring of the nearby p-terphenyl molecules about the molecular axis. The p-terphenyl molecules in a domain wall between two twins or near lattice defects may have lowered barriers to such central-ring tunneling motions. A theoretical study of the spectral diffusion trajectories [48–50] has allowed determination of specific defects that can produce this behavior, attesting to the power of SMS in probing details of the local nanoenvironment.

Spectral shifts of single-molecule lineshapes are common in many systems, appearing not only for certain crystalline hosts, but also for essentially all polymers studied (*vide infra*), and even for polycrystalline Shpol'skii matrices [51]. Moreover, spectral shifting has been observed on a much larger frequency scale for single molecules on surfaces at room temperature for both near-field [52] and far-field studies [53]. This behavior is turning out to be a ubiquitous feature of single-molecule experiments [14], giving us a new window into local dynamical behavior.

2.7 Single Molecules in Polymers: Light-Induced Spectral Shifts

The experiments described so far were performed with single impurity molecules doped into a crystalline matrix. Amorphous systems such as glasses or polymers have a number of interesting physical properties at low temperatures which are quite different from those for crystalline materials, in particular a complex, multidimensional potential energy surface [54]. According to the two-level system (TLS) model, the material can be approximated by a distribution of asymmetric double-well potentials in which only the two lowest energy levels of each double well are important. The effect of these TLS's on thermal and optical properties has been an area of extensive study. In 1992, perylene doped into a poly(ethylene) matrix became the first polymeric system to allow SMS [46,55]. Spectral diffusion manifested itself in this system in two different ways: as discontinuous jumps in frequency space on the scale of several hundred MHz similar to pentacene in p-terphenyl discussed above, and in addition, the observed linewidths varied from molecule to molecule due to fast shifts of small amplitude.

Light-driven shifts in absorption frequency were also observed, in which the rate of the process clearly increased with increases in laser intensity. This effect may be called "spectral hole-burning" by analogy with the earlier hole-

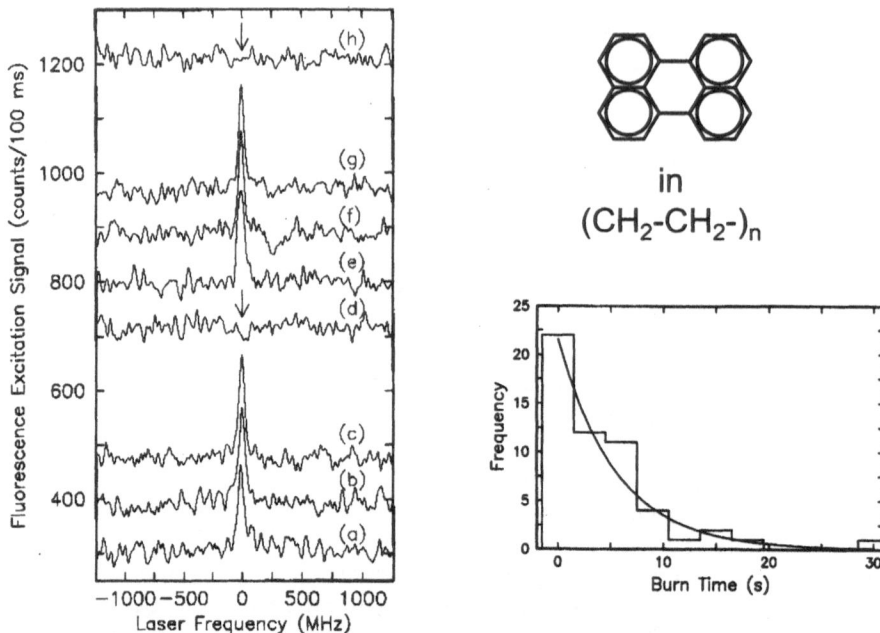

Fig. 2.6. Left side: Illustration of light-induced reversible frequency shifting for a single molecule of perylene in poly(ethylene) at 1.4 K. Right side: Histogram of waiting times before the spectral shift (hole-burning), from 53 events for the same single perylene molecule. For details, see [46,55]

burning literature [33], however, since here only one molecule is in resonance with the laser, the absorption line simply disappeared. An example is presented in Fig. 2.6, left side. Traces (a), (b), and (c) show three scans of one perylene molecule. After trace (c) the laser was tuned into resonance with the molecule until the fluorescence dropped (hole-burning). Trace (d) was then acquired, which showed that the resonance frequency shifted by more than ± 1.25 GHz as a result of the light-induced change in the nearby environment. Surprisingly, this effect was reversible for some molecules: a further scan some minutes later (trace (e)) showed that the molecule returned to the original absorption frequency. After trace (g) the molecule was burnt again and the whole sequence could be repeated many times.

The possibility to burn one and the same molecule several times was used to measure the kinetics of this process [55]. By measuring a large number of burning events for one perylene molecule in the poly(ethylene) host it was found that the various burning times are exponentially distributed, suggesting that the underlying process obeys Poisson statistics and is Markovian (Fig. 2.6 right side). It is fairly clear that the spectral shift is caused by changes in the states of one or more TLSs coupled to the perylene electronic transition; however, the exact microscopic mechanism needs further study and may be related to the generation of molecular internal vibrational modes during fluorescence emission or to nonradiative decay of the excited state. Several single-molecule systems have shown light-induced shifting behavior, for example, terrylene in poly(ethylene) [56], and terrylene in a Shpol'skii matrix [57], and it is hoped that future detailed study of this effect will shed light on the microscopic mechanisms of nonphotochemical hole-burning.

In principle, single-molecule hole-burning is a controllable process which could allow modification of the transition frequency of any arbitrary chosen molecule in the polymer host. This leads naturally to the possibility of optical storage at the single-molecule level. Such an idea would require more careful control of the spectral locations of the various molecules (formatting) as well as measures to deal with the exponential waiting times. In spite of these problems, optical modification of single-molecule spectra not only provides a unique window into the photophysics and low-temperature dynamics of the amorphous state, but it also allows such novel optical storage schemes to be considered.

2.8 Correlation Properties of Single-Molecule Emission; Photon Antibunching

The stream of photons emitted by a single molecule contains information about the system encoded in the arrival times of the individual photons. Figure 2.7a schematically shows the time-domain behavior of the photon stream for a single molecule with a dark triplet state, here taken to be pentacene. While cycling through the singlet states $S_0 \rightarrow S_1 \rightarrow S_0$, photons are emitted until intersystem crossing occurs. Since the triplet yield is 0.5%, on average 200 photons are emitted in a "bunch" before a dark period which has an average length equal to the triplet lifetime, τ_T. The corresponding decay in the autocorrelation function of the emitted photons for pentacene in p-terphenyl is easily observed [2], and this phenomenon has been used to measure the changes in the triplet yield and triplet lifetime from molecule to molecule which occur as a result of distortions of the molecule by the local nanoenvironment [58]. Such correlation measurements can also extract information on wide time scales about the spectral shifting behavior which occurs in amorphous systems [47] (*vide supra*). Although this method gives access to many decades in time, the dynamical process must be stationary, that is, the dynamics must not change during the relatively long time (many s) needed to record enough photon arrivals to generate a valid autocorrelation.

By contrast, in the nanosecond time regime within a single bunch (Fig. 2.7b), the emitted photons from a single quantum system are expected to show antibunching, which means that the photons "space themselves out in time", that is, the probability for two photons to arrive at the detector at the same time is small. This is a uniquely quantum-mechanical effect [59],

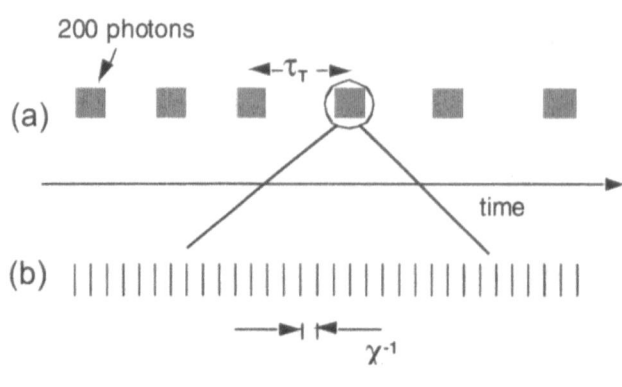

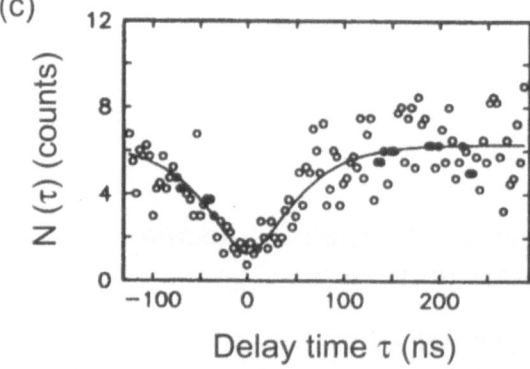

Fig. 2.7. Schematic of the temporal behavior of photon emission from a single molecule showing (**a**) bunching on the scale of the triplet lifetime and (**b**) antibunching on the scale of the inverse of the Rabi frequency. **c** Measured distribution of time delays between successive detected fluorescence photons for a single molecule of pentacene in p-terphenyl showing antibunching at $\tau = 0$. For details, see [61]

which was first observed for single Na atoms in a low-density beam [60]. For a single molecule, antibunching is easy to understand as follows. After photon emission, the molecule is definitely in the ground state and cannot emit a second photon immediately. A time on the order of the inverse of the Rabi frequency must elapse before the probability of emission of a second photon is appreciable.

To observe antibunching correlations, the second-order correlation function $g^{(2)}(t)$ is generally measured by determining the distribution of time delays $N(\tau)$ between the arrival of successive photons in a dual-beam detector. Photon antibunching in single-molecule emission was first observed for the pentacene in p-terphenyl model system [61], demonstrating that quantum optics experiments can be performed in solids and on molecules for the first time (Fig. 2.7c). The high-contrast dip at $\tau = 0$ is strong proof that the spectral feature in resonance is indeed that of a single molecule. This observation has opened the door to a variety of other quantum-optical experiments with single molecules [62], such as measurement of the AC Stark shift [63]. The convenient "trap" that the solid forms for a single molecule will allow further studies of the interactions of single molecules with the quantum radiation field.

2.9   Magnetic Resonance of a Single Molecular Spin

Historically, the standard methods of electron paramagnetic resonance and nuclear magnetic resonance have been limited in sensitivity to about 10^8 electron spins and about 10^{15} nuclear spins, respectively, due chiefly to the weak interaction between the individual spins and the magnetic fields used to excite the transition. The power of SMS is that only one molecule is in resonance with the laser; hence the detection of the effect of secondary perturbing fields on the optical emission can lead to observation of weak resonances. For example, the magnetic resonance transition of a single molecular spin was observed at IBM [64] and at Bordeaux [65] using the pentacene in p-terphenyl model system and a combination of SMS and optically detected magnetic resonance (ODMR). In essence, ODMR allows higher sensitivity because the weak spin transition is effectively coupled to a much stronger optical transition with oscillator strength near unity. The method involves selecting a single molecule with the laser as shown in Fig. 2.8 (left side) and monitoring the intensity of optical emission (here, fluorescence from the first excited singlet state S_1) as the frequency of a microwave signal is scanned over the frequency range of the triplet spin sublevels T_x, T_y, T_z. Since the emission rate is dependent upon the overall lifetime of the triplet (bottleneck) state, the emission rate is affected when the microwave frequency is resonant with transitions among the triplet spin sublevels.

Figure 2.8 (right side) shows examples of the 1480 MHz magnetic resonance transition among the T_x-T_z triplet spin sublevels at 1.5 K for a single molecule of pentacene in p-terphenyl, where the signal plotted is the change in the fluorescence emission rate as a function of the applied microwave frequency [64]. Traces (c)–(g) show the single-molecule lineshapes for four different molecules. An interesting observation is that the onset of the transition varies from molecule to molecule, in a fashion similar to the difference in onsets for the two inhomogeneously broadened site origins. The lineshape of the microwave transition for a single spin is broadened by hyperfine interactions induced by the large number of different configurations possible for the nearby proton nuclear spins. This occurs because many different configurations of the proton nuclear spins in the molecule are sampled on the time scale of the measurement of the triplet state transition. In contrast, in the large N experiment, an ensemble average is measured rather than a time average.

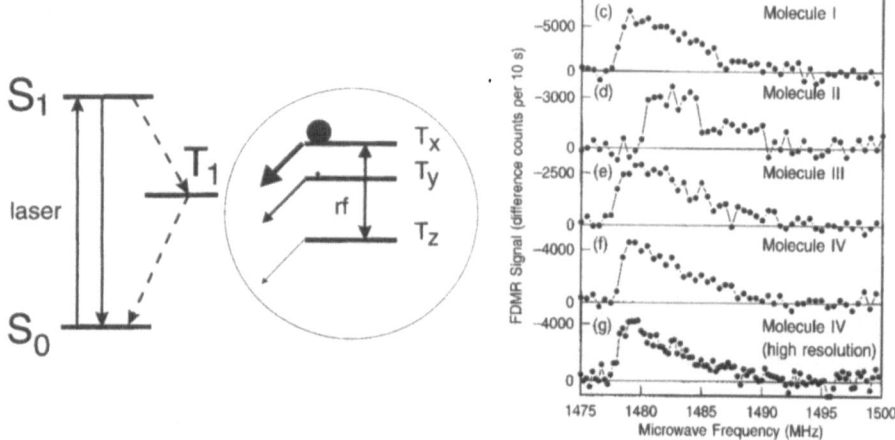

Fig. 2.8. Left: Energy levels relevant for fluorescence detected magnetic resonance of a single molecular spin. Right: Reductions in fluorescence as a function of microwave frequency for four different single molecules of pentacene in p-terphenyl. For details, see [64]

These observations have opened the way for a variety of new studies of magnetic interactions in solids at the level of a single molecular spin, such as the use of external magnetic fields and deuteration to reduce proton spin flips and hence the hyperfine broadening [66]. In future work, the properties of amorphous organic materials may be studied in greater detail, as the selection of a single molecular spin removes all orientational anisotropy as well as all inhomogeneous broadening. Imaging on the spatial scale of a single molecule may be possible with a sufficiently large magnetic field gradient. The power of magnetic resonance in general in the study of fine and hyperfine interactions, local structure, and molecular bonding can now be enhanced with these first demonstrations of useful sensitivity in the single-spin regime.

2.10   Vibrational Modes of Single Molecules

The majority of SMS studies utilize the total fluorescence excitation technique where all long-wavelength-shifted photons passing through a long-pass filter are recorded. With the use of a grating spectrograph and a CCD array detector (see Fig. 2.9, left side), vibrationally resolved emission spectra from single molecules both in crystals [67] and in polymers [68] have been obtained for the first time. Such experiments may also be regarded as resonance Raman studies, since the continuous-wave laser is in resonance with the 0–0 electronic transition, and even-parity vibrational modes of the ground state are detected by measuring the shift between the laser wavelength and the wavelength of the emission peak. The ability to examine vibronic or vibrational features of individual absorbers can generate specific details about the

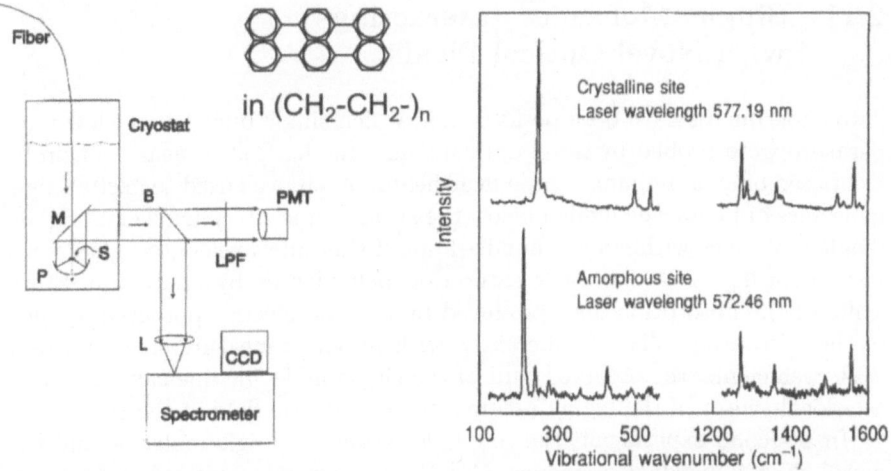

Fig. 2.9. Left: Optical configuration for exciting a single molecule at the tip of an optical fiber, collecting the fluorescence, and detecting both the spectrally dispersed emission on a CCD and the total fluorescence passing through a long-pass filter (LPF). Right: Emission signals from two different single molecules of terrylene in poly(ethylene) at 1.6 K showing the frequencies of even vibrational modes of the ground state. For details, see [68]

identity of the absorber and the nature of the interactions with the nanoenvironment producing shifts or intensity changes in the vibrational spectrum.

Figure 2.9, right side, shows typical dispersed fluorescence spectra for two different single molecules of terrylene in poly(ethylene) at 1.6 K. In addition to small ($\sim$cm^{-1}) shifts and intensity changes of various modes from molecule to molecule, two rather different classes of spectra were observed [68] as shown in the upper and lower parts of the figure. After considering various possibilities, a model was proposed in which the upper type of spectrum resulted from a terrylene molecule near or inside the crystalline region of the polymer, while the lower spectrum resulted from a single terrylene located in an amorphous region. Such results demonstrate the additional spectroscopic detail that can be obtained from individual molecules and used to probe truly local aspects of the structure of amorphous solids. For example, the lowest frequency mode in the figure is a long-axis ring expansion of the molecule; the shift to lower energy in the amorphous site can be understood as resulting from the reduced local density (greater free volume) compared to the crystalline site. Such measurements have stimulated new theoretical calculations of the spectral changes which result from specific local distortions for comparison with the single-molecule spectra [69].

2.11 Single Molecules Interacting with Novel Optical Fields

Two experiments have been performed in which single molecules at low temperature were probed by novel optical fields. In the first, a near-field probe composed of an aluminum-coated near-field fiber tip was used to excite single molecules of pentacene hidden below the surface of a p-terphenyl crystal [70]. Single molecules within a few hundred nm of this subwavelength light source were identified either by early saturation behavior or by analysis of Stark shifts of the absorption lines produced by a static electric potential applied to the Al coating. Viewed differently, with previous measurements of Stark shift coefficients, the observed shift of the single-molecule line can be viewed as a local sensor of the highly inhomogeneous electric field of the tip.

In a second experiment, the coupling between a single molecule and the morphology-dependent resonances (MDRs) of dielectric microspheres was explored [71]. A thin sublimed crystal of p-terphenyl doped with pentacene and terrylene was placed in optical contact with a small microsphere, and the high-Q ($\sim 10^6$) resonances of the microsphere were excited by the evanescent field from a nearby prism. By scanning the laser frequency and detecting the emission from the sample via a low temperature wide-field microscope, situations could be found where single-molecule absorption lines were coincident with MDRs of the sphere. Single-molecule linewidths were measured in the hope of detecting an enhanced or suppressed linewidth signifying modifications of the spontaneous emission rate by cavity quantum electrodynamic effects. Although anomalous linewidths were observed, it was not possible to conclusively assign the source to modification of the density of photon states by the MDR, due to interference from spectral diffusion effects. It remains an active research goal to couple the narrow optical absorption of a single molecule or ion trapped in a solid to a high-Q cavity in order to observe strong coupling of the molecule and the light field.

2.12 Room Temperature SMS: Diffusion of Single Molecules in Gels

As described in the introduction, several techniques are now available for imaging single molecules at room temperature, including scanning near-field optical microscopy, confocal microscopy, epifluorescence, and total internal reflection microscopy. The work of the Moerner group at room temperature has concentrated on the last three methods. Milestones are listed in Table 2.2, several of which will now be summarized.

For experiments hoping to obtain biologically relevant information, aqueous environments are required. Although the observation of single fluorophores diffusing through a focal volume had been accomplished by several

Table 2.2. Room temperature and single-biomolecule milestones: Moerner group

Experiment	Refs.
Diffusion of single fluorophores in poly(acrylamide) gels	[32][75]
Blinking and switching of single copies of green fluorescent protein	[74][81]
Imaging z-oriented molecules with total internal reflection microscopy	[77]
Cameleon – single-pair FRET between two GFP mutants in a [Ca++]-sensitive protein	[84]
Dynamics of a single-molecule pH sensor	[85]
Single-molecule source of single photons on demand	[86]

groups by the mid-1990's, more information can be obtained when the molecules are at least partially immobilized and the same single molecule is studied for a long time. In water, a small fluorophore can move a mean-squared distance of ~ 300 μm^2 in one second due to Brownian motion. This problem can be solved by attaching the system of interest to a surface, but this must be done carefully in order to avoid denaturation. Although useful information was obtained from dye-labeled lipids [72] in lipid bilayers and motor proteins bound on their natural counterparts [73], it is necessary for other systems to develop appropriate immobilizing aqueous environments.

In 1996, partial immobilization of single molecules was achieved by using the water-filled pores of poly(acrylamide) gels, a technique that has been demonstrated for organic dye molecules [32] as well as for green fluorescent protein [74] (see the next section). These poly(acrylamide) gels, which are widely used for separation techniques, are not only suitable as immobilizing matrix for SMS but also can be studied themselves by using the diffusion of embedded dye molecules as a probe for the local gel environment as shown in Fig. 2.10. The left side of the figure shows the total internal reflection (TIR) microscopy setup required for wide-field imaging. The method operates by illuminating a thin slice (~ 125 nm thick) of the sample with the evanescent light field produced by total internal reflection of the pumping laser beam at the interface between the upper cover slip C and the sample S. This technique has the advantage of pumping only a thin pancake-shaped volume of the aqueous sample to reduce background signals. Emission from single molecules is collected by a microscope objective and imaged onto a high-speed intensified CCD camera.

Excitation and detection of single nile red dye molecules in poly(acrylamide) gels by TIR microscopy allowed measurement of the three-dimensional trajectories of the spatial motion for the first time [32]. An example for the case of rhodamine 6G dye molecules in a gel [75] is shown in Fig. 2.10, right side. For each 100 ms time step provided by the integration time of the camera, the x and y transverse position information was simply provided by the position on the two-dimensional detector (with the natural diffraction-limited resolution of ~ 200 nm). The z-position information was obtained in

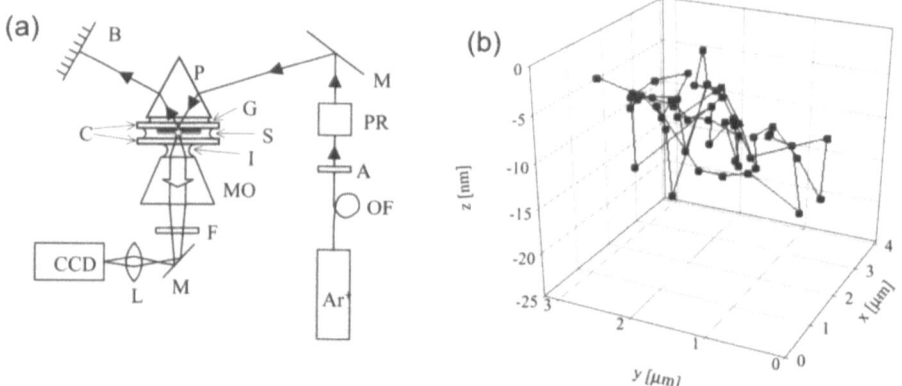

Fig. 2.10. Left: Experimental configuration for total-internal-reflection microscopy of single molecules in ambient samples. For details, see [32,75]. Right: Three-dimensional trajectory of a single molecule of rhodamine 6G in a poly(acrylamide) gel at room temperature, 100 ms per point. For details, see [75]

a novel way: with the reasonable assumption of fast rotation of the small fluorophores in the gel, the brightness of the molecule was used to determine its axial position in the exponentially-decaying evanescent field. The molecules still undergo Brownian motion, but with reduced diffusion coefficients that could be determined for various gel concentrations by analyzing the time-dependence of the mean square displacement in the x–y transverse plane [75]. The crosslinked meshwork of the poly(acrylamide) gel matrix was able to decrease the diffusion coefficient for small organic dye molecules by one to three orders of magnitude, depending on the gel composition. The relation between gel concentration and diffusion coefficient confirmed estimates for the dependence of the mean pore size on concentration that were derived from molecular-sieve chromatography of proteins [76].

2.13   Orientations of Single Molecules by Ring Patterns

For many years, it has been well-known that useful information from the optical microscopy of specimens can be obtained by altering the polarization of the pumping light, or by resolving the polarization of the emitted light. However, due to the transverse nature of a propagating electromagnetic wave, only pump or emission components in the x–y plane could be easily examined. It was demonstrated recently that the highly inhomogeneous electric field of a near-field optical tip can provide z-axis electric fields [29], but this method requires specialized optical apparatus.

The use of TIR excitation to pump single molecules provides a new and convenient way to access optical emission dipole moments aligned along the optic (z) axis of the microscope. The apparatus involves using a standard oil-immersion inverted microscope to view single molecules through some thickness of water (Fig. 2.10, left side). When imaging an aqueous sample of typical thickness of several microns with TIR, the focal plane is located at the interface between the upper cover slip and the water. However, oil-immersion objectives provide lowest aberration for focal planes located at the top of the lower cover slip. The extra propagation distance through water and the ensuing index mismatch relative to glass introduce aberrations for focal planes near the upper cover slip.

This seemingly arcane fact has fascinating consequences when the emission source is a single molecule. Z-axis oriented molecules emit light predominantly at polar angles near $\pi/2$, and the light propagating at these high angles is aberrated the most, resulting in a ring-shaped structure at the image plane [77]. At the same time, molecules with emission dipoles in the x–y plane do not show a ring structure. Due to its comparative simplicity, this technique should be useful in determining the three-dimensional orientation of single-molecule labels in a wide variety of biophysical situations in which oil-immersion objectives are used to observe aqueous samples [78].

2.14   Optical Study of Single Copies of the Green Fluorescent Protein

The Green Fluorescent Protein (GFP) and its mutants are currently of great importance in molecular and cellular biology [79]. GFP is a small (238 amino acids), water-soluble protein isolated from the jellyfish *Aequoria victoria* which has a β-barrel structure as shown in Fig. 2.11a. The strongly absorbing and fluorescing chromophore of GFP located inside the barrel is formed spontaneously (in the presence of oxygen) from three amino acids in the native protein chain. No external cofactor (as in most other fluorescent proteins) is necessary. In cell biology, GFP is currently widely used as an indicator for gene expression or as a fluorescent label for a large variety of proteins. Since the quantum yield of the emission is fairly high and the absorption can be pumped by convenient Ar+ laser lines, single copies of GFP mutants may easily be observed using SMS techniques. For example, a confocal microscopy image of the "10C" mutant at low concentration in a poly(acrylamide) gel is shown in Fig. 2.11b [80]. In this scanning method, the image is built up point by point as the small focal spot is translated over the sample, and the emission is imaged through an aperture before detection on a small-area photon counting detector.

Research in the Moerner group has yielded the first example of a room temperature single-molecule optical switch [74] and the first details of the photophysical character of GFP on the single-copy level. By studying two different red-shifted GFP mutants (S65G/S72A/T203Y denoted "T203Y" and S65G/S72A/T203F denoted "T203F") which differ only by the presence of a hydroxyl group near the chromophore, slight differences in the photo-

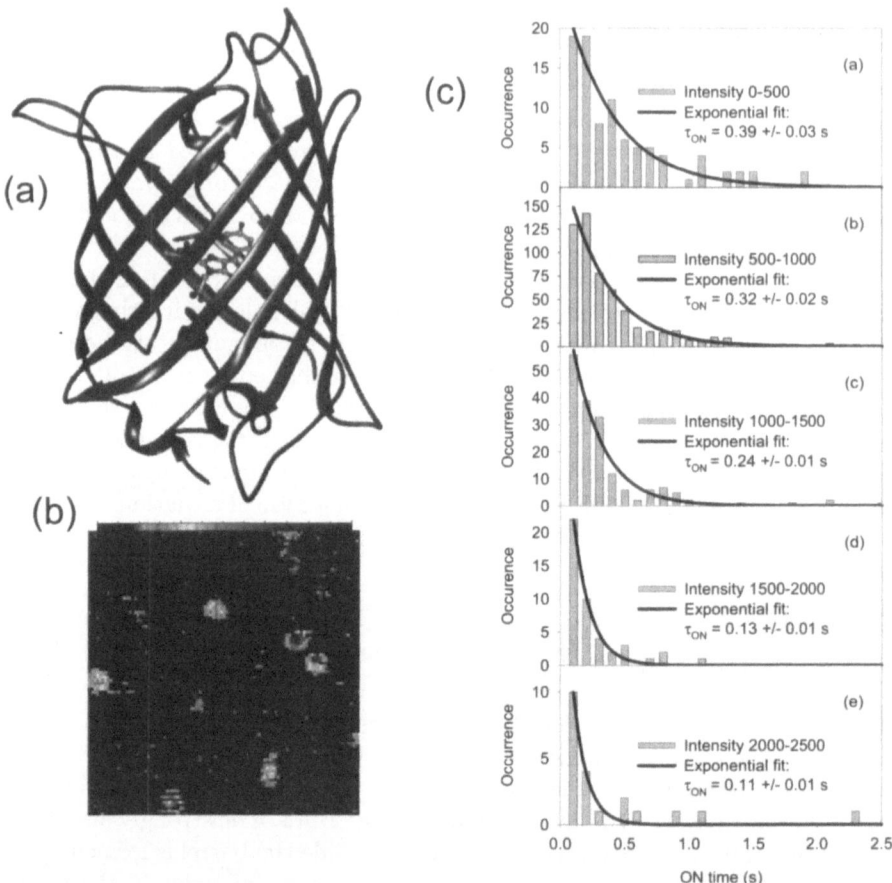

Fig. 2.11. a Structure of the green fluorescent protein [88]. **b** Confocal fluorescence image of single copies of the mutant 10C embedded in a poly(acrylamide) matrix, 10 μm × 10 μm range, 100 points × 100 points, 10 ms counting time per point. For details, see [80]. **c** Emission-intensity-binned histograms of the on times of the fluorescence of EGFP in agarose (pH 8) (at excitation intensities of 5 kW/cm^2, 1.5 kW/cm^2, and 0.5 kW/cm^2). The molecules were binned in 5 emission intensity classes (intensity in arbitrary units). For details, see [81]

physical properties were observed. In particular, a fascinating and unexpected blinking behavior appeared [74], discernable only on the single-molecule level (see the molecule at the lower edge of Fig. 2.11b for example). This blinking behavior likely results from transformations between at least two states of the chromophore, only one of which is capable of being excited by the 488 nm pumping laser and producing fluorescence. Additionally, a much longer-lived dark state is accessible through excited state processes. Thermally stable in the dark for many minutes, this long-lived dark state can be excited at 405 nm to regenerate the original fluorescent state.

Recently these studies have been extended to other mutants of GFP ('EGFP' and '10C') and to different environments (different pH, different matrix (agarose)) [81]. Rather than measuring the autocorrelation of the entire time trace, the 'on-time' and 'off-time' distributions were separately extracted. Mutant variation showed that the blinking behavior is very similar for all mutants studied in the phenolate ion class, and is very similar in both gel hosts. Furthermore, changing the buffer pH from 6 to 10 has no effect on the on-time distributions. This observation strongly suggests that the mechanism of blinking is not proton transfer, as was suggested earlier. However, changing the excitation intensity had a strong effect on the distribution of 'on-times': the higher the pumping intensity, the less time the fluorescence of the molecule is 'on' (Fig. 2.11c). This clearly shows that the termination of the emission is photo-induced. In contrast, no evidence was found for intensity dependence of the 'off-times', suggesting that the conversion from the 'off-' to the 'on-state' is spontaneous. A more likely explanation for the blinking is a reversible switching between different conformational states of the chromophore (e.g., isomerization) and/or the rest of the protein. From the single-molecule blinking experiments, the probability of termination of emission per photon absorbed was determined and found to be in agreement with the bulk bleaching quantum yield ($8(\pm2) \times 10^{-6}$), thus suggesting that the two processes are related. These studies, along with separate work on dynamics of the emission at shorter time scales with FCS [82], show that the photophysical dynamics of the GFP emission are quite complex.

2.15   Single-Molecule Concentration Sensors

SMS can be used to explore the dynamics of environment-dependent fluorescent molecules that are usually used as concentration reporters in biological media. One study concerns the "cameleon YC2.1" complex, whose structure is based on a cyan-emitting GFP (CFP) separated from a yellow-emitting GFP (YFP) by the calmodulin Ca^{2+}-binding protein and a calmodulin-binding peptide (M13) (see Fig. 2.12, left side). This complex was designed by A. Miyawaki and R.Y. Tsien to allow sensing of calcium concentration in cells by fluorescence resonant energy transfer (FRET). If Ca^{2+} ions are bound, the construct forms a more compact shape, leading to a higher efficiency of excitation transfer from the donor CFP to the acceptor YFP. The degree of FRET in cameleon is therefore a sensitive ratiometric reporter of the concentration of Ca^{2+} in solution and cells [83].

Analysis of single-molecule signals from the cameleon YC2.1 complex diluted in aqueous agarose gels allowed retrieval of several interesting features of the energy transfer between the donor and acceptor mutants of the construct, as a function of the calcium concentration in the medium [84]. The

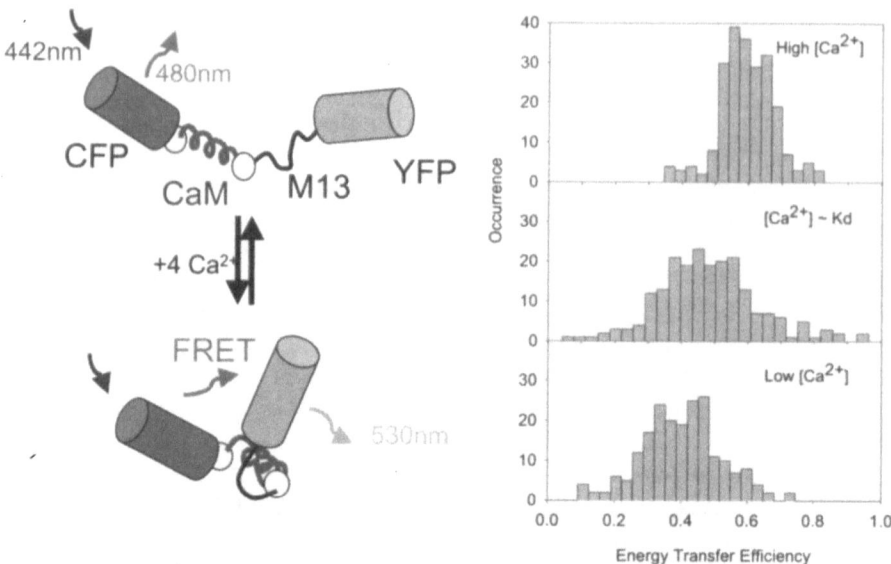

Fig. 2.12. Left: Schematic of the structure of cameleon and the change in FRET upon binding and unbinding of Ca^{2+} ions. Right: Histograms of the energy transfer efficiency measured for single molecules of cameleon in agarose gels, at three different Ca^{2+} ion concentrations. For details, see [84]

energy transfer efficiency distribution deduced from single-molecule confocal fluorescence signals shows an increased width at the Ca^{2+} dissociation constant concentration (Fig. 2.12, right side). This observation is consistent with the ligand binding kinetics, whose time scale at intermediate calcium concentration is close to our measurement time scale (20–200 ms). The complex dynamics of the fluctuations were examined using a combination of autocorrelation and cross-correlation in conjunction with polarization information. Beside reorientation fluctuations of the two dipoles that seem to occur slowly compared to the emission time scale but fast compared to the integration time, slower variations in the energy transfer between the two GFP mutants were observed. Both negative and positive cross-correlations in the donor and acceptor emission signals were present, the former related to the energy transfer process, and the latter caused by other perturbations of the donor and acceptor emission.

In separate measurements, the system Dextran-SNARF-1, which is pH-sensitive, was explored at the single molecule level by confocal fluorescence microscopy [85]. A variation of the pH in the medium induces a change of the ratio between the emission intensities at 580 nm and 640 nm, with much faster binding kinetics than the measurement time scale. The distribution of this ratio measured on single copies of this chromophore immobilized in agarose showed however an increased width close to the pK_a. Because the binding kinetics are fast, inhomogeneities in either the Dextran or in the gel environment may be responsible for this observation.

2.16   Summary

The research described here shows how the high signal-to-noise ratios now available in optical single-molecule measurements allow for a wide variety of experiments in many fields. In all cases, the main benefit of single-molecule isolation is the unraveling of ensemble averages. New insight is thus gained into the structure and dynamics of condensed or living matter at nanometer scales, not only when inhomogeneities cause different molecules to behave in wildly different ways, but also for identical nanosystems which fluctuate in time. Whereas time-resolved measurements on an ensemble demand synchronization of all subsystems, single-molecule experiments yield time-resolved information automatically.

The impact of single-molecule spectroscopy has been felt in physics and chemistry, and the application to biological problems is currently under intense investigation. In quantum optics, the isolation of a single quantum system greatly simplifies the description and understanding of its interaction with light. In a recent study, a single terrylene molecule in a p-terphenyl crystal at room temperature was used to create a highly nonclassical triggered source of single photons, one at a time, upon demand [86]. In the area of physical chemistry, single molecules at low temperatures have provided a wealth of new information about such fundamental processes as tunneling and activation in molecular crystals and polymers. Single molecules will also help solve problems in physical chemistry at ambient conditions, such as surface-enhanced Raman scattering, nanocrystal photophysics [87], adsorption and desorption of molecules on surfaces, and dynamics in polymers and in other complex environments. The scope and potential of single-molecule methods in biology are quite broad, as both molecular heterogeneity and single-molecule operations are characteristic of biomolecular systems. The domains of biophysical chemistry which could benefit from single-molecule investigations range from protein folding and colocalization to the interaction between single biomacromolecules, from the mechanisms of molecular motors to the sequencing of single nucleic acid strands. Finally, much work in modern science concerns manipulation of atoms and molecules at the nanoscale, with one long-term goal being the actual production of man-made molecular-scale devices and computing machines at some time in the future. Optical probing of single molecules may be a natural readout scheme which may be useful when interconnection problems and density issues prevent conventional wiring.

Acknowledgments. The author warmly thanks many members of the Moerner single-molecule spectroscopy group for their crucial contributions to the work through early 2000 reported here, in particular: T.P. Carter, L. Kador,

W.P. Ambrose, Th. Basché, P. Tchénio, A.B. Myers, J. Köhler, D.J. Norris, R.M. Dickson, S. Kummer, E.J.G. Peterman, S. Brasselet, J. Deich, and B. Lounis. Stimulating collaborations with U.P. Wild, M. Orrit, and J. Schmidt are gratefully acknowledged. This work has been supported in part by grants from the U.S. Office of Naval Research and the National Science Foundation, and by support from the IBM Research Division, the University of California San Diego, and Stanford University.

References

1. W. E. Moerner and L. Kador, "Optical detection and spectroscopy of single molecules in a solid," Phys. Rev. Lett. **62**, 2535–38 (1989)
2. M. Orrit and J. Bernard, "Single pentacene molecules detected by fluorescence excitation in a p-terphenyl crystal," Phys. Rev. Lett. **65**, 2716–19 (1990)
3. W. E. Moerner and T. Basché, "Optical spectroscopy of single inpurity molecules in solids," Angew. Chemie Int. Ed. Engl. **32**, 457 (1993)
4. W. E. Moerner, "Examining nanoenvironments in solids on the scale of a single, isolated molecule," Science **265**, 46–53 (1994)
5. W. E. Moerner, "High-resolution optical spectroscopy of single molecules in solids," Acc. Chem. Res. **29**, 563 (1996)
6. M. Orrit et al., "Optical spectroscopy of single molecules in solids," in: *Progress in Optics*, E. Wolf (ed.), pp. 61–144 (Elsevier, 1996)
7. T. Basché et al., *Single Molecule Optical Detection, Imaging, and Spectroscopy* (Verlag-Chemie, Munich, 1997)
8. S. Nie and R. N. Zare, Ann. Rev. Biophys. Biomol. Struct. **26**, 567–596 (1997)
9. T. Plakhotnik, E. A. Donley, and U. P. Wild, "Single-molecule spectroscopy," Ann. Rev. Phys. Chem. **48**, 181–212 (1996)
10. X. S. Xie and J. K. Trautman, "Optical studies of single molecules at room temperature," Ann. Rev. Phys. Chem. **49**, 441.-480 (1998)
11. W. E. Moerner and M. Orrit, "Illuminating single molecules in condensed matter," Science **283**, 1670–1676 (1999)
12. S. Weiss, "Fluorescence spectroscopy of single biomolecules," Science **283**, 1676–1683 (1999)
13. W. P. Ambrose et al., "Single molecule fluorescence spectroscopy at room temperature," Chem. Rev. **99**(10), 2929–2956 (1999)
14. W. E. Moerner, "Those blinking single molecules," Science **277**, 1059 (1997)
15. W. M. Itano, J. C. Bergquist, and D. J. Wineland, Science, **237**, 612 (1987)
16. F. Diedrich et al., IEEE J. Quant. Elect. **24**, 1314 (1988)
17. H. Dehmelt, W. Paul, and N. F. Ramsey, Rev. Mod. Phys. **2**, 525 (1990)
18. G. Binnig and H. Rohrer, Rev. Mod. Phys. **59**, 615 (1987)
19. G. Binnig, C. F. Quate, and C. Gerber, Phys. Rev. Lett. **56**, 930 (1986)
20. B. Sakmann and E. Neher, *Single Channel Recording* (New York, Plenum Press, 1995)
21. T. T. Perkins, D. E. Smith, and S. Chu, "Single polymer dynamics in an elongational flow," Science **276**, 2016.-2021 (1997)
22. S. M. Block, "Kinesin: What gives?," Cell **93**, 5–8 (1998)
23. D. L. Magde, E. L. Elson, and W. W. Webb, Biopolymers **13**, 29 (1974)

24. S. Wennmalm, L. Edman, and R. Rigler, "Conformational 'fluctuations in single DNA molecules," Proc. Nat. Acad. Sci. (USA) **94**, 10641–10646 (1997)
25. M. Eigen and R. Rigler, "Sorting single molecules: Application to diagnostics and evolutionary biotechnology," Proc. Natl. Acad. Sci. USA **91**, 5740–5747 (1994)
26. A. M. Stoneham, Rev. Mod. Phys. **41**, 82 (1969)
27. K. K. Rebane, *Impurity Spectra of Solids*, p. 99 (New York, Plenum, 1970)
28. E. B. Shera et al., "Detection of single fluorescent molecules," Chem. Phys. Lett. **174**, 553 (1990)
29. E. Betzig and R. J. Chichester, "Single molecules observed by near-field scanning optical microscopy," Science **262**, 1422–1428 (1993)
30. S. Nie, D. T. Chiu, and R. N. Zare, "Probing individual molecules with confocal fluorescence microscopy," Science **266**, 1018–1021 (1994)
31. J. K. Trautman and J. J. Macklin, "Time-resolved spectroscopy of single molecules using near-field and far-field optics," Chem. Phys. **205**, 221–229 (1996)
32. R. M. Dickson et al., "Three dimensional imaging of single molecules in pores of poly(acrylamide) gels," Science **274**(5289), 966–969 (1996)
33. W. E. Moerner (ed.), *Persistent Spectral Hole-Burning: Science and Applications*. Topics in Current Physics **44** (Springer, Berlin, 1988)
34. W. E. Moerner and T. P. Carter, "Statistical fine structure in inhomogeneously broadened absorption lines," Phys. Rev. Lett. **59**, 2705 (1987)
35. G. C. Bjorklund, Opt. Lett. **5**, 15 (1980)
36. W. E. Moerner, "Persistent spectral hole-burning: Photon-gating and fundamental statistical limits," in: *Polymers for Microelectronics, Science, and Technology*, Y. Tabata, I. Mita, and S. Nonogaki (eds.) (Kodansha Scientific, Tokyo, 1990)
37. L. Kador, D. E. Horne, and W. E. Moerner, J. Phys. Chem. **94**, 1237 (1990)
38. G. C. Bjorklund et al., Appl. Phys. B **32**, 145 (1983)
39. L. Kador et al., "Absorption spectroscopy on single molecules in solids," J. Chem. Phys. **111**, 8755–8758 (1999)
40. W. E. Moerner and W. P. Ambrose, "Comment on 'single pentacene molecules detected by fluorescence excitation in a p-terphenyl crystal'," Phys. Rev. Lett. **66**, 1376 (1991)
41. F. G. Patterson et al., Chem. Phys. **84**, 51 (1984)
42. H. de Vries and D. A. Wiersma, J. Chem. Phys. **70**, 5807 (1979)
43. W. P. Ambrose and W. E. Moerner, "Fluorescence spectroscopy and spectral diffusion of single impurity molecules in a crystal," Nature **349**, 225 (1991)
44. W. P. Ambrose, T. Basché, and W. E. Moerner, "Detection and spectroscopy of single pentacene molecules in a p-terphenyl crystal by means of fluorescence excitation," J. Chem. Phys. **95**, 7150 (1991)
45. J. Friedrich and D. Haarer, in: *Optical Spectroscopy of Glasses*, I. Zschokke (ed.), p. 149 (Reidel, Dordrecht, 1986)
46. T. Basché and W. E. Moerner, "Optical modification of a single impurity molecule in a solid," Nature **355**, 335 (1992)
47. A. Zumbusch et al., Phys. Rev. Lett. **70**, 3584 (1993)
48. P. D. Reilly and J. L. Skinner, "Spectral diffusion of single molecule fluorescence: A probe of low-frequency localized excitations in disordered crystals," Phys. Rev. Lett. **71**, 4257–4260 (1993)
49. P. D. Reilly and J. L. Skinner, J. Chem. Phys. **102**, 1540 (1995)

50. E. Geva and J. L. Skinner, "Theory of single-molecule optical line-shape distributions in low-temperature glasses," J. Phys. Chem. B **101**, 8920–8932 (1997)
51. Plakhotnik, T., et al., "Single-molecule spectroscopy in Shpol'skii matrices," Chimia **48**, 31 (1994)
52. J. K. Trautman et al., "Near-field spectroscopy of single molecules at room temperature," Nature **369**, 40–42 (1994)
53. X. S. Xie, "Single-molecule spectroscopy and dynamics at room temperature, Acc. Chem. Res. **29**(12), 598 (1996)
54. W. A. Phillips (ed.), *Amorphous Solids: Low-Temperature Properties*, Topics in Current Physics **24** (Springer, Berlin, 1981)
55. T. Basché, W.P. Ambrose, and W. E. Moerner, "Optical spectra and kinetics of single impurity molecules in a polymer: Spectral diffusion and persistent spectral hole-burning," J. Opt. Soc. Am. B **9**, 829 (1992)
56. P. Tchénio, A. B. Myers, and W. E. Moerner, "Optical studies of single terrylene molecules in polyethylene," J. Lumin **56**, 1 (1993)
57. W. E. Moerner et al., "Optical probing of single molecules of terrylene in a Shpolskii matrix – A two-state single-molecule switch," J. Phys. Chem. **98**, 7382–7389 (1994)
58. J. Bernard et al., "Photon bunching in the fluorescence from single molecules: A probe for intersystem crossing," J. Chem. Phys. **98**, 850 (1993)
59. R. Loudon, *The Quantum Theory of Light*, 2nd ed., pp. 226–249 (Clarendon, Oxford, 1983)
60. H. J. Kimble, M. Dagenais, and L. Mandel, "Photon anti-bunching in resonance fluorescence," Phys. Rev. Lett. **39**, 691 (1977)
61. T. Basché et al., "Photon antibunching in the fluorescence of a single dye molecule trapped in a solid," Phys. Rev. Lett. **69**, 1516–1519 (1992)
62. W. E. Moerner, R. M. Dickson, and D. J. Norris, "Single-molecule spectroscopy and quantum optics in solids," Adv. Atom. Molec. Opt. Phys. **38**, 193–236 (1997)
63. P. Tamarat et al., Phys. Rev. Lett. **75**, 1514 (1995)
64. J. Köhler et al., "Magnetic resonance of a single molecular spin, Nature **363**, 242–244 (1993)
65. J. Wrachtrup et al., Nature **363**, 244 (1993)
66. J. Köhler et al., "Single molecule electron paramagnetic resonance spectroscopy: Hyperfine splitting owing to a single nucleus," Science **268**, 1457–1460 (1995)
67. P. Tchénio, A. B. Myers, and W. E. Moerner, "Dispersed fluorescence spectra of single molecules of pentacene in p-terphenyl," J. Phys. Chem. **97**, 2491 (1993)
68. P. Tchénio, A. B. Myers, and W. E. Moerner, "Vibrational analysis of dispersed fluorescence from single molecules of terrylene in polyethylene, Chem. Phys. Lett. **213**, 325 (1993)
69. A. B. Myers et al., "Vibronic apectroscopy of individual molecules in solids," J. Phys. Chem. **98**, 10377 (1994)
70. W. E. Moerner et al., "Near-rield optical spectroscopy of individual molecules in solids," Phys. Rev. Lett. **73**, 2764 (1994)
71. D. J. Norris, M. Kuwata-Gonokami, and W. E. Moerner, "Excitation of a single molecule on the surface of a spherical microcavity," Appl. Phys. Lett. **71**, 297 (1997)
72. T. Schmidt et al., "Imaging of single molecule diffusion," Proc. Nat. Acad. Sci. USA **93**, 2926–2929 (1996)

73. A. D. Mehta et al., "Single molecule biomechanics using optical methods," Science **283**, 1689–1695 (1999)
74. R. M. Dickson et al., "Blinking and switching behavior of individual green fluorescent protein molecules," Nature **388**, 355 (1997)
75. S. Kummer, R. M. Dickson, and W. E. Moerner, "Probing single molecules in poly(acrylamide) gels," Proc. Soc. Photo-Opt. Instrum. Engr. **3273**, 165–173 (1998)
76. J. S. Fawcett and C. J. O. R. Morris, "Molecular-sieve chromatography of proteins on granulated polyacrylamide gels," Sep. Sci. **1**, 9-26 (1966)
77. R. M. Dickson, D. J. Norris, and W. E. Moerner, "Simultaneous imaging of individual molecules aligned both parallel and perpendicular to the optic axis," Phys. Rev. Lett. **81**, 5322–5325 (1998)
78. A. P. Bartko, "Imaging three-dimensional single molecule orientations," J. Phys. Chem. B **103**, 11237–11241 (1999)
79. R. Y. Tsien, "The green fluorescent protein," Ann. Rev. Biochem. **67**, 509–544 (1998)
80. W. E. Moerner et al., "Optical methods for exploring dynamics of single copies of green fluorescent protein," Cytometry **36**, 232–238 (1999)
81. E. J. G. Peterman, S. Brasselet, and W. E. Moerner, "The fluorescence dynamics of single molecules of green fluorescent protein," J. Phys. Chem. A **103**, 10553–10560 (1999)
82. P. Schwille et al., "Fluorescence correlation spectroscopy reveals fast optical excitation-driven intermolecular dynamics of yellow fluorescent proteins," Proc. Nat. Acad. Sci. (USA) **97**, 151–156 (2000)
83. A. Miyawaki et al., "Fluorescent indicators for Ca++ based on green fluorescent proteins and calmodulin," Nature **388**, 882 (1997)
84. S. Brasselet et al., "Single-molecule fluorescence resonant energy transfer in calcium-concentration-dependent cameleon," J. Phys. Chem. B **104**, 3676–3682 (2000)
85. S. Brasselet and W. E. Moerner, "Fluorescence behavior of single-molecule pH sensors," Single Molecules **1**, 15–21 (2000)
86. B. Lounis and W. E. Moerner, "Single photons on demand from a single molecule at room temperature," Nature **407**, 491–493 (2000)
87. M. Nirmal et al., "Fluorescence intermittency in single cadmium selenide nanocrystals," Nature **383**, 802–804 (1996)
88. M. Ormo et al., "Crystal structure of the Aequorea victoria green fluorescent protein," Science **273**, 1392–1395 (1996)

3   The Electronic Structure of Single Photosynthetic Pigment-Protein Complexes

A. M. van Oijen, M. Ketelaars, J. Köhler, T. J. Aartsma, and J. Schmidt

The development of optical techniques to study single molecules in the condensed phase [1] opened the way for the investigation of molecular interactions on a truly microscopic scale. These single-molecule measurements reveal the distribution of molecular properties in inhomogeneous systems, properties that are normally obscured by ensemble averaging. In the study presented here, single-molecule techniques are used to investigate the electronic structure of antenna complexes of photosynthetic purple bacteria. The initial event in bacterial photosynthesis is the absorption of a photon by a light-harvesting antenna system, which is followed by a rapid and highly efficient transfer of the energy to the reaction center (RC), where charge separation takes place and the energy becomes available as chemical energy. In most purple bacteria, the photosynthetic membranes contain two types of light-harvesting (LH) complexes, the LH1 and the LH2 complex. LH1 is known to directly surround the RC, whereas LH2 is not in direct contact with the RC but transfers the energy to the RC via the LH1 complex [2,3]. The high-resolution X-ray structure of the LH2 complex of *Rps. acidophila* [4], along with lower-resolution structural information for LH1 [5], showed a remarkable symmetry in the arrangement of the light-absorbing pigments in their protein matrix. This LH2 complex consists of nine copies of a pair of proteins (α and β) arranged in a ring structure with C_9 symmetry, where each $\alpha\beta$ unit binds three BChl a and (presumably) two carotenoid molecules. The arrangement of the pigments is indicated in Fig. 3.1, where only the bacteriochlorin rings of the BChl a molecules are shown for clarity. A striking feature of the organization of the 27 BChl a molecules is their separation in two parallel rings. One ring consists of a group of 18 closely packed BChl a molecules, with their bacteriochlorin planes parallel to the symmetry axis, absorbing at 850 nm (B850). The other ring comprises nine well-separated BChl a molecules absorbing at 800 nm (B800). The molecules in this B800 ring have their bacteriochlorin planes perpendicular to the symmetry axis of the complex. Upon excitation, energy is transferred from B800 to B850 molecules in 1 to 2 ps [6–8], while energy transfer among the B850 molecules is an order of magnitude faster [9–11]. The transfer of energy from LH2 to LH1 and subsequently to the RC occurs in vivo on a time scale of 30–40 ps,

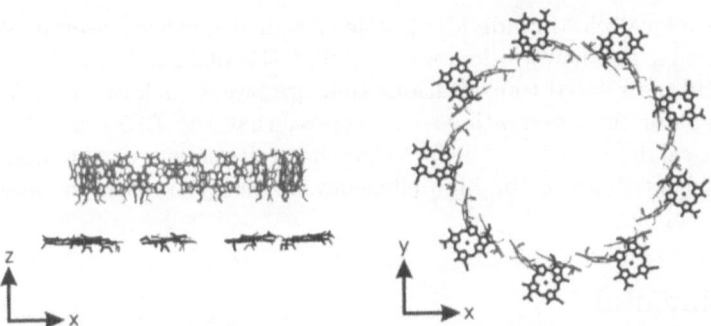

Fig. 3.1. Geometrical arrangement of the 27 BChl a molecules of the LH2 complex of *Rps. acidophila* obtained by x-ray crystallography. The phytol chains of the BChl a molecules are omitted for clarity. The data have been taken from the Brookhaven protein data bank (www.pdb.bnl.gov)

i.e., very fast compared to the fluorescence decay of B850 in isolated LH2, which has a time constant of 1 ns.

From intermolecular distances as determined from the X-ray structure it can be concluded that the dipolar interaction between neighboring BChl a molecules in the B800 ring will be significantly weaker than between the B850 molecules. The size of the transition dipole–dipole interaction strength between neighboring molecules compared to the variations of the site energy of the same molecules governs the extent of delocalization of the excited states of the LH2 complex. For the B800 BChl a molecules the dipole–dipole interaction strength is estimated to be -24 cm^{-1} (12), i.e., much smaller than the variation in site energy of about 200 cm^{-1}. The latter is estimated from the width of the ensemble absorption line. In contrast, for the B850 band the dipolar coupling strength is estimated to be about 300 cm^{-1} [12]. These values suggest that in the case of the B800 ring the excitation energy is largely localized on individual BChl a molecules [13], whereas for B850 one expects that the excitation is coherently distributed at least over a significant part of the ring [6,10,12–18].

The light-harvesting 2 complex provides a key to understand the details of the highly efficient energy transfer process in the light-harvesting complexes [19]. It represents one of the rare cases of pigment-protein complexes of this size, of which the geometric structure is known in detail. As yet, the problem in spectroscopic studies of this complex is the difficulty in determining unambiguously the various parameters which govern the energy-transfer process, such as the dipolar interaction between adjacent BChl a molecules, the actual site energy of the individual pigments and their spread in energy, and the strength of the electron–phonon interaction.

In this report we present an overview of the results of a series of single-molecule experiments on the B800 and B850 absorption bands of LH2. With regard to the B800 band we focus in particular on the question of whether the excited states are localized on individual molecules or delocalized over small groups of pigments. Further we demonstrate that the observed transitions in the B850 band are related to excitations that are largely delocalized over the ring. To explain the observations we propose that the B850 ring has undergone elliptical deformation. The fact that the excitations are so strongly delocalized may contribute to the high efficiency of energy transfer in these photosynthetic complexes.

3.1 Experimental

Thin polymer films with isolated LH2 were prepared by adding 1% polyvinyl alcohol (PVA; molecular weight 12500) to a solution of 5×10^{-11} M LH2 from *Rps. acidophila* (strain 10050) in buffer (0.1% LDAO/10 mM Tris/1 mM EDTA/pH 8.0) which was then spin coated on a LiF substrate. By dropping 10 µl of solution on the substrate and spinning it at 500 rpm for 15 seconds followed by 2000 rpm for 60 seconds, high quality films could be produced with an estimated thicknes of less than 1 µm. In order to perform fluorescence microscopy and fluorescence-excitation spectroscopy the sample was illuminated with the light from a cw tunable Ti:Sapphire laser featuring a spectral bandwidth of 1 cm^{-1}. Because the experiments are carried out in liquid helium only a single aspheric lens (numerical aperture 0.55, working distance 0.85 mm), mounted close to the sample, served as microscope objective.

To obtain wide-field images of parts of the sample, an area of 100 × 100 µm^2 was illuminated through a rear window of the cryostat. The fluorescence emitted by the LH2 complexes, with a wavelength of 890 nm, was collected by the aspheric lens and imaged on a CCD camera after passing appropriate filters to block residual laser light. The field of view was approximately 50 × 50 µm^2 with a lateral spatial resolution of 0.9 µm. The concentration of the LH2 complexes in the film, 50 pM, was chosen such that the average distance between individual complexes was much larger than 0.9 µm allowing for spatial selection of a single complex. The substrate which supported the film could be moved *in situ* in the lateral direction with respect to the wide-field illumination and the aspheric lens, to be able to probe different regions of the film.

In order to perform detailed experiments on a specific LH2 complex, the microscope was switched to the confocal mode. In this mode the sample was illuminated through the aspheric lens resulting in an excitation volume of about 1µm^3. By ensuring that this volume coincided with the position of one of the complexes observed with the CCD camera the fluorescence of a single LH2 complex, collected by the same lens, was focussed confocally on an avalanche photodiode. For all experiments described in this report, the detection occurred in a spectral window of 20 nm (FWHM) centered around 890 nm. By scanning the laser frequency fluorescence-excitation spectra of single LH2 complexes could be obtained with high signal-to-noise ratios.

3.2   Results and Discussion

An example of a wide-field fluorescence image of a $60 \times 40 \ \mu m^2$ region of the film is given in Fig. 3.2. The sample was illuminated at an intensity of $125 \ W/cm^2$ during 25 seconds. During accumulation of such an image, the excitation wavelength was scanned continuously between 798 and 801 nm at a speed of 3 nm/s. This leads to a significant increase in the contrast and overall quality of the images, because the effect of sudden light-induced spectral changes was minimized. To make sure that the single dots in the fluorescence images correspond to single LH2 assemblies, we checked that the number of dots per unit of surface is consistent with the number calculated with a concentration of 50 pM and a film thickness of 1 μm. The probability of finding one complex in the excitation volume is about 3%. The probability to find two complexes in this volume is already less than 0.1%. Furthermore, the number of dots observed in an image scales linearly with the LH2 concentration of the sample.

In Fig. 3.3 the fluorescence-excitation spectra of several single LH2 complexes and of an ensemble of LH2 complexes are compared. The ensemble spectrum features two broad structureless bands around 800 nm and 860 nm, corresponding to absorption of the B800 and B850 pigments, respectively. When observing the single complexes, the ensemble averaging in these bands is removed and remarkable spectral features become visible. The striking differences between the two absorption bands can be rationalized by considering the intermolecular interaction strength J between neighboring BChl a molecules in the rings and the spread in their transition energies Δ. J is mainly determined by the intermolecular distance and the relative orientation of the molecular dipole moments. Variations in site energies Δ can often

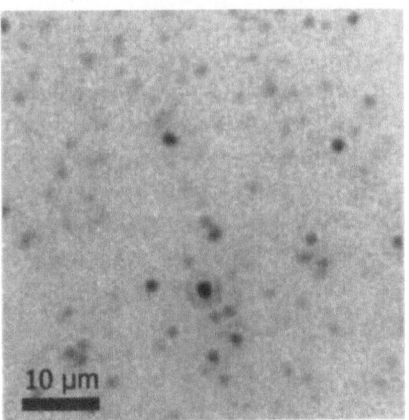

Fig. 3.2. Wide-field image of a sample region of about $40 \times 60 \ \mu m^2$. The black dots correspond to diffraction-limited images of the fluorescence of single LH2 complexes of *Rps. acidophila*

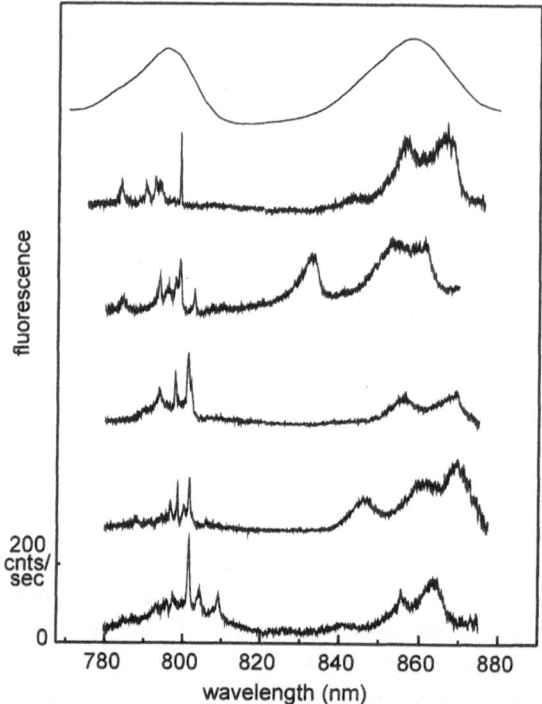

Fig. 3.3. Comparison of fluorescence-excitation spectra for an ensemble of LH2 complexes (top trace) and several individual LH2 complexes at 1.2 K. The vertical scale applies to the bottom spectrum; all other spectra are offset for clarity. cps stands for counts per second. The spectra were recorded with an excitation intensity of 20 W/cm^2

be attributed to structural variations in the local environment of the BChl a molecules, resulting in changes in the electrostatic interactions with the surrounding protein. If the ratio J/Δ is small, one would predict that the excitations are mainly localized on individual BChl a molecules. It is expected that this is the case for the B800 band. If the coupling strength J between the BChl a is much larger than Δ, the description should be in terms of delocalized excited state wave functions with relatively short energy relaxation times. This situation is expected to apply to the B850 band. In the following we will first discuss the spectra of the B800 band and then switch to those observed in the B850 band.

3.2.1   The B800 Band

As can be seen in Fig. 3.3, the B800 band of an individual LH2 complex consists of several relatively narrow spectral lines. As mentioned above, the dipolar coupling between the BChl a molecules in the B800 ring is predicted

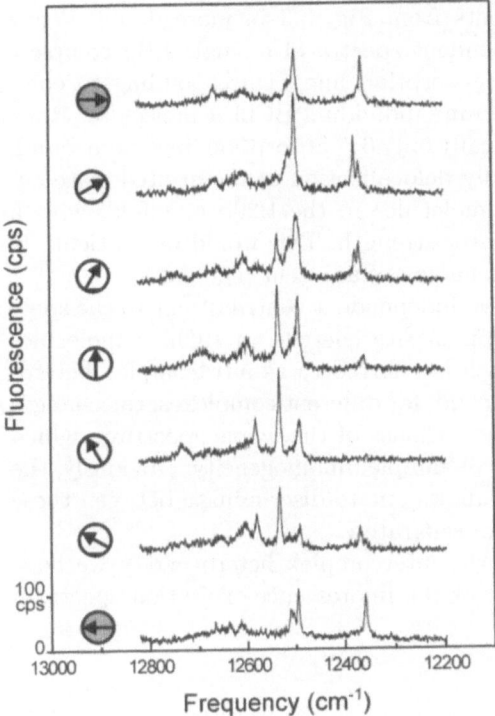

Fig. 3.4. Dependence of the B800 fluorescence-excitation spectrum of a single LH2 complex from *Rps. acidophila* on the polarization of the incident radiation. The polarization vector has been changed in steps of 30° from one spectrum to the next. The vertical scale is valid for the lowest trace; all other are displaced for clarity. The acquisition time for every trace was 25 min at an illumination intensity of 20 W/cm²

to be small with respect to the variation in site energy and it is expected that the excitation energy is mainly localized on individual B800 BChl *a*. Therefore, the narrow lines around 800 nm can be attributed to absorption by individual BChl *a* molecules in the B800 ring. This interpretation is corroborated by the strong dependence of the relative intensities of these lines on the polarization of the incident radiation. In Fig. 3.4 seven fluorescence-excitation spectra of the B800 band of an individual LH2 complex are shown, each taken with a different orientation of the polarization vector of the exciting laser light. It is seen that the relative intensities of the absorption lines change appreciably upon changing the polarization. This is what one would expect if the excitations are localized on the individual BChl *a* molecules because their Q_y transition dipole moments are arranged in a circular manner and, as a result of this, all have different orientations.

To verify the point of view that the diagonal disorder leads to a localization of the excitation mainly on one B800 molecule, we have examined the polarization-dependent measurements from Fig. 3.3 in more detail. When combining all the polarization-dependent spectra of a single LH2 complex, one would expect to observe nine absorption lines, corresponding to completely localized excitations on the nine individual BChl a molecules. However, in these combined spectra typically only 6–7 absorption lines were found, supporting the idea that occasionally delocalization of the excited state occurs between neighboring BChl a molecules in the B800 manifold, with a concomitant redistribution of oscillator strength. This would be particularly significant in the case of (near) degeneracy of adjacent pigments.

In general we have to consider two independent contributions to the spectral distribution. First, the variation in site energies of BChl a molecules within the same LH2 complex, which is referred to as intracomplex heterogeneity or diagonal disorder. And second, for different complexes, the changes in the spectral position of the center of mass of the whole spectrum, which is called intercomplex heterogeneity or sample inhomogeneity. Obviously, the study of individual LH2 complexes allows one to discriminate between these two contributions and to study them separately.

In order to find a measure for the intercomplex heterogeneity we have defined the spectral mean value, $\bar{\nu}$, of the fluorescence-excitation spectrum of a single LH2 complex by

$$
\bar{\nu} = \frac{\sum\limits_i I(i) \cdot \nu(i)}{\sum\limits_i I(i)} \ , \tag{3.1}
$$

where $I(i)$ denotes the fluorescence intensity at datapoint i, $\nu(i)$ the spectral position corresponding to datapoint i, and the sum runs over all datapoints of the spectrum. The respective histogram for $\bar{\nu}$ obtained from the spectra of 46 complexes is depicted in Fig. 3.5a, and has a width of about 120 cm^{-1}.

The intracomplex heterogeneity or diagonal disorder is extracted from the data by calculating the standard deviations σ_ν of the intensity distributions in the individual spectra

$$
\sigma_\nu = [\overline{\nu^2} - \bar{\nu}^2]^{1/2} \ , \tag{3.2}
$$

where $\overline{\nu^2}$, is given by

$$
\overline{\nu^2} = \frac{\sum\limits_i I(i) \cdot [\nu(i)]^2}{\sum\limits_i I(i)} \ . \tag{3.3}
$$

The result is shown in Fig. 3.5b. The distribution for σ_ν is centered at a value of about 55 cm^{-1}. Because the diagonal disorder is commonly defined as the full width at half maximum of the distribution of site energies this value has to be multiplied by a factor of 2.36 to obtain a value of 130 cm^{-1} for the diagonal disorder. Clearly, an ensemble spectrum reflects the convolution of

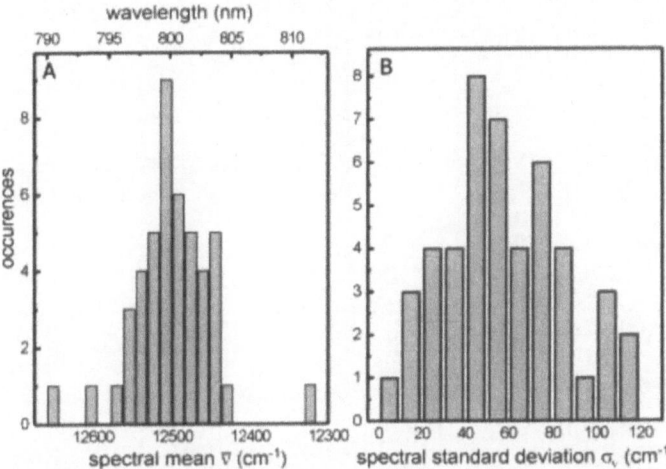

Fig. 3.5. a Distribution of the spectral mean of the B800 fluorescence-excitation spectrum for 46 LH2 complexes featuring the amount of intercomplex heterogeneity. **b** Distribution of the standard deviations for the spread of absorption lines in the individual fluorescence-excitation spectra for the same 46 LH2 complexes

both contributions to the heterogeneity. From our data we expect for the B800 band a total inhomogeneous linewidth of about 180 cm^{-1}, in excellent agreement with the results from bulk spectra of LH2 of *Rps. acidophila* taken at 1.2 K.

To study the photostability of the complexes and the effect of spectral diffusion we followed the evolution of the B800 fluorescence-excitation spectra of single LH2 complexes (see Fig. 3.6). In this figure the first four spectra were recorded consecutively at intervals of 10 minutes; it can be seen that only minor spectral changes of the excitation spectra occur. In contrast to the work of Bopp et al. [20], who observed photobleaching under ambient conditions after a few tens of seconds for similar excitation energies, our data demonstrate that such effects are very small at cryogenic temperatures. None of the 46 complexes studied showed photobleaching on a timescale of hours.

Tuning the laser into resonance with one of the B800 absorptions of a single LH2 complex and recording the total emitted fluorescence as a function of time, strong fluctuations on a timescale of seconds were observed (see Fig. 3.7). The fluorescence time trace becomes more erratic when the excitation intensity is increased, indicating that the "blinking" behavior is light-induced. In order to study these effects in more detail, fluorescence-excitation spectra were recorded by scanning the laser wavelength at a speed of 3 nm per second. This yields a temporal resolution of 10 ms per data point, corresponding to a spectral separation between two data points of 0.5 cm^{-1}. The actual spectral resolution of the scans is then determined by the spectral bandwidth of the Ti:Sapphire laser. The upper panel on the left-

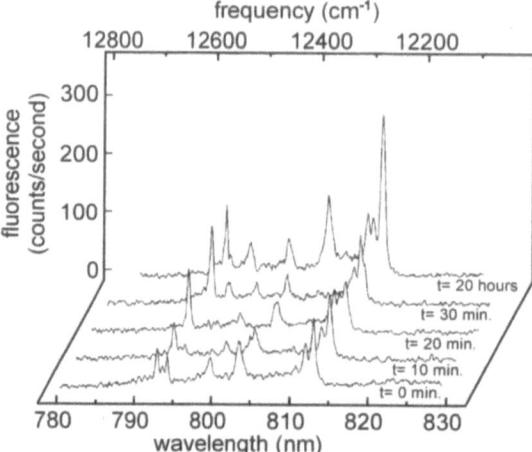

Fig. 3.6. Sequence of fluorescence-excitation spectra in the B800 band of an individual LH2 complex. The first four spectra were obtained at intervals of 10 minutes; the last spectrum was recorded 20 hours later, including a period of 10 hours of continuous illumination at 80 W/cm^2

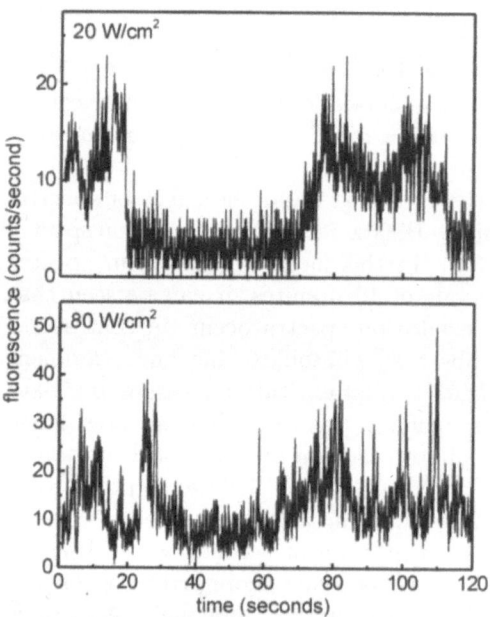

Fig. 3.7. Time dependence of the fluorescence intensity when the excitation frequency is tuned into resonance with a particular absorption of a single complex in the B800 spectrum for excitation intensities of 20 W/cm^2 (upper panel) and 80 W/cm^2 (lower panel)

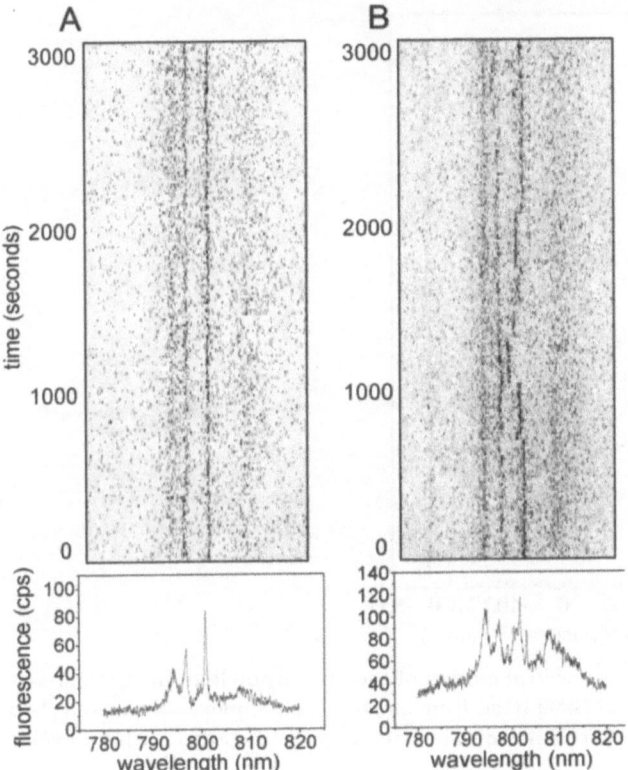

Fig. 3.8. a Stack of 200 fluorescence-excitation scans through the B800 band recorded at a scan speed of 3 nm/s and an excitation intensity of 20 W/cm^2 (upper panel). **b** Stack of 200 fluorescence-excitation scans recorded at a scan speed of 80 W/cm^2 (upper panel). In both cases the fluorescence intensity is indicated by the gray scale, and the average of the 200 spectra is displayed in the lower panel

hand side of Fig. 3.8 shows the result of a sequence of 200 of such fast scans with an excitation intensity of 20 W/cm^2. Each horizontal trace represents a single spectrum in which the fluorescence intensity is given by the shade of gray. Successive spectra are displaced vertically along the time axis. The lower panel on the same side of the figure displays the fluorescence-excitation spectrum that results when all 200 independent scans are averaged in computer memory. Apparently, the variation in time of the spectral positions of the absorptions are restricted to a small spectral range for this illumination condition. The situation changes drastically when the excitation intensity is increased to 80 W/cm^2 as illustrated on the right-hand side of the figure. As is evident from the collection of fast scans, sudden spectral jumps of the absorptions occur and the averaged fluorescence-excitation spectrum shows significant broadening of the spectral lines due to spectral diffusion effects. Interestingly, it is possible to switch between these two situations reversibly

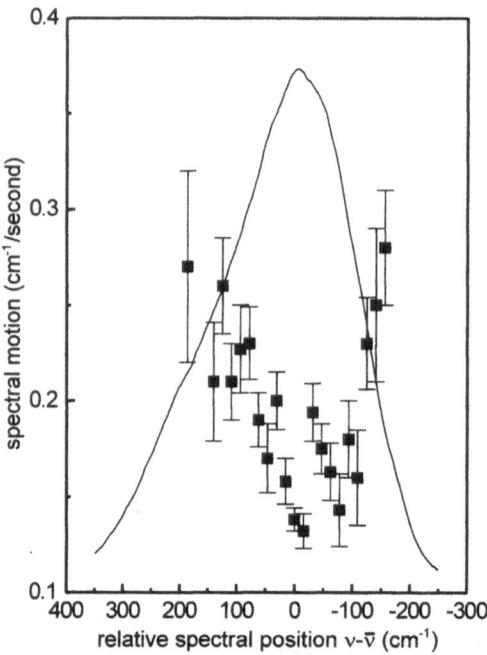

Fig. 3.9. The amount of spectral motion of the absorption lines in the fluorescence-excitation spectra of the B800 band from single LH2 complexes with respect to the separation from the spectral mean. The spectra were recorded with an excitation intensity of 20 W/cm^2

by changing the excitation intensity. We think that the spectral diffusion is caused by internal conversion, B800–B850 energy transfer, intersystem crossing, and subsequent dissipation of vibrational energy in the complex. This energy is dumped as heat in the protein surrounding the pigments, inducing conformational changes which in turn give rise to changes of the absorption frequencies of the chromophores.

To analyze the statistics of the spectral jumps we have fitted every transition in the B800 band in every single sweep with a Lorentzian. From the fits the spectral position of each transition was determined as a function of time. This yields the absolute spectral distance covered per unit time for each particular transition (spectral motion in cm^{-1}/s). Figure 3.9 shows the amount of spectral motion for the absorption lines of single BChl a molecules as a function of the spectral position in the B800 band. To exclude intercomplex heterogeneity and to be able to compare the data from different LH2 complexes the spectral position is given with respect to the spectral mean, $\bar{\nu}$, of the complex under study. Remarkably, the spectral motion of a particular absorption is correlated with its spectral position within the B800 band. The spectral diffusion increases towards the wings of the spectral distribution of absorption, as can be seen from the ensemble spectrum which is

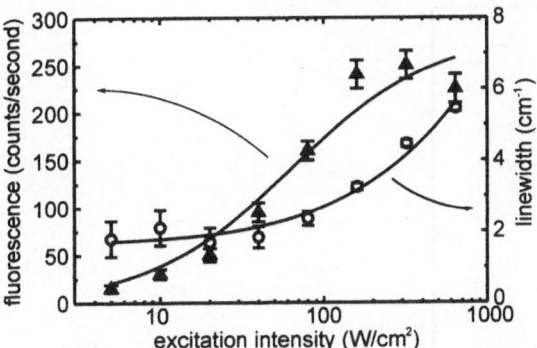

Fig. 3.10. The fluorescence emission rate for a particular B800 transition in units of detected photons per second (black triangles, left vertical scale) and the homogeneous linewidth (open circles, right vertical scale) versus the excitation intensity

included in Fig. 3.9 for illustration. Apparently, the probability for conformational changes in the immediate environment is larger for pigments which show absorption frequencies with large deviations from the mean.

The homogeneous linewidth of the individual absorption lines in the B800 band could not be obtained directly due to spectral diffusion effects. However, the previously described method of data acquisition allowed us to diminish the influence of the spectral motions on the observed linewidths in the following way. First every single transition in each fast data scan was fitted by a Lorentzian from which the peak position for each absorption was obtained. Second, for a particular transition the separate scans were shifted in the spectral position such that the fitted peak positions for that transition coincided. In the third step the shifted raw spectra were averaged and the width of the absorption line under study could be determined after deconvolution with the laser linewidth, devoid of artificial line broadening caused by spectral diffusion. This procedure was repeated for all absorption lines in a single complex spectrum. Given the scan speed of the laser, all light-induced spectral movement on a timescale slower than 50 ms could be suppressed. Figure 3.10 shows the dependence of the linewidth, Γ, and the emission rate, R, for a particular B800 transition on the excitation power. The data could be fitted by the well-known expressions for the saturation behavior of two-level systems [21]

$$R(I) = R_\infty \frac{I/I_s}{(1 + I/I_s)} , \tag{3.4}$$

$$\Gamma(I) = \Gamma(0)\sqrt{1 + I/I_s} , \tag{3.5}$$

where R_∞ is the fully saturated emission rate, I_s is the saturation intensity, and $\Gamma(0)$ is the homogeneous linewidth. For the data shown we obtain $R_\infty = 280$ detected photons per second, which corresponds to roughly

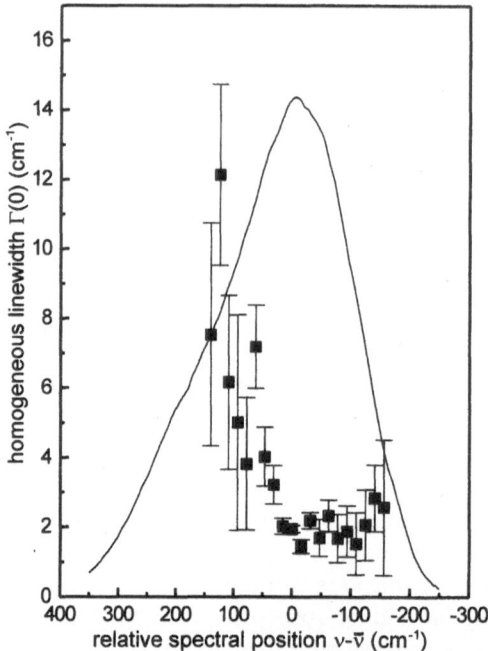

Fig. 3.11. The dependence of the homogeneous linewidth of the observed absorption lines on the spectral position in the B800 band with respect to the spectral mean. The spectra were recorded with an excitation intensity of 20 W/cm^2

600,000 emitted photons per second. The homogeneous linewidth is determined to be $\Gamma(0) = 1.7 \pm 0.2$ cm^{-1}, resulting in an excited state lifetime of 3.2 ps.

The homogeneous linewidth of a particular absorption shows a strong dependence on spectral position within the B800 band, as shown in Fig. 3.11. To exclude intercomplex heterogeneity we have plotted the inhomogeneous linewidth as a function of the spectral separation from the spectral mean rather than as a function of the absolute spectral position. The homogeneous linewidth decreases from about 10 cm^{-1} on the blue side of the B800 band to less than 2 cm^{-1} in the center and to the red side of this band.

A similar dependence of the homogeneous linewidth on the absolute spectral position has been found by hole-burning studies on other light-harvesting pigment-protein complexes of purple bacteria [8,22]. Based on these ensemble data it was concluded that for all B800 states B800 → B850 interband excitation energy transfer is effective, whereas for the energetically higher B800 states also, energy transfer occurs to lower-lying B800 molecules, driven by a Förster-like process [23,24]. The different pathways result in variations of the lifetimes of the respective states over the region of the inhomogeneous linewidth. The justification for applying first-order perturbation theory, upon which the Förster model is based, lies in the assumption that the dipole–dipole interaction between neighboring pigments is very small compared to their difference in transition energy. The energy-transfer rates in the Förster picture are then determined by the spectral overlap between donor and acceptor molecules. However, pronounced phonon sidebands are absent in our spectra and the zero-phonon lines are spectrally distributed over a region about 25 times their homogeneous linewidth, both observations implying very small spectral overlap between donor and acceptor. Wu and coworkers have shown that under such conditions Förster transfer would occur with a transfer time of tens of picoseconds [8], i.e., much slower than observed in hole-burning experiments. This leads us to the conclusion that Förster processes cannot explain the observed B800 intraband energy transfer. Moreover, a comparison of the value of 130 cm^{-1} for the diagonal disorder with the interaction strength, -24 cm^{-1}, in the point-dipole approximation, shows that there is no question of the weak coupling between neighbouring BChl a molecules.

From experiments and modeling on the FMO antenna complex from green sulfur bacteria it is known that similar ratios of the diagonal disorder and the interaction strength lead to a slight delocalization of the excitation over 2–3 pigments [25]. In this situation it has been shown that the dynamical properties of the excited states are governed by exciton–phonon interactions and that the lifetimes of individual levels are determined by vibronic relaxation to the lower states in the exciton manifold. We believe that a similar description applies to the dynamic properties of the levels in the B800 band of the LH2 complex, i.e., that the observed increase in the homogeneous linewidth towards the high-energy side of the B800 absorption band is caused by vibronic relaxation rather than a Förster-type energy transfer

3.2.2 The B850 Band

In the B850 band, the interaction strength between BChl a molecules is determined to be about 300 cm^{-1} [12], i.e., considerably larger than the disorder (estimated to be about 125 cm^{-1}). Therefore, we have to consider excitonic interactions in order to understand the optical spectra. As a starting point, we calculated the excited-state manifold of a cylindrically symmetric B850 assembly with zero disorder. Of the two nondegenerate (denoted $k = 0$ and $k = 9$) and eight pairwise degenerate ($k = \pm1$, $k = \pm2$, ..., $k = \pm8$) exciton states, only the low-energy degenerate pair $k = \pm1$ will carry appreciable oscillator strength (Fig. 3.12 left). Upon introducing diagonal disorder in the ring the pairwise degeneracies will be lifted, and the oscillator strength will be redistributed over adjacent exciton states [13] (Fig. 3.12 right). The transition dipole moments associated with the $k = \pm1$ states will have orthogonal polarizations. This orthogonality is maintained when disorder is introduced, assuming that the diagonal disorder is dominated by variations in electrostatic interactions and possibly intermolecular distances, rather than by changes in the orientations of the BChl a molecules.

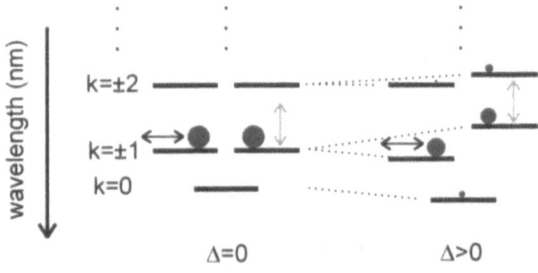

Fig. 3.12. Schematic representation of the energy-level scheme of the lowest states in the excited-state manifold of the B850 ring in LH2 of *Rps. acidophila*. Compared are the relative positions of the lowest levels in the presence (left) and in the absence (right) of ninefold rotational symmetry. The gray circles indicate the initial population of a given excited state, and the arrows indicate the relative orientation of the transition dipole moments in the plane of the ring

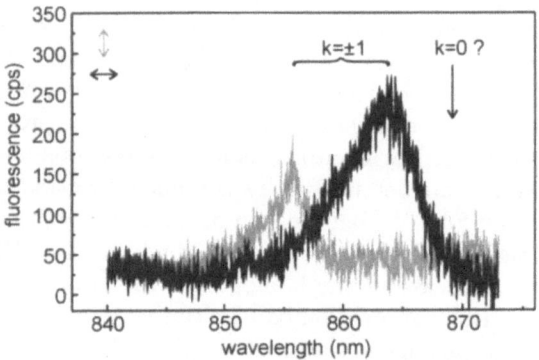

Fig. 3.13. Fluorescence-excitation spectrum of the long-wavelength region of the B850 band of an individual LH2 complex of *Rps. acidophila* for mutually orthogonal polarized excitation

In all spectra of the single LH2 complexes that we observed, the B850 band consisted of two broad absorption lines at ∼860 nm, sometimes accompanied by a weaker third transition at the higher energy side. These observations can be explained in terms of the exciton model. The two absorptions correspond to transitions to the $k = \pm 1$ states which have their degeneracies lifted. By performing polarization-dependent experiments on these two bands [26], the orthogonality of the associated transition dipole moments predicted by the exciton model could be ascertained (Fig. 3.13). This orthogonality was observed in all individual LH2 complexes that we studied and is a strong argument in favor of a high degree of delocalization of the excitations. The observed homogeneous linewidth of ∼50 cm^{-1} for the transitions to the $k = \pm 1$ states is consistent with anisotropy decay times of ∼100 fs found in pump–probe experiments [11]. The extent of delocalization will decrease, and the

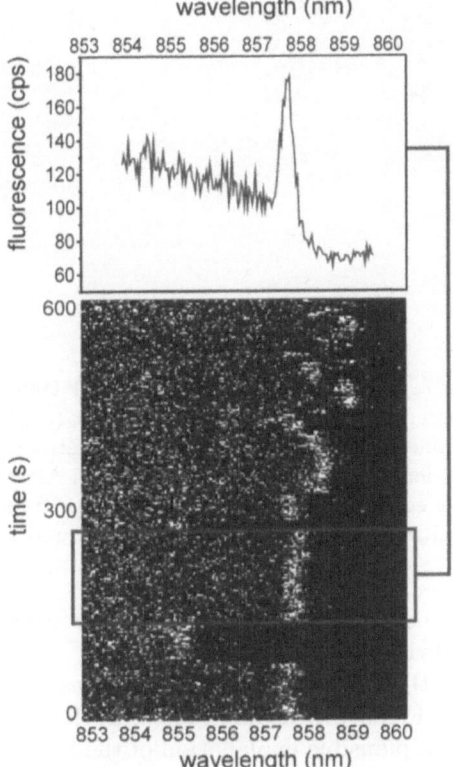

Fig. 3.14. Fluorescence-excitation spectrum of the red wing of the long-wavelength absorption of the B850 band of a single LH2 complex of *Rps. acidophila*. In the bottom panel, a stack of 200 consecutively recorded spectra (3 s per scan) is shown where the fluorescence intensity is given in gray scale. The spectrum in the top panel corresponds to an average of only those scans that are covered by the box. For this particular complex, the whole set of lines in the B850 band is shifted towards higher energies in comparison to the complex shown in Fig. 3.13

dynamical properties will change at higher temperatures, where mixing of the exciton states by vibronic coupling will occur [27].

Another observation supporting the excitonic level scheme is the detection of the lowest exciton state $k = 0$. By repeatedly scanning the excitation wavelength quickly through the low energy side of the $k = \pm 1$ pair and following the spectral features through time, the presence of the spectrally rapidly diffusing lowest exciton state could be observed in a fraction of the complexes studied (Fig. 3.14). The low intensity of the $k = 0$ transition, which in principle is symmetry-forbidden, and spectral diffusion on a time scale faster than that of the experiment, may explain the absence of this lowest exciton transition in most of the complexes. The linewidth of the $k = 0$ state should be ~ 0.005 cm^{-1}, as determined by the 1.1-ns fluorescence lifetime

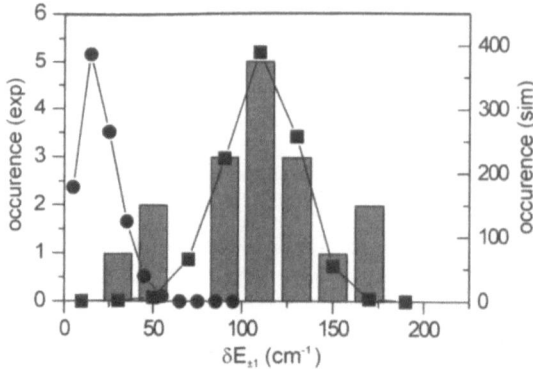

Fig. 3.15. The distribution of the energy separations $\delta E \pm_1$ of the $k = \pm 1$ transitions in the B850 band. The histogram represents the experimental data (exp) for all complexes studied. The values obtained from the Monte Carlo simulations, assuming only a disorder of 125 cm^{-1}, are depicted by the solid black circles. After an additional elliptical deformation with an eccentricity of $\varepsilon = 0.52$ is introduced in the simulations (sim), one obtains the data represented by the solid black squares

of the system, but the observed value of ~ 5 cm^{-1} is mainly determined by residual spectral diffusion and the bandwidth of the excitation source.

The energy splitting $\delta E \pm_1$ between the $k = +1$ and $k = -1$ states was measured for all complexes investigated (Fig. 3.15). As mentioned previously, the presence of the disorder Δ provides a plausible explanation of the lifting of the degeneracy of these exciton states. To check this point of view Monte Carlo simulations were performed on the basis of the crystal structure of LH2 of *Rps. acidophila*. Only nearest-neighbor interactions were taken into account in the point-dipole approximation. Random diagonal disorder (FWHM = 125 cm^{-1}) was introduced for both the undistorted and distorted rings, assuming that it is centered at the same transition energy for every B850 pigment. In the case of the distorted ring, the individual pigments were located on an ellipse, whereas the long-wavelength (Q_y) transition dipoles retained the alignment seen in the crystal structure. The results of these simulations show that the observed average $\delta E \pm_1$ of ~ 110 cm^{-1} cannot be explained by taking into account only random disorder (as depicted in Fig. 3.15) for a simulated Δ with a full width at half maximum (FWHM) of 125 cm^{-1}. Even an unreasonably large value of $\Delta \approx 500$ cm^{-1} for the width of the distribution of site energies did not result in an energy separation between the $k = \pm 1$ exciton states, as observed experimentally. To exclude the possibility that these abnormally large splittings are caused by an anisotropic environment in the polymer matrix, we repeated the experiment on single LH2 complexes in a glycerol matrix, which resulted in similar values of $\delta E \pm_1$. A Jahn–Teller-like deformation in the excited state causing the observed splitting is unlikely in view of the unrealistically large values needed for the electronic–nuclear coupling strength.

Although random disorder can give rise to large variations in the absorption wavelengths of the B850 bands (Fig. 3.3), the observed splitting $\delta E_{\pm 1}$ can only be explained in terms of largely correlated disorder, such as a static symmetric distortion of the protein complex in the ground state. In the case of an elliptical deformation of the ring in first approximation only the $k = \pm 1$ exciton states split. This is in agreement with the absence in the spectra of a splitting and a polarization effect of the higher-lying k states. The eigenfunctions of the $k = \pm 1$ states then belong to the long and short axes of the ellipse and hence exhibit orthogonal polarization of their transition moments, as observed experimentally. Our simulations show that the observed splittings can be explained by assuming an eccentricity ε of the ring of 0.52, corresponding to a ratio of the long and short radius of 0.85, and a random disorder of 125 cm^{-1} (Fig. 3.15); $\varepsilon = (1 - a^2/b^2)^{1/2}$, where a and b are the length of the short and long axes, respectively. An explanation for the symmetry lowering in the LH2 complex from ninefold in the crystals used for resolving the x-ray structure to the twofold symmetry observed in our experiments may be found in the extremely dense packing of LH2 in the x-ray crystals, causing a stabilization of the structure. In our case of completely isolated complexes these stabilizing forces are absent and the complex is more susceptible to deformation. What the symmetry properties of the LH2 complex are in a natural environment, surrounded by a limited number of LH2 complexes in the photosynthetic membrane, is therefore an intriguing question and deserves further study.

3.3   Conclusion

This work demonstrates that single-molecule spectroscopy is a powerful tool to reveal in detail the factors determining the electronic structure of pigment-protein complexes and, more generally, of molecular aggregates. Various manifestations of disorder can be probed directly, providing valuable information for the theoretical modeling of energy-transfer processes in these systems, a better understanding of the structure of these biologically important systems, and an understanding of how these systems function.

Acknowledgment. This work is supported by the "Stichting voor Fundamenteel Onderzoek der Materie (FOM)" with financial aid from the "Nederlandse Organisatie voor Wetenschappelijk Onderzoek (NWO)." One of us (J. K.) was fellow of the Heisenberg Program of the "Deutsche Forschungsgemeinschaft (DFG)."

References

1. W. E. Moerner and M. Orrit, "Illuminating single molecules in condensed matter, Science **283**, 1670–1676 (1999)

2. M. Z. Papiz, S. M. Prince, A. M. Hawthornthwaite-Lawless, G. McDermott, A. A. Freer, N. W. Isaacs, and R. J. Cogdell, "A model for the photosynthetic apparatus of purple bacteria," Trends in Plant Science **1**, 198–206 (1996)

3. X. Hu, T. Ritz, A. Damjanovic, and K. Schulten, "Pigment organization and transfer of electronic excitation in the photosynthetic unit of purple bacteria," J. Phys. Chem. B **101**, 3854–3871 (1997)

4. G. McDermott, S. M. Prince, A. A. Freer, A.M. Hawthornthwaite-Lawless, M. Z. Papiz, R. J. Cogdell, and N. W. Isaacs, "Crystal structure of an integral membrane light-harvesting complex from photosynthetic bacteria," Nature **374**, 517–521 (1995)

5. S. Karrasch, P. A. Bullough, and R. Ghosh, "The 8.5 angstrom projection map of the light-harvesting complex I from Rhodospirillum rubrum reveals a ring composed of 16 subunits," EMBO J. **14**, 631–638 (1995)

6. J. T. M. Kennis, A. M. Streltsov, H. Permentier, T. J. Aartsma, and J. Amesz, "Exciton coherence and energy transfer in the LH2 complex of *Rhodopseudomonas acidophila* at low temperature," J. Phys. Chem. B **101**, 8369–8374 (1997)

7. R. Monshouwer, I. Ortiz de Zarate, V. F. Mourik, and R. van-Grondelle, "Low-intensity pump-probe spectroscopy on the B800 to B850 transfer in the light harvesting 2 complex of *Rhodobacter sphaeroides*," Chem. Phys. Lett. **246**, 341–346 (1995)

8. H.-M. Wu, S. Savikhin, N. R. S. Reddy, R. Jankowiak, R. J. Cogdell, W. S. Struve, and G. J. Small, "Femtosecond and hole-burning studies of B800's excitation energy relaxation dynamics in the LH2 antenna complex of *Rhodopseudomonas acidophila* (Strain 10050)," J. Phys. Chem. **100**, 12022–12033 (1996)

9. M. Chachisvilis, O. Kuhn, T. Pullerits, and V. Sundstrom, "Excitons in photosynthetic purple bacteria: wavelike motion or incoherent hopping?" J. Phys. Chem. B **101**, 7275–7283 (1997)

10. R. Jimenez, S. N. Dikshit, S. E. Bradford, and G. R. Fleming, "Electronic excitation transfer in the LH2 complex of *Rhodobacter sphaeroides*," J. Phys. Chem. **100**, 6825–6834 (1996)

11. S. I. E. Vulto, J. T. M. Kennis, A. M. Streltsov, J. Amesz, and T. J. Aartsma, "Energy relaxation within the B850 absorption band of the isolated light-harvesting complex LH2 from Rhodopseudomonas acidophila at low temperature," J. Phys. Chem. B **103**, 878–883 [1999)

12. Sauer, K., R. J. Cogdell, S. M. Prince, A. A. Freer, N. W. Isaacs, and H. Scheer, "Structure-based calculations of the optical spectra of the LH2 bacteriochlorophyll-protein complex from *Rhodopseudomonas acidophila*," Photochemistry and Photobiology **64**, 564–576 (1996)

13. R. G. Alden, E. Johnson, V. Nagarajan, W. W. Parson, C. J. Law, and R. J. Cogdell, "Calculations of spectroscopic properties of the LH2 bacteriochlorophyll-protein antenna complex from *Rhodopseudomonas acidophila*," J. Phys. Chem. B **101**, 4667–4680 (1997)

14. J. T. M.,Kennis, A. M. Streltsov, T. J. Aartsma, T. Nozawa, and J. Amesz, "Energy transfer and exciton coupling in isolated B800-850 complexes of the photosynthetic purple sulfer bacterium *Chromatium tepidum*. The effect of structural symmetry on bacteriochlorophyll excited states," J. Phys. Chem. B **100**, 2438–2442 (1996)

15. R. Monshouwer, M. Abrahamsson, F. van-Mourik, and R. van-Grondelle, "Superradiance and exciton delocalisation in bacterial photosynthetic light-harvesting systems," J. Phys. Chem. B **101**, 7241–7248 (1997)

16. V. I. Novoderezhkin and A. P. Razjivin, "Exciton dynamics in circular aggregates: Application to antenna of photosynthetic purple bacteria," Biophys. J. **68**, 1089–1100 (1995)

17. T. Pullerits, M. Chachisvilis, and V. Sundstrom, "Exciton delocalization length in the B850 antenna of *Rhodobacter sphaeroides*," J. Phys. Chem. **100**, 10787–10792 (1996)

18. A. P. Shreve, J. K. Trautman, H. A. Frank, T. G. Owens, and A. C. Albrecht, "Femtosecond energy-transfer processes in the B800-850 light-harvesting complex of Rhodobacter sphaeroides 2.4.1," Biochim. et Biophys. Acta **1058**, 280–288 (1991)

19. V. Sundstrom, T. Pullerits, and R. van-Grondelle, "Photosynthetic light-harvesting: Reconciling dynamics and structure of purple bacterial LH2 reveals function of photosynthetic unit," J. Phys. Chem. B **103**, 2327–2346 (1999)

20. M. A. Bopp, J. Yiwei, L. Li, R. J. Cogdell, and R. M. Hochstrasser, "Fluorescence and photobleaching dynamics of single light-harvesting complexes," Proc. Natl. Acad. Sci. USA **94**, 10630–10635 (1997)

21. W. P. Ambrose, T. Basché, and W. E. Moerner, "Detection and spectroscopy of single pentacene molecules in a p-terphenyl crystal by means of fluorescence excitation, J. Chem. Phys. **95**, 7150–7162 (1991)

22. C. De Caro, R. Visscher, R. van-Grondelle, and S. Volker, "Inter- and intraband energy transfer in LH2-antenna complexes of purple bacteria. A fluorescence line-narrowing and hole-burning study," J. Phys. Chem. **98**, 10584–10590 (1994)

23. T. Joo, Y. Jia, J.-Y. Yu, D. M. Jonas, and G. R. Fleming, "Dynamics in isolated bacterial light harvesting antenna (LH2) of *Rhodobacter sphaeroides* at room temperature," J. Phys. Chem. **100**, 2399–2409 (1996)

24. J. T. M. Kennis, A. M. Streltsov, S. I. E. Vulto, T. J. Aartsma, T. Nozawa, and J. Amesz, "Femtosecond dynamics in isolated LH2 complexes of various species of purple bacteria," J. Phys. Chem. B **101**, 7827–7834 (1997)

25. S. I. E. Vulto, M. A. de-Baat, S. Neerken, F. R. Nowak, H. van-Amerongen, J. Amesz, and T. J. Aartsma, "Excited state dynamics in FMO antenna complexes from photosynthetic green sulfur bacteria: A kinetic model," J. Phys. Chem. B **103**, 8153–8161 (1999)

26. A. M. van-Oijen, M. Ketelaars, J. Köhler, T. J. Aartsma, and J. Schmidt, "Unraveling the electronic structure of individual photosynthetic pigment-protein complexes," Science **285**, 400–402 (1999)

27. J. A. Leegwater, "Coherent versus incoherent energy transfer and trapping in photosynthetic antenna complexes," J. Phys. Chem. **100**, 14403–14409 (1996)

4 Single-Molecule Optical Switching: A Mechanistic Study of Nonphotochemical Hole-Burning

F. Kulzer and T. Basché

Persistent spectral hole-burning of dopant chromophores embedded in solid matrices has proven to be a sensitive high-resolution spectroscopic tool to investigate structural and dynamic properties of amorphous and crystalline hosts at low temperature [1]. A commonly encountered mechanism of hole-formation is the nonphotochemical process, for which it is assumed that the frequency selective laser excitation and the subsequent relaxation of guest and host eventually leads to a change of configurational degrees of freedom in the nearby environment of the photo-excited centers or in the impurities themselves (or both) [2]. However, detailed knowledge about the microscopic mechanism of the nonphotochemical process is rare. Methyl group spin conversion [3] and rearrangement of hydrogen bond networks [1] belong to the few mechanisms known in the literature. In the present work we want to introduce a model system which allows the reproducible observation of nonphotochemical hole-burning at the single molecule level, a phenomenon which amounts to the controlled optical manipulation of an isolated chromophore. We will illustrate how a number of experimental techniques available in single molecule spectroscopy can be combined to obtain ample information about the underlying hole-burning mechanism. Then we will introduce a theoretical approach [4] to elucidate the microscopic nature of the configurational degrees of freedom responsible for the formation of the photoproduct: the results of recent molecular dynamics simulations do indeed admit a detailed mechanistic scenario for the hole-burning process in our model system. We thus hope to demonstrate how studies at the single molecule level can serve to improve our understanding of the structural dynamics of solids at low temperatures. To do so, we will start by giving a brief overview of the basic concepts of low-temperature single-molecule spectroscopy.

Single-molecule spectroscopy (SMS) has become a well established experimental technique in the past 10 years, with applications in such diverse fields as quantum optics, solid state physics, material sciences and biophysical research [5–8]. The key issue in single-molecule detection in the condensed phase is being able to selectively address just one absorber and detect its signal above the background resulting from the matrix. To achieve this goal, it is important to illuminate a suitably small excitation volume because of background reduction, and to collect the emitted fluorescence with efficient optics. With far-field optics the minimum dimensions of the sample volume are determined by the diffraction limit. An important technique of that variety is confocal microscopy, which uses a high numerical aperture microscope objective to both focus the excitation light and collect the emitted fluorescence [9]. For the experiments presented below the resolution achieved by far-field illumination turned out to be quite sufficient, since mixed molecular crystals of terrylene in p-terphenyl with an extremely low concentration of chromophores can easily be prepared by sublimation [10]. If the average number of absorbers in the excitation volume becomes significantly smaller than one, then most, if not all, absorbers can be excited individually when the laser spot is scanned across the sample.

The procedure described above amounts to a spatial isolation of chromophores and its applicability is independent of temperature. An additional effect can be utilized to single out individual absorbers, but this approach is only feasible in the low temperature regime: defects in the structure of the solid matrix and the resulting variations in the local environment of the guest molecules will lead to a distribution of transition frequencies, and the absorption band is said to be inhomogeneously broadened. At liquid helium temperatures, when coupling to low-energy excitation modes (phonons) of the host lattice is of minor importance, the chromophores can have homogeneous line-widths which are several orders of magnitude smaller than the range of the inhomogeneous distribution of their transition frequencies. Under these conditions the number of absorbers in the excitation volume can be significantly larger then one without losing the ability to address single molecules. Scanning a single-frequency dye laser across the inhomogeneous band then makes it possible to distinguish individual chromophores by their absorption frequency [11,12].

The present paper focuses on terrylene ($C_{30}H_{16}$, see Fig. 4.2) in p-terphenyl, a system which exhibits many favorable properties for single-molecule studies both in a cryogenic environment and at room temperature [10,13–15]. Therefore we will dedicate the next section to a brief description of the structure of the p-terphenyl matrix and its temperature-dependent phase transition.

4.1   The Structure of Crystalline p-Terphenyl

The room temperature modification of pure p-terphenyl crystal has monoclinic symmetry and the corresponding elementary cell contains two inequivalent molecules, see Fig. 4.1a. The two outer phenyl rings of each p-terphenyl molecule lie in one plane and the long axes of all molecules are aligned parallel to the crystallographic c-axis. Matrix molecules in alternating rows are tilted $\pm 32.7°$ relative to the b-axis. The central phenyl rings of all molecules perform librational movements in a symmetric double-well potential with

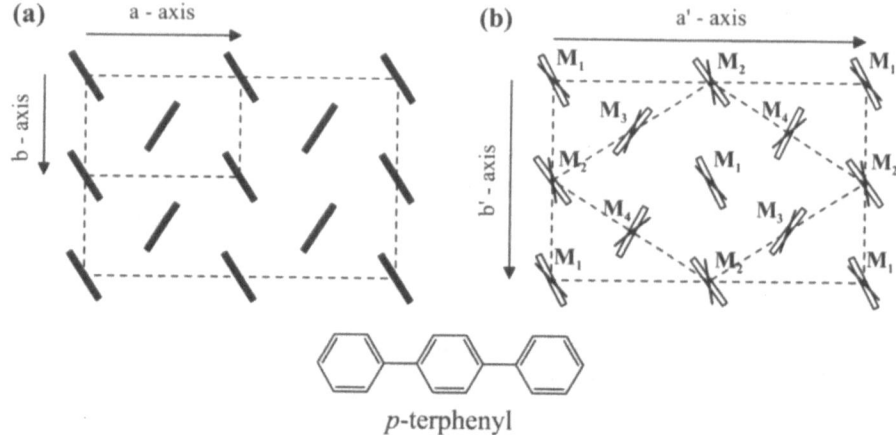

p-terphenyl

Fig. 4.1. a The monoclinic room temperature modification of p-terphenyl crystal, as seen in projection onto the crystallographic (a,b)-plane. The elementary cell contains two inequivalent molecules, both of which are aligned with their long molecular axes parallel to the crystallographic c-axis. The figure shows a larger section consisting of 2×2 elementary cells to facilitate comparison with the low-temperature modification. (adapted from [16].) **b** Schematic representation of the corresponding low-temperature ($T < 193$ K) modification. The elementary cell now contains four nonequivalent p-terphenyl molecules denoted M_1–M_4. The figure depicts a larger, pseudo-monoclinic cell; the true triclinic elementary cell is indicated by the inner dashed rhomboid. The open bars show the orientation of the two outer phenyl rings of each molecule, while the shorter solid lines indicate the direction of the central ring. There are two possible structures which form separate domains at low temperatures; the one not shown here is related to Fig. 4.1b by simple mirror symmetry (adapted from [17])

minima at a tilt angle of $\pm 13°$ against the plane of the outer rings, so that the molecules appear to be pseudo-planar in X-ray structures [16].

This librational movement occurs in an asymmetric double-well potential below 193 K, which leads to the formation of small clusters of antiferromagnetic central ring ordering. Slow cooling below the temperature of this disorder–order phase transition will allow the clusters to grow into macroscopic domains [17] which can be observed by polarization microscopy [18]. There are two inequivalent domain structures which can be transformed into each other by simple symmetry operations.

Recently a microscopic all-atom model of p-terphenyl has been developed by Bordat and Brown [19], which is in close correspondence with structural and dynamical properties of the real system, as deduced from X-ray, neutron diffraction and scattering and NMR data. Their model also correctly predicts the disorder–order transition around 190 K, the ring flipping dynamics, and the critical slowing near the phase transition. These results will

become important for the understanding of an intriguing single-molecule optical switching phenomenon which will be discussed in the next section.

4.2 Single-Molecule Optical Switching of Terrylene in p-Terphenyl

An important criterion for the suitability of a given host/guest system for
frequency selective SMS is the structural compatibility of the guest chromophores and the host lattice, since the former have to replace one or more
matrix molecules in a "well-defined" and stable way. In spite of the considerable size difference between terrylene and p-terphenyl (cf. Figs. 4.1 and
4.2) there obviously exist four distinct crystallographic insertion sites in p-
terphenyl, giving rise to four well-separated purely electronic origins at liquid
helium temperatures, which are denoted sites X_1–X_4 [10,13].

Single molecules can be studied in all four sites, but the photophysical properties of the chromophores depend strongly on the site investigated
[10,13]. The characteristic difference between the sites is the stability of the
single molecule absorption frequencies under prolonged resonant excitation.
Absorbers from site X_2 generally exhibit excellent photostability, meaning
that they can be exposed to high excitation intensities for a long time (hours
at least) without any changes of their resonance frequency, achieving high
saturation count rates. These favorable features of site X_2 have made possible a number of interesting experiments [20–25]. Molecules from X_4 generally
show comparable photostability and count rates, but those in the wings of
that site's inhomogeneous distribution are sometimes prone to spectral diffusion (small spontaneous frequency jumps) [26].

The situation is quite different as far as chromophores belonging to X_1
or X_3 are concerned. Single molecule fluorescence excitation spectra can be
recorded in these two sites as well, provided that only a moderate excitation
intensity is applied. When the laser wavelength is held fixed at the absorption maximum of a given chromophore, however, this will inevitably induce

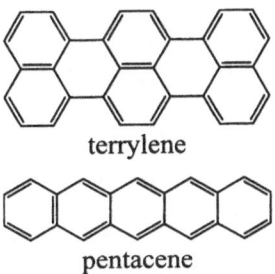

terrylene

pentacene

Fig. 4.2. Chemical formulae of the dopant chromophores terrylene and pentacene,
respectively. The present work is concerned with terrylene in p-terphenyl, but at
some points a comparison to pentacene will become relevant

a change in its absorption frequency, thus shifting it out of resonance. It is remarkable that these light-induced frequency shifts do not involve an irreversible photodestruction of the chromophores (photobleaching), but constitute the single-molecule analogue of nonphotochemical hole-burning. It can be demonstrated that the structural changes triggered by resonant irradiation are furthermore fully reversible. This can be seen directly in site X_3, where all molecules which have been burned away return to their old spectral position spontaneously (thermally induced) within a few minutes – even at $T = 1.4\,\mathrm{K}$ – and the process can be repeated over and over. The most promising properties for mechanistic investigations are found in site X_1 where the photoproducts are stable at temperatures below $10\,\mathrm{K}$, which allows detailed spectroscopic investigations [27,28]. The next two subsections summarize the investigations conducted so far and introduce a detailed mechanistic model for the single-molecule optical switching process.

4.2.1 Experimental Results

Optical switching of a single-molecule resonance has already been demonstrated for two other chromophore/matrix systems [29–32]. However, conclusive mechanistic investigations of the light-driven process were always severely hindered by spontaneous spectral dynamics, nonuniform behavior of the chromophores and often unknown spectral positions of the photoproducts. Terrylene molecules in the X_1 site of p-terphenyl on the other hand allow optical manipulation at the single-molecule level with unprecedented control and reproducibility [27]. All molecules exhibit an identical and fully reversible light-driven behavior which is not disturbed by any spontaneous frequency jumps at $T = 1.4\,\mathrm{K}$. This system additionally provides the opportunity for a detailed spectroscopic characterization of the photoproducts. The enormous and reproducible jump widths (more than 10 000 times the single-molecule line-width) indicate substantial and well-defined rearrangements of the insertion geometry. An illustration of this behavior is given in Fig. 4.3.

In the beginning (trace a) we find both molecules at their original spectral positions in X_1 and the photoproduct site (XY) is empty. Prolonged resonant excitation of either chromophore eventually induces a single-molecule hole-burning event, which manifests itself by a sudden drop of the fluorescence count rate to the background level as the molecule is shifted out of resonance. All such spectral jumps are indicated by grey arrows in Fig. 4.3. The marked (*) molecule is sent to the XY site first (b) and increases its absorption frequency by $843\,\mathrm{GHz}$ in the process (this corresponds to almost 1 nm; note the huge axis break in Fig. 4.3). The photoproduct state is stable at low temperatures ($T < 10\,\mathrm{K}$) and does not change spontaneously, but any prolonged excitation in XY will bring the molecule back exactly to its original position in X_1; this can be seen in (c) with the other molecule serving as a frequency marker during this first light-driven jump cycle. This marker molecule can be switched to XY just as easily (d) and this is reversible as well (not shown in Fig. 4.3). When both chromophores are brought to the photosite (see trace e) we find that their spectral pattern in X_1 is reproduced in XY, which proves that both molecules exhibit exactly the same jump width. The light-driven nature of the frequency jumps can easily be verified by determining the intensity-dependent distribution of waiting times prior to spectral jumps [33].

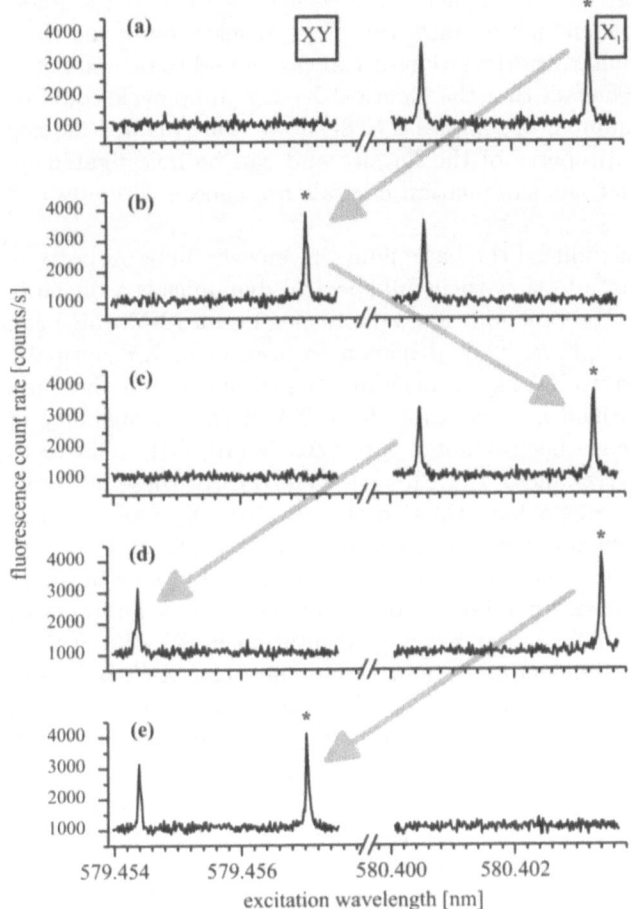

Fig. 4.3. Reversible optical switching of two terrylene molecules in p-terphenyl at $T = 1.4\,\text{K}$. Both absorbers undergo light-driven jump cycles between the X_1 site and the XY photosite; every switching processes is triggered by light irradiation and is symbolized here by a grey arrow. The lines corresponding to one of the chromophores have all been marked with an asterisk for identification purposes. The content of the five traces is described in the text. Note the huge break in the wavelength axis (adapted from [27])

The identical behavior of all X_1 molecules marks a clear distinction between terrylene in p-terphenyl and the other known cases of single-molecule hole-burning. In "low concentration" samples (about 10^{-11} mol/mol) all X_1 molecules exhibit identical reversible jumps over $\Delta\nu = 843 \pm 2$ GHz, with a variation of 500 MHz at most for molecules in the same crystal [27]. An increasing variation in the jump widths is found for samples with a higher guest concentration, as each terrylene molecule itself intrinsically constitutes a defect in the ideal p-terphenyl structure. However, as long as the guest concentration is kept low enough to allow the spectral selection of the chromophores at the band center, the deviations are always found to be well below a 0.5% threshold [28]. The fact that the identical X_1-XY jump cycle could be reproduced for quite a number of samples [27,28] shows that this well-defined behavior is an intrinsic property of the X_1 site and can be investigated reproducibly, provided that the sublimation crystals are chosen carefully and handled properly.

Now that we have introduced the basic light-driven switching property of X_1 molecules, we have to add that their full spectral dynamics is a bit more involved, meaning that there are additional photosites besides XY which can be visited by the chromophores [27]. Resonant excitation in XY normally sends the absorbers back to X_1, as has been mentioned above, but there are three more photosites which are accessible from XY with a probability of about 15% and which have been denoted XY' ($\Delta\nu = 1109$ GHz relative to X_1), XY'' ($\Delta\nu = 1401$ GHz) and XY''' ($\Delta\nu = 1644$ GHz). All these positions are highly reproducible over a long time so that a given X_1 molecule can be observed at its various spectral positions for several weeks at least and one never "loses" a single molecule signal due to irreversible hole-burning. Typically, the chromophores undergo five to ten of the X_1-XY jump cycles before they burn away to XY' after resonant excitation in XY. From then on, one finds reversible light-induced jump cycles between XY'-XY'' and XY''-XY''' but no direct transitions XY'''-XY' and no light driven jumps back to XY or X_1 [27]. It appears that when driving a molecule from X_1 to XY''' this process always occurs successively via XY, XY', and XY'' without skipping any of the intermediate positions. The molecules can always be "reset" to their X_1 spectral position by raising the temperature above 20 K. Thus X_1 seems to be the most stable spectral position and one never finds any molecules in the photosites (in low concentration crystals) without prior burning in X_1.

The excellent reproducibility of the single-molecule optical switching process described above makes it possible to employ a combination of secondary spectroscopies to explore the underlying nonphotochemical hole-burning mechanism. The fundamental question arises whether the structural changes responsible for the spectral jumps are restricted to the host lattice or if they include distortions of the chromophores themselves. From dispersed fluorescence spectra of the photoproducts [27] it was found that the spectral jumps are most probably correlated to a significant change in the structural relaxation of the host lattice which accompanies the repeated $S_1 \leftrightarrow S_0$ transitions of the chromophores. Therefore it seems most probable that the mechanism of the single-molecule hole-burning process involves conformational rearrangements in the immediate vicinity of the terrylene molecules or of the absorbers themselves.

The most valuable information, however, could be obtained from DC Stark effect measurements on the original and the photoproduct state. Application of an external electric field induces a shift of the molecular absorption frequencies, and a quantitative analysis of this behavior yields information about molecular dipole moments and polarizabilities. The feasibility of DC Stark effect measurements at the single molecule level was first demonstrated for pentacene in p-terphenyl [34] and for terrylene in polyethylene [35]. Since then the use of this technique was extended to other organic host/guest-systems [36] as well as to single semiconductor nanocrystals [37].

When we record the fluorescence excitation spectrum of a single chromophore and determine its absorption frequency, we measure the difference in energy between the ground state (S_0) and the first excited singlet state (S_1). Hence the shift of the absorption line in an external electric field is related to the dipole moment and polarizability *differences* of these two states. However, care must be exercised in taking account of the dielectric properties of the matrix. A unified theory of the electronic Stark effect in molecular crystals, which takes into account the host–guest interactions within a point-dipole model, has been formulated by R. W. Munn [38], but a quantitative application to a given system requires a substantial amount of additional experimental work in order to obtain all pertinent parameters. For practical purposes a slightly simplified approach can be utilized to interpret Stark experiments: Its basic idea is to neglect the feedback effect of the induced guest dipole moment on the host matrix (the reaction field), only taking into account how the externally applied field is transformed by the dielectric medium. The measurements on terrylene in p-terphenyl discussed below were all conducted by applying a homogeneous electric field E^0 along the c-axis of the p-terphenyl crystal. The Stark effect response $\Delta\nu$ of the absorbers can then be described in good approximation by the following equation, which is a simplified scalar form of the full vector/tensor expression [39]:

$$h \cdot \Delta\nu = -\Delta\mu \cdot (f \cdot E^0) - \frac{1}{2} \cdot \Delta\alpha \cdot (f \cdot E^0)^2 \,, \qquad (4.1)$$

where $\Delta\mu$ and $\Delta\alpha$ correspond to the differences in dipole moment and polarizability, respectively, as seen in projection onto the direction of the applied electric field. These parameters can thus be readily obtained by recording the frequency shift $\Delta\nu$ as a function of the external electric field E^0, fitting a second-order polynomial to the data, and interpreting the linear and quadratic coefficient according to (4.1). The factor f, which relates the externally applied field to the field strength experienced in the dielectric host matrix, can

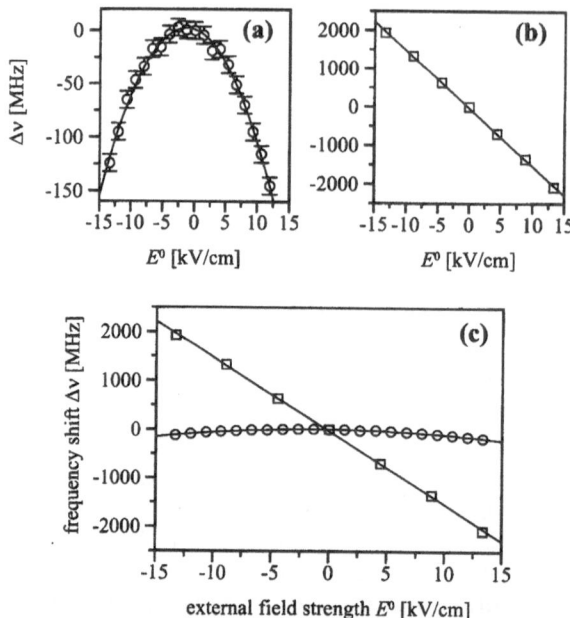

Fig. 4.4. Stark effect measurement on a single terrylene molecule at sites X_1 and XY. Panel **a** depicts the data corresponding to the X_1 site. The open circles with error bars mark the observed spectral shifts $\Delta\nu$ whose dependence on E^0 can be described by a second-order polynomial (*solid line*). In **b** we see the results of the same measurement for the XY photosite. The open squares indicate the Stark shifts which now can be described by a linear function. Panel **c** simply shows the data from **a** and **b** in one graph to facilitate comparison (adapted from [28])

be obtained from the field transformation tensor, which has been calculated for monoclinic p-terphenyl [40] (here $f = 1.254$).

A detailed investigation [28] showed that the reaction to the external field is dramatically different for the two spectral positions of the molecule. Predominantly quadratic behavior with only a small linear contribution is found in X_1; the Stark effect in XY on the other hand is characterized by a pronounced linear dependence and the maximum spectral displacement achieved is about ten times greater in the photosite. The field-dependent frequency shifts of one molecule at the X_1 and XY position can be seen in Fig. 4.4.

Application of (4.1) yields $\Delta\mu = +(3.5 \pm 0.2)$ mD (unit conversion factor: 10^{-3} Debye $= 3.33564 \cdot 10^{-33}$ Cm) and $\Delta\alpha = +(2.12 \pm 0.03)$ mD/(kV/cm) for this molecule in X_1 [28]. The increase of the dipole moment difference was found to be quite impressive when the XY photosite was investigated, and the observed Stark shifts could be fitted with a simple linear function in that case [28]. One obtains $\Delta\mu = +(237 \pm 2)$ mD, and we can only estimate an upper boundary for the polarizability difference, namely $|\Delta\alpha| \leq 2.5$ mD/(kV/cm).

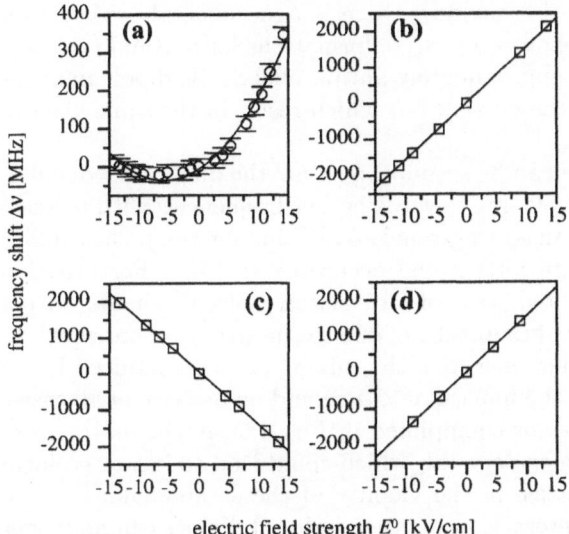

Fig. 4.5. Changes in the single-molecule Stark effect for consecutive visits in the photosite XY ($T = 1.4\,$K). Panel **a** shows the initial measurement in X_1, where predominantly quadratic behavior was found, corresponding to $\Delta\mu = (-16.3 \pm 0.6)\,$mD and $\Delta\alpha = (-2.3 \pm 0.1)\,$mD/kV/cm). Panels **b–d** depict the results of three measurements in XY, separated by light-induced visits in X_1. Linear fits to the data from panels **b** and **d** yield $\Delta\mu = (-247 \pm 2)\,$mD, while for panel **c** one obtains $\Delta\mu = (+231 \pm 2)\,$mD (adapted from [28])

So far 15 X_1 molecules from three different samples have been subjected to the measurement described above, and the same substantial increase in the dipole moment difference was always found when the chromophores were sent to the photosite XY [28]. The mean absolute value and the empirical standard deviation of $\Delta\mu$ in X_1 were found to be $|\Delta\mu| = (18 \pm 7)\,$mD, as compared to $|\Delta\mu| = (237 \pm 12)\,$mD obtained for the same 15 absorbers in XY. The fact that results discussed here were uniform for three different samples, in which the well-defined nature of the X_1–XY jumps was verified for all observed X_1 molecules [28], validates the assumption that the induced dipole moment changes found in these experiments constitute an intrinsic property of the configurational changes responsible for the spectral jumps.

An interesting observation can be made when absorbers are sent to photosite XY repeatedly: the chromophores are found to exhibit almost the same magnitude but different signs of the Stark shift $\Delta\nu$ (and correspondingly of $\Delta\mu$) for consecutive visits in XY. This behavior is illustrated in Fig. 4.5.

It is well known from hole-burning spectroscopy in crystalline matrices that chromophores with degenerate absorption frequencies can belong to subsets with different orientations of the dipole moments, leading to a splitting of spectral holes in external electric fields [42]. The novel result of the SMS experiments reported above is that now there is experimental evidence for the existence of two "XY" photoproduct conformations for one and the same absorber, which give rise to approximately antiparallel S_1–S_0 dipole moment differences (with respect to the c-axis), but which result in the same absorption wavelength (± 0.001 nm).

An additional peculiarity can be seen in Fig. 4.5a: the quadratic contribution clearly shifts the resonance line towards higher frequencies with increasing external field strength, suggesting a negative value for the polarizability difference $\Delta\alpha$ if the data are interpreted according to (4.1). Perturbation theory predicts that the excited state of an isolated molecule should have a higher polarizability than the ground state, due to the stronger contributions of electronic states with higher energy in the mixing of states induced by an external field. Consequently the molecular $\Delta\alpha$ should be positive in all cases. An explanation for the behavior exemplified in Fig. 4.5a might be the existence of structural defects in the crystal, which could lead to increased local polarizability of the host lattice in the vicinity of these "anomalous" chromophores. In that case the interaction between induced dipoles can no longer be neglected as is done in the simplified approach on which (4.1) is based. A more sophisticated description of the internal electric field in the crystal at an atomic level [43], as well as additional experimental work, will be needed to extract the effective molecular polarizabilities. However, one important piece of structural information can always be obtained from single-molecule Stark experiments, and that is whether a given insertion site is centrosymmetric or not (quadratic versus linear response).

The above presentation of the Stark effect measurements concludes our summary of experimental results on the single-molecule optical switching phenomenon. We wanted to illustrate the relative ease with which a combination of spectroscopic techniques can be employed to investigate the underlying nonphotochemical hole-burning mechanism in great detail. With the combined empirical evidence we are now able to outline a detailed mechanistic scenario, which was obtained by recent theoretical efforts [4].

4.2.2 Mechanistic Model

As a basis for making reasonable assumptions about the mechanism of the light-induced spectral jumps of terrylene in p-terphenyl, we need to have a detailed model of the crystalline insertion geometries corresponding to the spectral sites X_1–X_4. The difference in size between terrylene (1.36×0.67 nm^2) and p-terphenyl (1.36×0.48 nm^2) suggests that each terrylene molecule might have to replace more than one p-terphenyl in order to achieve a stable substitution geometry. The first molecular packing calculations [13] indeed seemed to favor a 1:2 substitutional scheme for the main sites, but unfortunately these calculations did not allow for structural relaxation other than translational and rotational movements of "stiff" molecules with fixed conformations. Therefore no conclusions about secondary minima in the potential energy surface – which constitute possible structures of photoproducts – could be reached. Quite recently, however, a very elaborate molecular dynamics study on terrylene in p-terphenyl has been reported by Bordat and Brown [4]. As a first step an all-atom empirical potential for the p-terphenyl matrix was developed, which correctly predicts the crystal structure, the phenyl ring flipping dynamics and the disorder–order phase transition [19]. The dopant chromophores pentacene and terrylene (see Fig. 4.2) were then modeled by adjusting the standard harmonic potentials for aromatic hydrocarbons [44] to reproduce their respective low-energy vibrational frequencies. Thus the full set of force-field parameters necessary for molecular dynamics simulations on the doped p-terphenyl crystals was obtained.

The procedure outlined above was first applied to pentacene [45] which shows four well-defined, purely electronic origins (commonly referred to as sites O_1–O_4) in low-temperature p-terphenyl, just like terrylene. Pentacene can easily be envisioned to form stable insertion sites in p-terphenyl by 1:1 substitution since there is an almost perfect match in the molecular dimensions of this guest and the host. The observed spectral sites O_1–O_4 can then be readily explained to arise from the respective substitution of one of the four inequivalent molecules which form the host lattice. However, it has proven difficult to establish a rigorous one-to-one correlation between the insertion sites $\{M_1, M_2, M_3, M_4\}$ (see Fig. 4.1b) and the spectral sites $\{O_1, O_2, O_3, O_4\}$ from empirical evidence alone, although most of the possible 24 permutations could be shown to be incompatible with specific experimental results [46–49].

The detailed computational investigations on the pentacene inclusion sites proceeded roughly as follows [45]. A sufficiently large section of the host lattice was modeled with the central p-terphenyl molecule being replaced by pentacene; this was done separately for all four 1:1 insertion sites. These initial structures were then allowed to relax under constant external pressure to find the corresponding minimum-energy conformations for low temperatures. As these molecular dynamics calculations made it possible to couple the modeled minicrystal to an adjustable heat bath, it was possible to study annealing processes at different temperatures. Furthermore, thermal heating–quenching cycles could be simulated to probe the potential energy hypersurface for secondary minima, the existence of which would indicate possible photoproduct conformations.

The results of the calculations outlined in the last paragraph were found to be in good agreement with experimental evidence and sufficient to unambiguously single out one of the remaining possible combinations as the most probable assignment of the substitutional sites to the individual electronic origins O_1–O_4. It is especially noteworthy for the subsequent comparison to terrylene that the simulations did not predict any stable photoproducts, which is in accordance with the absence of well-defined light-induced frequency jumps in the system pentacene/p-terphenyl.

The encouraging results on pentacene in p-terphenyl indicate that the treatment outlined above might be equally useful to address the question of the terrylene inclusion sites in the same matrix. First of all the problem of how many host molecules are replaced by terrylene had to be taken care of. To this end all possible structures stemming from either 1:1 or 1:2 substitution were subjected to the molecular dynamics simulation and the minimum-energy conformation was determined in each case [4]. These refined calculations now clearly indicated that a 1:2 substitution leads to energetically unfavorable structures with an exceptionally high degree of nonplanar distortions in the terrylene geometry. The 1:2 inclusion sites were furthermore found to be non-centrosymmetric, which is incompatible with the observed quadratic Stark effect [28] (this particular piece of experimental evidence had not been available when the 1:2 insertion model of [13] was proposed). Bordat and Brown could thus conclusively argue in favor of a 1:1 substitutional model, according to which the spectral sites X_1–X_4 of terrylene result form a direct replacement of the four structurally inequivalent matrix molecules M_1–M_4. The increased size of terrylene has to be accommodated by introducing significant distortions of the host lattice in the vicinity of each dopant chromophore, as can be seen in Fig. 4.6, which shows one of the four resulting inclusion geometries.

The first thing which is noteworthy about the computed structures is that they are all centrosymmetric and thus are in accordance with the observation of a primarily quadratic Stark effect in X_1, X_2 and X_4 [28,33]. Furthermore, the nonplanar terrylene geometry which was postulated in order to explain the occurrence of certain additional vibrational bands [13] is predicted by the model as well. It seems that terrylene has to "mimic" the host in order to fit the crystal. It can be seen in Fig. 4.6 that the staggered conformation of the three naphthalene subunits somewhat resembles that of the phenyl rings of an M_2 matrix molecule. Most important with regard to the single-molecule optical switching phenomenon, however, is that the simulations provide a way to distinguish between inclusion conformations of different photostability. When virtual heating–quenching cycles were performed, it was found that two of the substitutional sites (M_1 and M_4) would always recover the same respective minimum-energy structure, while the other two sites exhibited a number of metastable secondary minima in the potential energy hypersurface. This result corresponds to the observed photostability of the spectral sites: X_2 and X_4 are stable, while X_1 and X_3 are prone to nonphotochemical hole-burning. A detailed investigation of the calculated structures, in combination with the results of polarization-dependent measurements [26], has indeed made it possible to establish an unambiguous correlation between the inclusion conformations and the main spectral sites X_1–X_4 [4].

The mechanism of light-induced frequency jumps can be elucidated by the structures of secondary minima found for insertion site M_3, which is thought to correspond to the X_1 spectral site [4]. The model predicts relative energies for certain secondary minima which are in accordance with the relative

b'   c-axis

a'

Fig. 4.6. An insertion site of terrylene in low-temperature p-terphenyl crystal, as predicted by molecular dynamics simulations [4]. The figure shows the minimum-energy structure for the substitution of an M_1 matrix molecule by the terrylene guest and can be compared directly to the triclinic elementary cell of the pure host crystal as shown in Fig. 4.1b. The axes of the pseudo-monoclinic elementary cell are indicated at the bottom to facilitate orientation

widths of some of the observed spectral jumps; it furthermore provides an explanation for the alternating behavior found in the slope of the linear Stark response of the XY photosite. The theoretical results are quite elaborate [4] and their detailed discussion goes far beyond the scope of the present article, so we will conclude with only a simplified presentation of the key concepts involved.

The formation of the various photoproduct structures proceeds along complex paths in a multidimensional conformational space, with the necessary activation energy provided by the prolonged optical excitation of the guest. It turns out, however, that as a first approximation the X_1–XY–XY''' transition can be described in terms of fairly simple localized degrees of freedom, namely flip–flop motions of the central phenyl ring of matrix molecules in the near vicinity of the chromophore. This has already been proposed in the initial, largely intuitive models of the optical switching process [27,28], and is corroborated by the molecular dynamics simulations [4]. An illustration of this simplified scenario can be found in Fig. 4.7.

It can be seen that the aforementioned central ring flip can occur on either side of the terrylene molecule, leading to two possible "XY" conformations,

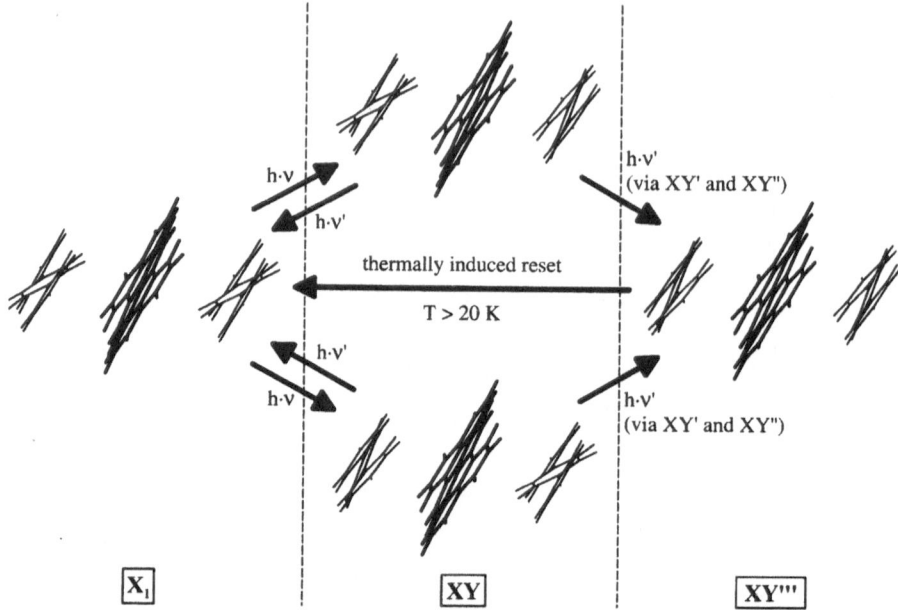

Fig. 4.7. A simplified mechanistic scenario for the X_1–XY–XY$'''$ spectral jumps of terrylene in p-terphenyl. The conformational rearrangements are, as a first order approximation, described by two localized degrees of freedom in the second "shell" around the impurity molecule. These are central ring flips of either of the two p-terphenyl molecules shown here (molecular sizes and relative positions not scaled to proportions), which are thought to be triggered by the prolonged optical excitation of the guest chromophore. The resulting two possible conformations for the primary photosite XY exhibit the same spectral shift, but can be distinguished by their reaction to an external electric field. The XY$'''$ conformation is achieved when both ring flips have occurred. Hence the spectral separation of X_1–XY$'''$ ($\Delta\nu = 1644\,\text{GHz}$) is roughly equal to twice the jump width X_1–XY ($\Delta\nu = 843\,\text{GHz}$)

which are no longer centrosymmetric and hence exhibit a predominantly linear response to an external electric field. The slope of the Stark plots can indeed be expected to have alternating signs for different visits in XY (see Fig. 4.5), depending on which of the two conformations is realized at that time. The fact that a simultaneous ring flip of both p-terphenyl molecules is several orders of magnitude less probable than either individual flip also explains why the XY$'''$ state can only be reached indirectly from X_1, via the intermediate photosites. The respective shift in the terrylene absorption frequency can be modeled by dipolar coupling of the p-terphenyl molecules to the electronic transition of the chromophore. The two matrix molecules in question are symmetry-equivalent in the X_1 conformation due to the existing center of inversion, which is occupied by terrylene, so they can be expected to exhibit the same strength of dipolar coupling to the absorber. This explains both why the two conformations bring about the same shift in absorption frequency, leading to only one observed photosite (i.e. XY), and why the difference in absorption frequency between X_1 and XY''' ($\Delta\nu = 1644\,\text{GHz}$) is roughly equal to twice the jump width X_1–XY ($\Delta\nu = 843\,\text{GHz}$).

With this we conclude the present work. We have tried to illustrate how low-temperature single-molecule spectroscopy can be employed in combination with advanced computational efforts to obtain a detailed, microscopic understanding of conformational dynamics in solids and of the optical manipulation of host/guest systems on a nanoscopic scale. Such efforts might be relevant for understanding the mechanism of nonphotochemical hole-burning, for finding new ways to employ single impurity molecules as truly local probes and for developing novel concepts of high-density optical data storage.

Acknowledgment. We are much obliged to P. Bordat, R. Brown, and M. Orrit for generously providing us with unpublished results and for valuable discussions. Financial support of the Deutsche Forschungsgemeinschaft (Ba 1405/6–1) and the Fonds der Chemischen Industrie is gratefully acknowledged.

References

1. W. E. Moerner (ed), *Persistent Spectral Hole-Burning: Science and Applications* (Springer, Berlin, 1989)
2. J. M. Hayes, R.P. Stout, and G.J. Small, J. Chem. Phys. **74**, 4266 (1981)
3. C. von Borcyzskowski, A. Oppenländer, H. P. Trommsdorff, and J.C. Vial, Phys. Rev. Lett. **65**, 3277 (1990)
4. P. Bordat, R. Brown, Chem. Phys. Lett. **331**, 439 (2000)
5. T. Basché, W. E. Moerner, M. Orrit, and U. P. Wild (eds), *Single-Molecule Optical Detection, Imaging and Spectroscopy* (Verlag Chemie, Weinheim, 1997)
6. J. L. Skinner, W. E. Moerner, J. Phys. Chem. **100**, 13251 (1996)
7. W. E. Moerner and M. Orrit, Science **283**, 1593 (1999)
8. P. Tamarat, A. Maali, B. Lounis, and M. Orrit, J. Phys. Chem. A **104**, 1 (2000)
9. S. Nie, D. T. Chiu, and R. N. Zare, Science **266**, 1018 (1994)
10. S. Kummer, T. Basché, and C. Bräuchle, Chem. Phys. Lett. **229**, 309 (1994); Chem. Phys. Lett. **232**, 414 (1995)
11. W. E. Moerner and L. Kador, Phys. Rev. Lett. **62**, 2535 (1989)
12. M. Orrit and J. Bernard, Phys. Rev. Lett. **65**, 2716 (1990)
13. S. Kummer, F. Kulzer, R. Kettner, T. Basché, C. Tietz, C. Glowatz, and C. Kryschi, J. Chem. Phys. **107**, 7673 (1997)
14. L. Fleury, B. Sick, G. Zumofen, B. Hecht, and U. P. Wild, Mol. Phys. **95**, 1333 (1998)
15. F. Kulzer, F. Koberling, T. Christ, A. Mews, and T. Basché, Chem. Phys. **247**, 23 (1999)
16. H. M. Rietveld, E. N. Maslen, and C. J. B. Clews, Acta Cryst. B **26**, 693 (1970)

17. J. L. Baudour, Y. Delugeard, and H. Cailleau, Acta Cryst B **150**, 32 (1976)
18. M. Sougoti, Ph. D. Thesis, Université des Rennes I, Rennes (1994)
19. P. Bordat and R. Brown, Chem. Phys. **246**, 323 (1999)
20. T. Basché, S. Kummer, and C. Bräuchle, Nature **373**, 132 (1995)
21. M. Vogel, A. Gruber, J. Wrachtrup, and C. von Boczyskowski, J. Phys. Chem. **99**, 14915 (1995)
22. P. Tamarat, B. Lounis, J. Bernard, M. Orrit, S. Kummer, R. Kettner, S. Mais, and T. Basché, Phys. Rev. Lett. **75**, 1514 (1995)
23. S. Kummer, S. Mais, and T. Basché, J. Chem. Phys **99**, 17078 (1995)
24. D. J. Norris, M. Kuwata-Gonokami, and W. E Moerner, Appl. Phys. Lett. **71**, 297 (1997)
25. A. C. J. Brouwer, E. J. J. Groenen, and J. Schmidt, Phys. Rev. Lett. **80**, 3944 (1998)
26. S. Kummer, Ph. D. Thesis, Universität München, München (1996)
27. F. Kulzer, S. Kummer, R. Matzke, C. Bräuchle, and T. Basché, Nature **387**, 688 (1997)
28. F. Kulzer, R. Matzke, C. Bräuchle, and T. Basché, J. Phys. Chem. A **103**, 2408 (1999)
29. T. Basché, W. P. Ambrose, and W. E. Moerner, J. Opt. Soc. Am. B **9**, 829 (1992)
30. T. Basché and W. E. Moerner, Nature **352**, 600 (1991)
31. L. Fleury, A. Zumbusch, M. Orrit, R. Brown, and J. Bernard, J. Lumin. **56**, 15 (1993)
32. W. E. Moerner, T. Plakhotnik, T. Irngartinger, M. Croci, V. Palm, and U. P. Wild, J. Phys. Chem. **98**, 7382 (1994)
33. F. Kulzer, Ph. D. thesis, Universität Mainz, Mainz (2000)
34. U. P. Wild, F. Güttler, M. Pirotta, and A. Renn, Chem. Phys. Lett. **193**, 451 (1992)
35. M. Orit, J. Bernard, and A. Zumbusch, Chem. Phys. Lett. **196**, 595 (1992)
36. M. Pirotta, A. Renn, and U. P. Wild, Helv. Phys. Acta **69**, 7 (1996)
37. S. A. Empedocles and M. G. Bawendi, Science **278**, 2114 (1997)
38. R. W. Munn, Chem. Phys. **76**, 243 (1983)
39. D. M. Hanson, J. S. Patel, I. C. Winkler, and S. A. Morrobel-Sosa, "Effects of Electric Fields on the Spectroscopic Properties of Molecular Solids," in: *Modern Problems in Condensed Matter Sciences: Spectroscopy and Excitation Dynamics of Molecular Systems Vol. 4*, V. M. Agranovich and R. M. Hochstrasser (eds.), pp. 621–679 (North–Holland, Amsterdam, 1983)
40. J. H. Meyling, P. J. Bounds, and R. W. Munn, Chem. Phys. Lett. **51**, 234 (1977)
41. J. Gerblinger, U. Bogner, and M. Maier, Chem. Phys. Lett. **141**, 31 (1987)
42. M. Maier, Appl. Phys. B **41**, 73 (1986)
43. B.E. Kohler and J.C. Woehl, J. Chem. Phys. **102**, 7773 (1995)
44. N. Neto, M. Scrocco, and S. Califano, Spectrochimica Acta **22**, 1981 (1966)
45. P. Bordat and R. Brown, Chem. Phys. Lett. **291**, 153 (1998)
46. C. Kryschi, H.C. Fleischhauer, and B. Wagner, Chem. Phys. **225**, 485 (1992)
47. F. Güttler, J. Sepioł, T. Plakhotnik, A. Mitterdorfer, A. Renn, and U. P. Wild, J. Lunim. **56**, 29 (1993)
48. F. Güttler, M. Croci, A. Renn, and U. P. Wild, Chem. Phys. Lett. **211**, 421 (1996)
49. J. Köhler, A. C. J. Brouwer, E. J. J. Groenen, and J. Schmidt, Chem. Phys. Lett. **250**, 137 (1996)

5 Triggered Emission of Single Photons by a Single Molecule

C. Brunel, P. Tamarat, B. Lounis, and M. Orrit

Intense laser light is often represented as a classical Maxwell wave. At low intensities, however, light absorption leads to discrete detection events in photon-counting detectors, or to shot noise in the photodetector current. Although these and many such observations can be interpreted in a semiclassical frame where matter is quantized and waves are still classical, subtler experiments have shown that light, as well as matter, is a quantum object, and that photons have physical reality [1]. The quantum nature of light entails Heisenberg uncertainty relations between two conjugate variables in the harmonic oscillator Hamiltonian of each mode. These so-called quadratures can be the phase and the amplitude of the field. In normal light, e.g. laser light, the noise is equally distributed on the two quadratures. But if the noise on one quadrature is reduced – at the cost of increased noise on the other quadrature – one gets a new state of radiation, called squeezed light [2].

There are several possible ways to squeeze light. The ideal example we consider here would be a very regular stream of photons, i.e. individual photons periodically coming out of a source. The corresponding light beam would have no amplitude- or photon-noise. With such a beam, and with a high-efficiency photodetector, it would be possible to measure extremely small absorptions, because the absence of even a single photon could be detected unambiguously. Indeed, any source delivering a single photon at a known instant could be used to measure weak absorptions, for example a setup producing twin photons [3], or a triggered source (a photon "gun") delivering single photons on command. Such a triggered source of single photons would also be very attractive for quantum cryptography [4]. Eavesdropping on a message involves a measurement, and therefore perturbs the quantum state of the light carrying the message. As a stream of single photons is a well-defined quantum state, the perturbation would be easy to detect in the received photon stream. Encoding and transmission of messages by single photons have already been demonstrated experimentally [5].

In the triggered source we are discussing here, each photon must be released within a short, given time interval after the trigger signal. This would be impossible with a continuously excited source of single photons. For example, a single atom, ion, or molecule excited with a cw laser beam emits single photons. The emission of photons one by one has already been demonstrated via the antibunching of fluorescence light [6–8], but this emission is random in time. Several other schemes have been used to generate single photons, or twin photons. For example, a cascade process in a calcium atom can produce a pair of correlated photons [9]. Another scheme is the parametric generation of twin photons by a laser pulse in the nonlinear crystal of an optical parametric oscillator (OPO). The photons appear at conjugate positions in the phase-matching output cone of the OPO [10,11]. In these two schemes, the efficiency of the generation must be kept low to avoid pile-up, and the emission process is still random. Even though each emission event is known with certainty via the detection of the twin photon, it is impossible to predict from which particular atom or during which particular pump laser pulse a photon pair will be emitted.

To obtain a single photon on command, a single emitting state must be prepared *with certainty*. In recent work on semiconductor quantum well structures, Yamamoto and colleagues used Coulomb blockade effects to let a single electron and a single hole recombine, thereby producing a single photon [12]. In the solution we discuss here, we want to prepare a single molecule in its excited state with certainty, as proposed earlier by de Martini and colleagues [13]. By spontaneous emission, the molecule will return to its ground state and emit a single photon within a few excited-state lifetimes T_1 (assuming that the fluorescence quantum yield is close to unity, which is the case of many aromatic molecules). We can imagine several processes to prepare a single molecule in its excited state. For example, with a short, resonant pi pulse, we could coherently bring the molecule into its excited state. Alternatively, we could use an incoherent pathway, by pumping the molecule at high laser intensity to an upper vibronic level, which would then relax to the emitting excited state. Both these solutions require pulsed lasers which were not available in our laboratory. Instead, we chose to sweep the frequency difference between molecule and laser rapidly through resonance, a process called rapid adiabatic passage [14] (explained in detail hereafter), to prepare the molecule in its excited state. The passage could be achieved by scanning the laser frequency very quickly with modulators or by chirping a laser pulse. Instead of sweeping the laser frequency, we decided to sweep the frequency of the molecule with an applied electric field, which can be varied very fast, by applying a variable voltage (here RF) to electrodes. In the present paper, we will describe in more detail experiments reported earlier [15,16], which demonstrate how a single molecule can be used as a source for triggered single photons.

Rapid adiabatic passage (RAP, also called adiabatic following [14]) can be easily described in the effective spin picture of magnetic resonance. The molecular Bloch vector, which represents the density matrix of the two-level molecule, precesses around a rotation vector, or an effective magnetic field.

In the rotating frame, the horizontal component of the rotation vector is the (resonant) Rabi frequency, proportional to the optical electric field of the laser, while its vertical component is the frequency difference between molecule and laser. When we sweep the molecular frequency through resonance with the laser, the Bloch vector keeps precessing around the effective magnetic field, and flips from the downward-pointing to the upward-pointing direction; the molecule ends up in its excited state. The adiabatic passage can also be seen in the dressed-atom picture [17]: upon slowly sweeping the detuning between laser and molecule, the dressed atom is driven adiabatically from the ground state with $n+1$ laser photons to the excited state with n laser photons [18].

The sweep must be slow enough to remain adiabatic, so that the Bloch vector accompanies the effective magnetic field as the molecule is swept through resonance with the laser. On the other hand, the sweep should not be too slow, because any relaxation event of population or coherence would misalign the Bloch vector and the effective field, leading to complex dynamics, with a final state different from the pure excited state. The conditions for ideal rapid adiabatic passage (RAP conditions) are [14]:

$$T_2 \gg T_{pass} \gg T_{Rabi} = 1/\Omega \tag{5.1}$$

where Ω is the Rabi frequency and T_2 is the coherence lifetime, equal to twice the excited state lifetime T_1 in the low-temperature experiments presented here. T_{pass} can be defined naively as "the time it takes the laser to cross the saturated resonance":

$$T_{pass} = \Gamma_S \cdot \frac{dt}{d\omega} \, , \text{ with } \Gamma_S = \Gamma\sqrt{1 + I/I_S} \, .$$

Γ_S is the saturated width, $\Gamma = T_1^{-1}$ the homogeneous width, I the laser intensity, and I_S the saturation intensity. $d\omega/dt$ is the rate of angular frequency scanning, whose physical meaning is considered in more detail hereafter. The RAP conditions require that the Rabi frequency be much larger than the inverse coherence lifetime T_2^{-1}, but the second inequality in (5.1) need not be very strong. Actually, the adiabatic passage can be good enough with T_{Rabi} of the same order of magnitude as T_{pass}. Therefore, a ratio of about 5 to 10 between Ω and Γ often suffices to ensure a good adiabatic passage. In our experiments, the Rabi frequency, which is proportional to the square root of the laser intensity, was limited to about 5Γ by the maximum admissible background.

The notion of a frequency varying with time is in principle self-contradictory. To follow a frequency as a function of time, the sweep must be slow enough. For a very fast passage through a frequency interval $\Delta\omega$, lasting some time $T_{pass} = \Delta\omega dt/d\omega$, the Fourier frequency uncertainty T_{pass}^{-1} would be larger than $\Delta\omega$. Therefore, the naive definition of the passage time is correct only if

$$T_{pass} \gg \sqrt{dt/d\omega} \, .$$

This condition is equivalent to the second inequality in (5.1), which states that the passage time, defined as the time to pass through the saturated resonance (with a width of about Ω) should be longer than the inverse of the Rabi frequency. If this condition is not fulfilled, broadening of the molecular frequency by its rapid sweep must be taken into account. Let us define an effective passage time by $(dt/d\omega)^{1/2} = T_C$. This effective sweep time T_C will be used in the following instead of $dt/d\omega$ to characterize the sweep rate. The RAP conditions (5.1) may also be rewritten as $T_2 \gg T_C \gg T_{Rabi}$.

The article is organized as follows. Section 5.1 gives experimental details. The theory and simulations are briefly discussed in Sect. 5.2. The results are given and discussed in Sect. 5.3.

5.1 Experimental

The proposed experiment requires a molecule with a large linear Stark effect (i.e., a large difference $\Delta\mu$ between static dipole moments in its excited and ground electronic states), because the zero-phonon line of the molecule must be swept quickly on a broad range of frequencies. We have previously studied the linear and quadratic Stark effect of dibenzanthanthrene (DBATT) molecules in a n-hexadecane Shpol'skii matrix [19]. A large fraction of molecules in this system present a very small $\Delta\mu$ (less than 5 mD), and are thought to have a nearly centrosymmetric crystal environment. However, some molecules do have a large Stark effect with $\Delta\mu$ of the order of 0.3 D. Such molecules probably sit close to crystal defects, or in the boundaries between crystalline grains. For the experiments presented here, we selected molecules in this second distribution, which gave us shifts of up to about 1 GHz for an electric field of 1 MV/m.

The sample, a drop of a dilute solution of DBATT in hexadecane (concentration well below 10^{-7} M), was placed on a thin glass plate glued on the flat face of the collecting paraboloid (see below). All experiments were done in pumped superfluid helium, at 1.8 K. The glass plate was fitted with two thick (more than 100 nm) evaporated aluminum electrodes, separated by a gap of 18 µm. Static voltages of up to 200 V could be applied without dielectric breakdown, but the RF voltages we applied were less than 40 V.

The optical setup is schematically presented in Fig. 5.1. The exciting laser beam is collimated and falls on an aspheric lens (Newfocus, model 5714, focal length 8 mm, NA = 0.28). The converging beam crosses the full silica paraboloid and focuses on the sample holder, held by capillarity with a little index-matching oil onto the flat front face of the paraboloid. The laser is focused to a nearly diffraction-limited spot on the sample by varying the distance between the lens and the sample through a mechanical deformation of the lens-holder. The parabolic mirror for the collection of molecular fluorescence is a silica parabolic lens custom-made by Sopelem SA [20]. Its convex side is coated with aluminum, except for the small hole (diameter

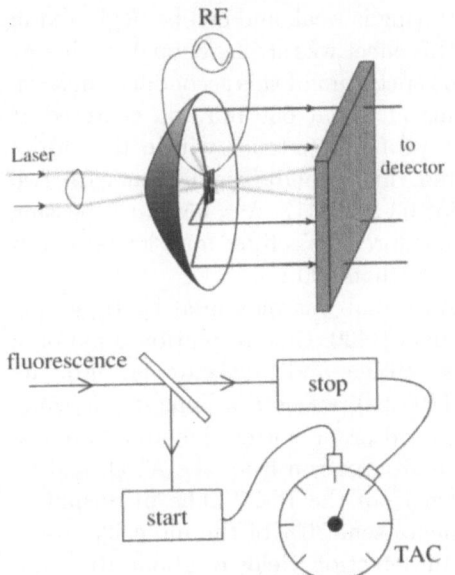

Fig. 5.1. Upper part: Schematic view of the optical setup, with the aspheric lens focusing the exciting laser, the full silica paraboloid collecting fluorescence emitted by the sample, and the filter blocking the laser light. The sample is placed in the 18 micron gap between the two aluminum electrodes. The RF ac voltage applied to the electrodes sweeps the molecular resonance frequency sinusoidally in time. The lower part shows a diagram of the start–stop experiment. The start detector is the avalanche photodiode, the stop detector is a photomultiplier tube. The delay between start and stop is measured by a time-to-amplitude converter (TAC)

1 mm) through which the exciting beam is sent. Because there is almost no index jump between the sample and the mirror, the collection optics is of the immersion type, and there are almost no losses from internal reflection. The smallest diameter of a spot obtained by focusing a parallel beam with the mirror is about 8 μm. The mirror is therefore not good enough to build a confocal microscope, but it works well enough to image the excited sample volume onto a detector as small as our avalanche photodiode. The exciting light is blocked out of the detected fluorescence by a notch filter (Kaiser, cutoff wavelength 592 nm) and by a colored glass filter (Schott, RG610).

For the intensity correlation measurement, we built a Hanbury-Brown and Twiss coincidence setup with two detectors (see Fig. 5.1): a photomultiplier tube, PMT (RCA, model C31034A-02), and a photon-counting avalanche photodiode, APD (EG&G, model SPCM-AQ-231, active area 200 μm in diameter). The fluorescence detected by the PMT was filtered spatially with a 0.2 mm pinhole and an 80-mm lens to reduce the background. In our first experiments [15], the photomultiplier detected some secondary photons emitted by the APD, which propagated back to the sample, and were scattered or reflected to the PMT. Such backscattering is weak and can be neglected in experiments on liquid solutions. Here, this effect was strong enough to distort the coincidence histograms severely. The spectrum of this secondary emission has two broad components, one starting at about 630 nm and centered at 730 nm, and a more intense one which probably extends within the gap of silicon, beyond 1000 nm. We eliminated this secondary emission with two filters, a short-pass mirror (Melles Griot 03-SWP412, $\lambda = 650$ nm), passing wavelengths shorter than 660 nm, and a colored glass filter for a better cutoff of the infrared light at wavelengths longer than 800 nm.

The temporal profile of the emission signal was measured by triggering a multichannel analyzer (Stanford, model SR430, time resolution 5 ns) on a rising edge synchronous with the Stark voltage, and by accumulating many photons. For a higher time resolution (300 ps), we used a home-made time-to-amplitude converter (TAC). The second-order correlation function was measured by starting the TAC with a photoelectron from the APD, and by stopping it with the next photoelectron from the PMT. The beamsplitter dividing the fluorescence into two branches sent 70% of the intensity to the PMT and 30% to the APD, with overall detection yields of about 10^{-3} and $2 \cdot 10^{-3}$, respectively. The PMT signal was delayed to give the correlation function for negative times. We made no attempt to measure absolute yields, since the optical path had to be readjusted every day.

5.2   Theory and Simulations

Earlier work has shown that optical Bloch equations describe high-resolution excitation spectra of single molecules very accurately. Many of the nonlinear optical effects observed earlier with atoms have also been observed with single molecules in solids at low temperatures: light-shift [21], hyper-Raman resonances [22], Rabi resonances [23], etc.. We simulated the dynamics of the molecule in the laser field with Bloch equations, after we had determined the various parameters by experimental measurements. We extracted the homogeneous linewidth and saturation intensity from a study of the optical saturation for each molecule we investigated: the width and intensity of each line were measured as functions of laser power. For DBATT in n-hexadecane below 3 K, dephasing by acoustic phonons is negligible ($T_2 = 2T_1$), and the only cause of coherence relaxation is the decay of the excited state, mainly by spontaneous emission. The internal conversion rate is not known precisely, but we deduced the radiative emission rate of DBATT from its absorption strength [24] with the Strickler–Berg formula. We found 80% of the measured fluorescence rate, which, allowing for experimental errors, implies that the quantum yield must be larger than about 70% at room temperature. It is probably close to unity at low temperatures. The amplitude of the linear Stark modulation of each single molecule line was first measured by applying a static voltage or a low-frequency ac voltage to the electrodes. As shown in

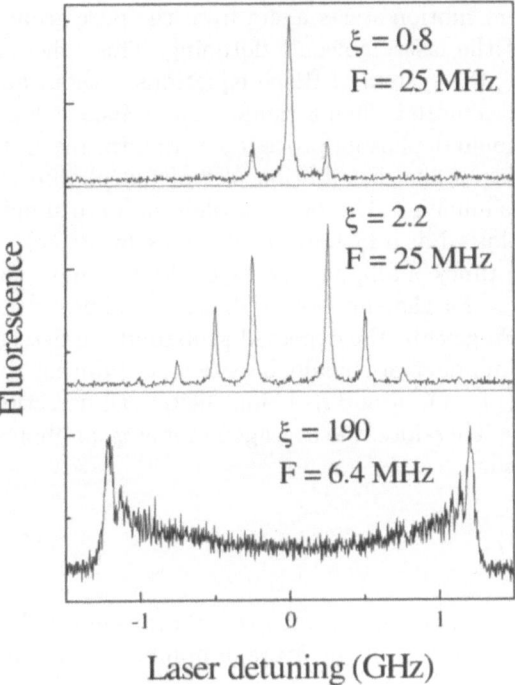

Fluorescence

Laser detuning (GHz)

Fig. 5.2. Examples of spectral shapes of single-molecule excitation lines when a sinusoidal RF voltage is applied to the electrodes. Top: For a weak modulation index ξ (i.e. the ratio of the frequency modulation amplitude to the RF frequency F), only two sidebands appear in addition to the carrier in the center. For larger ξ, (middle spectrum), more sidebands appear, and the carrier may vanish. Finally, for very large ξ (lower spectrum), the molecular frequency is spread over a wide frequency interval. Note the spectral fringes close to the turning points.

Fig. 5.2, the spectra at different modulation frequencies and for weak laser power are in excellent agreement with the simple theory of frequency modulation [23]. For a weak modulation frequency (Fig. 5.2c), the spectral profile is nearly the convolution of the homogeneous lineshape with the probability density of the frequency under sinusoidal scanning (note, however, the fringes close to the turning points, due to the finite RF frequency). Therefore, all the parameters needed for the simulation were known for each molecule.

The time-dependent differential system of optical Bloch equations was solved by a Runge–Kutta algorithm, with an integration step of $2 \cdot 10^{-3} T_1$, to yield the evolution of the density matrix during the sweep. The average time profile of the emission burst (see 5.1) was calculated from the time-dependent population of the excited state. To simulate our start–stop experiments, however, and to estimate the efficiency of our source, it was important to evaluate the full distribution of the number of photons emitted in each sweep. Optical Bloch equations only give average rates of photon emission. This distribution was obtained by introducing discrete spontaneous emission processes in quantum Monte Carlo simulations [25]. Since the period of the sweeps (i.e. half the RF period, typically a few hundreds of ns) was always much longer than the excited state lifetime (about 9 ns), we assumed that the molecule always starts from the ground state at the beginning of each sweep. After a spontaneous emission, relaxation brings the molecule back to the ground state very quickly, and a coherent motion starts again from the pure ground state, with the current value of the laser–molecule detuning. This coherent motion was simulated by solving the system of Bloch equations without any relaxation, and was suddenly interrupted when a random spontaneous emission process occurred. No additional dephasing processes were introduced. In order to obtain reliable statistics of the distribution of emitted photons, a large number of passages were simulated. The pairs of photons were simply counted for each delay τ, and plotted as a histogram to represent $g^{(2)}(\tau)$.

From the simulated emission times of all photons, we could determine the probabilities p(0), p(1), p(2), ... for the emission of $0, 1, 2, \ldots$ photons per sweep. These probabilities in turn gave us the expected probability of detecting a photon pair. The average number of coincidences detected during the same m-photon sweep is $\eta_1 \eta_2 m(m-1)$, η_1 and η_2 being the overall detection yields in the start and stop arms. Therefore, the average number q_0 of photon pairs detected within one sweep is:

$$q_0 = \eta_1 \eta_2 \sum_{m=2}^{\infty} m(m-1)p(m) \, , \tag{5.2}$$

whereas the average number q_1 of photon pairs with the photons emitted in two different sweeps is simply the square of the average number of photons detected in one sweep, i.e.:

$$q_1 = \eta_1 \eta_2 \left(\sum_{m=1}^{\infty} m\, p\,(m) \right)^2 . \tag{5.3}$$

The ratio q_0/q_1 deduced from simulations will be compared to the ratio of the experimental structures in 4.3.

Finally, we note that our start–stop measurement will give the distribution of the pairs of consecutive photons $C(\tau)$, instead of the distribution of all photon pairs $P(\tau)$, related to the correlation function and the average intensity $\langle I \rangle$ in counts/s by:

$$P(\tau) = \langle I \rangle \cdot g^{(2)}(\tau) \, . \tag{5.4}$$

Under steady state conditions, the following relation holds between their Laplace transforms $C_L(s)$ and $P_L(s)$, s being the conjugate variable of time [26,27]:

$$C_L(s) = P_L(s) / [1 + P_L(s)] . \tag{5.5}$$

Equations (5.4), (5.5) show that the distribution of consecutive pairs $C(\tau)$ is proportional to the correlation function for times well below $\langle I \rangle^{-1}$, i.e. for the whole range of experimental times if the detection quantum yield is low enough. Although these equations do not strictly apply to our present case because the excitation conditions are varying in time, they explain qualitatively that we can compare our simulations of $g^{(2)}(\tau)$ to measurements of $C(\tau)$ as long as the detection quantum yield is weak, i.e. as long as $\langle I \rangle$, and particularly the contribution to $\langle I \rangle$ from the background, is negligible. For the parameters of our experiment, the intensity ratio of the coincidence structures measured experimentally (see 5.3) was nearly insensitive to background.

5.3 Results and Discussion

5.3.1 Average Emission Signal

In a first experiment, we recorded the average time profile of the emission bursts synchronized with the applied RF voltage, which drives the molecule through its resonance with the fixed laser. The frequency of the laser was adjusted to be approximately in the middle of the swept interval, corresponding to zero applied voltage. The inset of Fig. 5.3 shows an averaged time trace of the fluorescence signal for many sweeps. The fluorescence bursts are emitted periodically, each half RF period. Figure 5.3 shows the time profile of one burst. It presents a short rise time (a few ns), and an exponential decay with a characteristic time of about 8–9 ns. This decay corresponds to the measured lifetime of the DBATT molecule [28], and is compatible with the homogeneous width of single DBATT molecules, about 20 MHz. Oscillations are superimposed on the decay. They should be absent for an ideal rapid adiabatic passage, but appear as soon as the RAP conditions are not perfectly

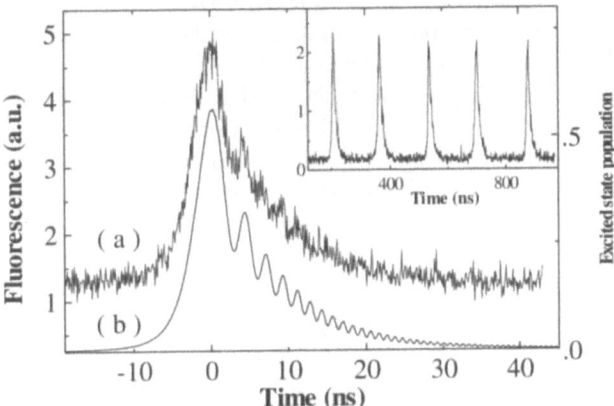

Fig. 5.3. The inset shows the average fluorescence signal triggered on the RF voltage and accumulated for a large number of sweeps. The emission occurs after the molecule has crossed resonance with the laser. The main plot shows the time profile of one of the bursts, with a fast rise and a nearly exponential decay, upon which oscillations are superimposed. The smooth line is a simulation with all parameters determined from experimental measurements

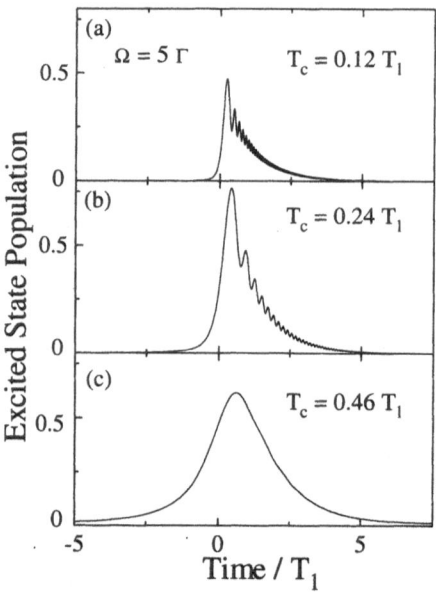

Fig. 5.4. Emission burst profiles calculated from optical Bloch equations for a given Rabi frequency and various sweep speeds (characterized by the effective sweep time T_C). Note that except for the very slow sweep (c), oscillations appear even for the best sweep time for RAP (0.24 T_1), and that the oscillation period is very sensitive to sweep time. The maximum population of the excited state is about 75% for curve (b), which corresponds approximately to the maximum probability of emission of a single photon (see Fig. 5.5)

fulfilled, in particular when the very strong inequality $\Omega \gg \Gamma$ does not apply. The data of Fig. 5.3 were obtained with a Rabi frequency $\Omega = 3\Gamma$, i.e. far from the ideal RAP condition $\Omega \gg \Gamma$. Figure 5.4 shows calculations of the burst profile for $\Omega = 5\Gamma$ (which was the largest Rabi frequency we could reach with a reasonable background) and for three values of the characteristic time $T_C = (dt/d\omega)^{1/2}$ (0.12, 0.24 and 0.46 T_1). Note that the oscillation is much faster than the Rabi oscillation at resonance, and that its frequency changes with time and strongly depends on the sweep rate.

Taking a closer look at the average time trace of the fluorescence in the insert of Fig. 5.3, one sees that the time interval between bursts is alternately slightly shorter and longer than half the RF period, whereas the delay between bursts n and $n + 2$ is always exactly equal to the RF period. This is because the laser frequency was not perfectly tuned to the middle of the scanned frequency interval.

The average time profile of the photon bursts shows that emission indeed takes place just after the molecule has crossed resonance. However, the average number of photons delivered in each burst is difficult to evaluate precisely, because we did not measure the absolute quantum yield of our detection accurately. Moreover, measuring an average number gives no hint about the statistical distribution of emitted photons. Determining which bursts produced no photon would require an ideal detector. But by using a pair of detectors, it is possible to obtain information about the number of bursts giving two or more photons. This amounts to measuring the second-order correlation function of the intensity, $g^{(2)}(\tau)$. Before explaining the measurements and their results, we examine theoretical simulations of this correlation function.

5.3.2 Theoretical Expectations for Photon Emission Probabilities

Quantum Monte Carlo simulations enabled us to calculate the probabilities for the emission of 0, 1, 2 and more photons in each sweep. Obviously, the faster the sweep, the lower the probability of excitation and of fluorescence. Figure 5.5 shows plots of $p(n)$, for $n = 0$ to 4, as functions of $T_C = (dt/d\omega)^{1/2}$, the effective scanning time defined in the introduction, for a given Rabi frequency (here 5Γ). For a very fast passage (T_C very short), the probability of excitation is very low, and zero-photon bursts are most probable.

Note, however, that the probability of one-photon bursts increases quadratically with T_C (i.e. linearly with the inverse sweep rate). From the naive definition of the passage time, one could have expected an excited state population increasing quadratically with the "time spent at resonance," i.e. quadratically with the inverse sweep rate. Instead, one obtains linear behavior because the molecular spectrum broadens for large sweep rates, with a broadening increasing like T_C^{-1}, while the saturated molecular width would remain independent of the sweep rate (see the discussion in Sect. 5.1).

When the scan rate is lowered, the probability of one-photon bursts reaches a maximum (here for $T_C = 0.25T_1$), then decreases for slower sweeps. The probability of emission of two photons then reaches its maximum, before decreasing for still slower sweeps, etc. For very slow sweeps, we would obtain a Poisson distribution, with an average number determined by the ratio of the passage time to the lifetime of the excited state. Note that

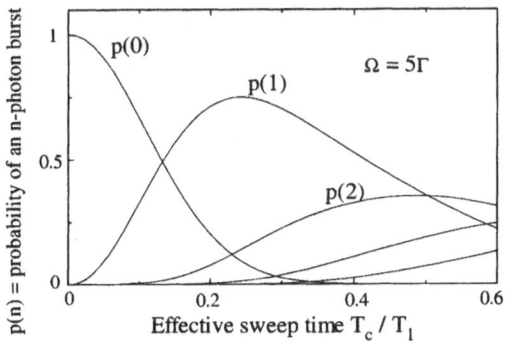

Fig. 5.5. Variations of the probabilities of emission of $0, 1, 2, \ldots$ photons as functions of effective sweep time (square root of the inverse frequency scanning rate), calculated from quantum Monte Carlo simulations and for the largest Rabi frequency we could achieve experimentally. A maximum probability of 75% for a one-photon burst is reached for $T_C/T_1 = 0.25$. For this value, the distribution of the number of emitted photons is very far from Poissonian

around the maximum of $p(1)$, the distribution is very far from Poissonian. The point of the adiabatic passage is to prepare the molecule in its excited state with certainty, i.e. to create a non-Poissonian distribution, ideally with $p(1) = 1$.

5.3.3 Start–Stop Measurements

Figure 5.6 shows the results of some start–stop measurements for different experimental conditions. As expected for the periodic emission of Fig. 5.3, the correlation function is periodic, with structures separated by half the RF period. In all three histograms the intensity of the central structure, around zero delay, is clearly reduced as compared to that of the lateral structures. The ratio of their areas roughly gives twice the probability of two-photon sweeps, compared to that of one-photon sweeps. If the passage is too fast (Fig. 5.6a), the ratio q_0/q_1 is very low, but there are much more zero-photon sweeps than one-photon sweeps, as Fig. 5.5 shows. On the other hand, if

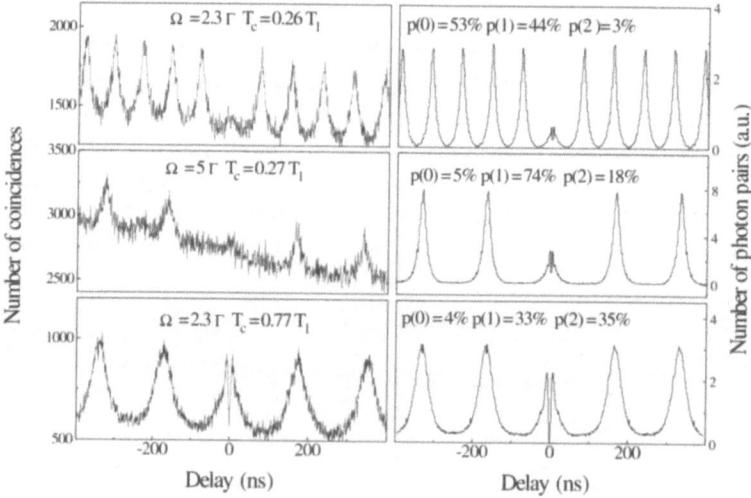

Fig. 5.6. Experimental histograms of consecutive photon pairs measured in the start–stop experiment (left curves), and simulated correlation functions (or distribution of photon pairs) for the same parameters, with the calculated probabilities of emission of 0, 1, and 2 photons. The upper curves are for a fast sweep: the number of two-photon sweeps is very low, but so is the number of one-photon sweeps as compared to zero-photon sweeps. The middle data correspond to the best conditions we achieved ($\Omega = 5\Gamma$), but at the price of a very high background. These curves correspond to the maximum of $p(1)$ in Fig. 5.5. The lower curves are for a slow sweep: the intensity of the central structure is comparable to that of the lateral ones. The antibunching dip is clearest in the lower curves because Rabi oscillations are slower (see Fig. 5.4)

the passage is too slow (Fig. 5.6c), the intensity of the central structure and the ratio q_0/q_1 grow, because the probability of emission of two or more photons increases. The highest probability of one-photon passage obtained in our experiments was 68% for $\Omega = 3\Gamma$ and 74% for $\Omega = 5\Gamma$. For the latter case, this corresponds to the maximum of $p(1)$ in Fig. 5.5. In some of the coincidence histograms (see Fig. 3 in [16]), the shapes of the first two lateral structures (and indeed that of all odd-numbered structures at larger delays) was different from those of the second ones (which are identical to that of all even-numbered structures, except the central one). Again, as discussed in Sect. 5.4.1, this happened whenever the laser was not tuned exactly to the center of the scanned frequency interval. The odd-numbered structures are split into two components, corresponding to delays slightly longer or shorter than half the RF period, whereas the even structures always turn up at integer numbers of whole RF periods.

The central structures of the histograms often show a clear dip for zero delay (see Fig. 5.6). This dip, due to photon antibunching [6–8,29,30], is a direct consequence of our exciting only one molecule. Considering the width of this dip for a cw excitation at the same Rabi frequency (several ns), we could have expected clearer antibunching structures. Experimental imperfections (binning, finite time resolution, time jitter of our experimental system) do not explain the weak contrast of the dip. Indeed, for fast frequency scanning, Fig. 5.4 shows what happens: for a fast passage, the Rabi oscillations are so fast that the dip's width can be less than 1 ns. The antibunching phenomenon has been discussed in detail in the literature, and much better data have been obtained with single molecules at cryogenic or room temperatures [8,29,30]. The aim of the present measurement was not to demonstrate or measure antibunching, but to show that the intensity of the central structure as a whole is dramatically reduced as compared to that of the lateral ones.

Finally, some of the experimental spectra of Fig. 5.6 show a clear slope in the background level when going from negative to positive times. This slope is due to the difference between the pair distribution as measured in the start–stop experiment and the true correlation function, where all photon pairs would be measured (as discussed in Sect. 5.3). Since the stop signal tends to occur more often at the beginning than at the end of acquisition, especially for strong background, a positive bias appears for short delays. The background not only limits the signal-to-noise ratio, it also prevents measurement of the pair distribution at long times. For practical uses of our single photon source, the background would have to be reduced dramatically, for example by reducing the laser intensity between passages through resonance, when the laser light has no effect on the molecule.

5.4   Conclusion and Outlook

The present article has demonstrated how a single molecule can be made to emit single photons on command.The experiment is not only feasible, but even relatively easy. Here, we have prepared the molecule in its excited state by a rapid adiabatic passage, i.e. by simply sweeping the molecular frequency through resonance with a fixed laser. In the best passage we could achieve, limited by the intensity of background fluorescence, we achieved 75% single-photon generation, with 25% either zero-photon or two-photon passages. This figure could be improved easily with a much higher Rabi frequency, using an optical shutter or a pulsed excitation laser to reduce the background. However, the system and preparation method presented here have several drawbacks. Keeping the molecule at cryogenic temperatures protects it from photobleaching, but is not very attractive for future applications. Continuously exciting the molecule produces a high background, but is useful only during a fraction of the time. It would be a big improvement to reduce the background between photon bursts using a pulsed excitation. For example, one could pump a four-level system such as a laser dye molecule to a vibronically excited level. After fast relaxation to the excited singlet state, the molecule would be able to emit only a single photon. Because the molecule could be pumped at high intensity, the probability of preparing the emitting state could approach unity. A four-level system would have the further advantage of spectrally separating laser and emission, while the resonant part of the emission (10 to 15 %) is lost in our present experiments. To improve the signal-to-background ratio, it would still be possible to gate the fluorescent emission with a fast optical shutter.

The rate of spontaneous emission could be enhanced if the molecule were placed in a cavity with high Purcell factor [31,32], or in a photonic band gap structure. This could improve the spectral purity of the emission. The emission spectrum of a single molecule at low temperature consists of several sharp lines (about 1 cm^{-1} broad) spread over several hundreds of cm^{-1}[28]. Selecting one of these lines with a resonant cavity would dramatically narrow the fluorescence spectrum.

One big advantage of the four-level scheme discussed above is that the source could operate at room temperature. The fluorescence spectrum would be much broader than at low temperature, but it could be somewhat narrowed by coupling to a cavity. Single dye molecules are easy to handle, but they are prone to photobleaching. Other small quantum systems with well-separated electronic levels, such as nanocrystals or quantum dots, could also be the core of a single-photon source at room temperature. The main problem to solve for practical applications will be to achieve an overall detection yield close to unity. Once efficient collection optics and high-yield photodetectors are available, single quantum systems will provide compact, cheap, and reliable sources of single photons.

References

1. R. Loudon, *The Quantum Theory of Light* (Clarendon, Oxford (England), 1983)
2. D. F. Walls and G. J. Millburn, *Quantum Optics* (Springer, Berlin, 1994)
3. J. G. Rarity, P. R. Tapster, and E. Jakeman, Opt. Commun. **62**, 201 (1987)
4. W. Tittel, G. Ribordy, and N. Gisin, Physics World **11**, 41 (1998)
5. P. D. Townsend, J. G. Rarity, and P. R. Tapster, Electron. Lett. **29**, 1291 (1993)
6. H. K. Kimble, M. Dagenais, and L. Mandel, Phys. Rev. A **18**, 201 (1978)
7. F. Diedrich and H. Walther, Phys. Rev. Lett. **58**, 203 (1987)
8. Th. Basché, W. E. Moerner, M. Orrit, and H. Talon, Phys. Rev. Lett. **69**, 1516 (1992)
9. P. Grangier and A. Aspect, Phys. Rev. Lett. **54**, 418 (1985)
10. A. Heidmann, R. J. Horowicz, S. Reynaud, E. Giacobino, and C. Fabre, Phys. Rev. Lett. **59**, 2555 (1987)
11. P. G. Kwiat, K. Mattle, H. Weinfurter, A. Zeilinger, A. V. Sergienko, and Y. Shih, Phys. Rev. Lett. **75**, 4337 (1995)
12. J. Kim, O. Benson, H. Kan, and Y. Yamamoto, Nature **397**, 500 (1999)
13. F. de Martini, G. di Giuseppe, and M. Marrocco, Phys. Rev. Lett. **76**, 900 (1996)
14. Y. R. Shen, *The Principles of Nonlinear Optics*, p. 399 (Wiley, New York, 1984)
15. C. Brunel, P. Tamarat, B. Lounis, J. Plantard, and M. Orrit, C. R. Acad. Sci. Paris **326**, 2679 (1998)
16. C. Brunel, B. Lounis, P. Tamarat, and M. Orrit, Phys. Rev. Lett. **83**, 2722 (1999)
17. C. Cohen-Tannoudji, J. Dupont-Roc, and G. Grynberg, *Atom–Photon Interactions* (Wiley, New York, 1992)
18. P. Tamarat, F. Jelezko, C. Brunel, A. Maali, B. Lounis, and M. Orrit, Chem. Phys. **245**, 121 (1999)
19. C. Brunel, P. Tamarat, B. Lounis, J. C. Woehl, and M. Orrit, J. Phys. Chem. **103**, 2429 (1999)
20. Y. Durand, J. C. Woehl, B. Viellerobe, W. Göhde, and M. Orrit, Rev. Sci. Instr. **70**, 1318 (1999)
21. P. Tamarat, B. Lounis, J. Bernard, M. Orrit, S. Kummer, R. Kettner, S. Mais, and T. Basché, Phys. Rev. Lett. **75**, 1514 (1995)
22. B. Lounis, F. Jelezko, and M. Orrit, Phys. Rev. Lett. **78**, 3673 (1997)
23. Ch. Brunel, B. Lounis, Ph. Tamarat, and M. Orrit, Phys. Rev. Lett. **81** (1998) 2679.
24. E. Clar, *Polycyclic Aromatic Hydrocarbons* (Academic Press/Springer, New York, 1995)
25. K. Moelmer, Y. Castin, and J. Dalibard, J. Opt. Soc. Am. B **10**, 524 (1993)
26. S. Reynaud, Ann. Phys. (Paris) **8**, 351 (1983)
27. H. Talon, Thèse de Doctorat, Université Bordeaux I (unpublished)
28. A.-M. Boiron, B. Lounis, and M. Orrit, J. Chem. Phys. **105**, 3969 (1996)
29. W. P. Ambrose, P. M. Goodwin, J. Enderlein, D. J. Semin, J. C. Martin, and R. A. Keller, Chem. Phys. Lett. **269**, 365 (1997)
30. U. Mets, J. Widengren, and R. Rigler, Chem. Phys. **218**, 191 (1997)
31. P. Goy, J.-M. Raimond, M. Gross, and S. Haroche, Phys. Rev. Lett. **50**, 1903 (1983)
32. J.-M. Gérard, B. Sermage, B. Gayral, B. Legrand, E. Costard, and V. Thierry-Mieg, Phys. Rev. Lett. **81**, 1110 (1998)

6 Photophysics of Conjugated Polymers Unmasked by Single Molecule Spectroscopy

J. Yu, D.-H. Hu, and P. F. Barbara

Conjugated polymers are linear polymer molecules with interacting π-bonds formed by the overlap of carbon p_z orbitals along the polymer backbone. Conjugated polymers have been extensively studied due to their excellent processability and favorable electronic and spectroscopic properties. The utilization of this type of material in various photonic applications (light emitting diodes, lasers, sensors, etc.) has been quite fruitful [1–6]. The investigation of Poly-p-phenylene-vinylene (PPV) type polymers such as MEH-PPV (Fig. 6.1) has been especially intense. The study of these materials has also led to many fundamental advances in the understanding of the electronic structure and dynamics of organic materials, including information on the optical- and charge- transport and localization in organic materials [7–9], and information on how polymer morphology [10–13], such as molecular conformation and aggregation, influences these processes.

The characterization of the photophysical properties of conjugated polymers in a bulk form, however, has often been proved difficult and sometimes confusing due to the inherent photophysical heterogeneities of this complex type of material. Not only is there a wide distribution of molecular weights for

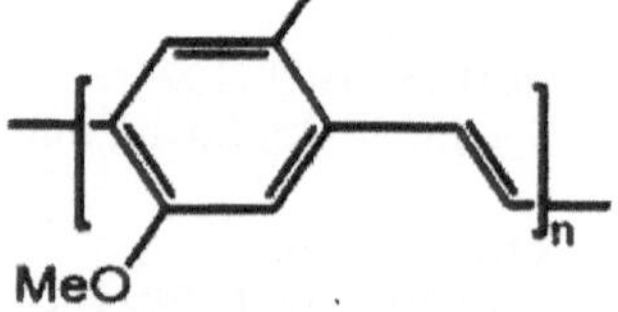

Fig. 6.1. The chemical structure of MEH-PPV

typical conjugated polymer sample, but also each polymer chain is a "multi-chromophoric" molecule [8,14,15]. It has been generally accepted that the optical coherent length of conjugated polymers is fairly limited, e.g., ~10–15 repeating units in PPV type polymers. This limitation is believed to originate from both conformational defects on the polymer backbone, i.e. chain bends, and electron–hole correlation. For a typical MEH-PPV sample that has an average Mw of 900,000, the absorption spectrum is due to an overlap of singlet–singlet absorption of more than a hundred weakly coupled, localized quasi-chromophores.

Single-molecule spectroscopy (SMS) allows for direct information on the molecular distribution of the physical quantities, rather than just their ensemble average values. This paper will focus on the research from our laboratory on applying single-molecule spectroscopy to the investigation of fluorescent conjugated polymer MEH-PPV, a conjugated polymer used extensively in light-emitting diodes and other photonic applications. In our experiments, MEH-PPV molecules were isolated at dilute concentration (about 0.2 molecules per μm^2) in a spin coated polymer thin film (polystyrene or poly-bisphenol-A-carbonate, 50 nm–1000 nm). Single-molecule fluorescence images, spectra, intensity kinetic traces, picosecond emission dynamics and other single molecules spectroscopic measurements were obtained with a home-built sample-scanning fluorescence confocal microscope.

6.1   Sample Preparation

One of the most common obstacles to single-molecule spectroscopy is irreversible photodestruction of the molecule of interest as a result of an oxidation reaction due to O_2. We present a relatively simple solution to this problem by making an O_2-depleted sample. After the polymer film was spin coated on glass substrate, the fresh-made sample was exposed to vacuum $(1.0 \times 10^{-7}$ mbar) for 30 minutes, and then coated with a 200-nm thick thermally evaporated Al overlayer in order to seal the film from atmospheric O_2. This procedure resulted in a more than three order of magnitude decrease in the rate of photodestruction and greatly improved signal-to-noise ratio due to longer averaging time. Under the extremely low oxygen concentration of our sample, MEH-PPV molecules exhibit a time-independent fluorescence intensity and spectral band shape (under low-intensity excitation).

6.2   Fluorescence Intensity Flickering

Typical single-molecule fluorescence intensity vs. irradiation time "transients" of MEH-PPV are shown in the Fig. 6.2, with each point corresponding to photon counts in a counting period of 10 ms. Molecules were continuously irradiated at 488 nm, which is close to the peak of the absorption spectrum. This should generate excitons at random locations along the polymer chain. The

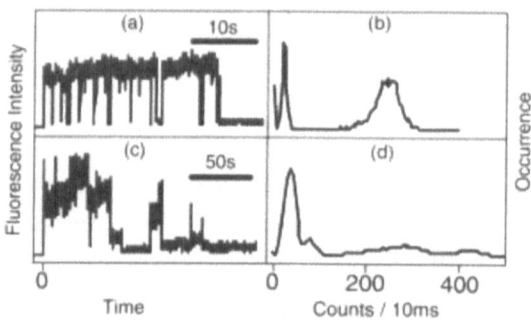

Fig. 6.2. Typical fluorescent transient of single MEH-PPV molecules (**a** and **c**), and the corresponding intensity histogram (**b** and **d**)

transients exhibit large-amplitude, "sudden" intensity fluctuations despite the large number of chromophores in one polymer molecule. We have varied host film thickness from less than 100 nm to 1 µm, and tried different types of polymers as the host matrix (e.g. polystyrene, polymethylmethacrylate and poly-bisphenol-A-carbonate). This general behavior of intensity jump persists for all the different experimental conditions. Many of the transients exhibit a small number (typically 3 or 4) quasi-discrete intensity levels. This is demonstrated by Figs. 6.2b and d, which are intensity histograms of the transients displayed Figs. 6.2a and c, respectively. This phenomenon is related to the fact that the singlet excitons in conjugated polymers are not spatially stationary but tend to move energetically downhill to other locations on the polymer chain. Analogous intensity jumps have also been reported for another conjugated polymer [7].

An individual intensity jump occurs within one counting period (typically 10 ms) and appears to be due to a single photochemical event. The time interval between jumps is on the millisecond to tens of second time scale. The transition from a higher intensity level to a lower level is assigned to the photochemical generation of the quencher defect, and from a lower to a higher level to the thermal repair of a quencher defect. The rate of forming the fluorescent quencher varies considerably for various samples, and is also a function of local O_2 concentration. The "aged" sample, which presumably has a higher O_2 concentration due to leakage of O_2 from the inperfection (pinholes, etc.) in the Al overlayer, shows an obvious higher rate of "flickering" than fresh-made samples. A possible chemical structure for the fluorescence quenchers is an O_2/polymer adduct with charge transfer character: $polymer^+/O_2^-$. The regular and reproducible events of quenching strongly suggest that the fluorescence quenchers are formed repeatedly at the same site on the polymer chain. In samples that do not have an Al overlayer, high O_2 concentrations led to the formation of multiple quenchers, resulting in complex spectral dynamics and rapid permanent photo-oxidation.

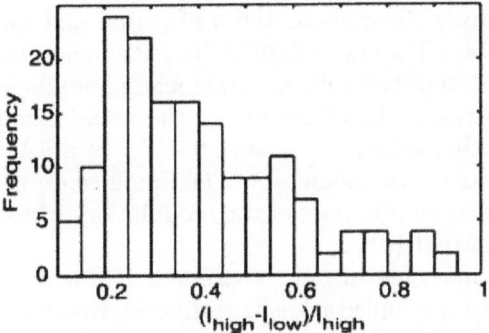

Fig. 6.3. A histogram of fluorescent quenching depths for different single MEH-PPV molecules. I_{high} denotes the fluorescence intensity of unquenched molecules and I_{low} is the intensity after a quenching event. Intensity jumps smaller than 10% (due primarily to noise) were discarded in constructing the histogram

The quenching efficiency of a single quencher defect can be estimated from the fluorescence intensity drop that occurs due to the transition from a higher to lower intensity level. The distribution of the quenching efficiency from ~150 molecules is plotted in Fig. 6.3. We can calculate the intramolecular exciton migration rate based on the quenching depth and the typical exciton lifetime for MEH-PPV of ~300 ps. Since the intensity drops were typically ~20% or greater, the intramolecular electronic energy migration length must correspond to at least ~20% of an average single-polymer molecule extended chain length. This corresponds to an electronic energy diffusion constant greater than 4 cm^2/s, based on a simple three-dimensional bimolecular quenching model. This rate is more than a factor of 40 greater than the bulk diffusion constant for singlet excitons in pure MEH-PPV films [16]. This result indicates that the diffusion of excitons in bulk film is limited by intermolecular exciton hopping processes.

6.3   Single Molecule Anisotropy

Due to their unique structure, conjugated polymers have absorption tensors that can be approximated by collections of transition dipoles along the local direction of the polymer chain [17]. Therefore employing single molecule polarization spectroscopy [18] to measure the polarization anisotropy of the fluorescence excitation of single MEH-PPV molecules gives direct information on the conformation of these molecules. The knowledge of conformation also gives further insight into the nature of optical excitation, since the optical properties of conjugated polymers are heavily influenced by the conformations of the polymer chain.

Individual MEH-PPV molecules were laterally separated in the polycarbonate film. Molecules are optically excited with focused linearly polarized light. We denote the light propagating direction as the **z** direction and the sample plane as the **x** and **y** directions. The emission of MEH-PPV molecules was collected along **z** direction and detected with an avalanche photodiode detector without polarization preference. The direction of the polarization (θ) of the excitation light was linearly modulated between 0 and π at a fixed frequency (usually 1 Hz) by an electro-optic modulator. The fluorescence intensity $\mathbf{I}(\theta)$ is synchronously recorded for 30 cycles of the modulation, which corresponds to $\sim 10^7$ excitations of MEH-PPV.

Absorption of one MEH-PPV molecule in our experiment is due to a collection of approximately 140 weakly coupled transition dipoles, whose directions are determined by the orientation of the local chain segments corresponding to the transition dipole. Thus the absorption properties of one MEH-PPV molecule are best described by an "absorption ellipsoid" in the molecular frame with three orthogonal absorption cross-sections, which we denote as $C_{x'}$, $C_{y'}$, and $C_{z'}$ [19], which generally do not coincide with the laboratory frame axes **x**, **y**, and **z**. Since the light propagation direction is along **z**, $\mathbf{I}(\theta)$ is the projection of this absorption ellipsoid in the **x**, **y** plane. Thus we have

$$I(\theta) \propto 1 + M \cos 2(\theta - \phi) , \tag{6.1}$$

where the modulation depth M is the anisotropy of the absorption ellipsoid projected on the **x**, **y** plane, and ϕ is the orientation of maximum absorption. $\mathbf{I}(\theta)$ is well fit by (6.1). The observed ϕ values are randomly distributed between 0 and π, as expected.

For the majority of the MEH-PPV molecules we studied, both the intensity and the phase of the $\mathbf{I}(\theta)$ from different averaging cycles remain constant within experimental error during this averaging process. The invariance of $\mathbf{I}(\theta)$ implies that the characteristic chain conformation of each molecule is fixed on the experimental time scale. Furthermore, time and wavelength resolved single-molecule spectroscopy demonstrates that $\mathbf{I}(\theta)$ is not distorted by spectral diffusion, blinking, or other photoinduced processes. Thus we believe that $\mathbf{I}(\theta)$ is a good parameter to analyze the polarization dependence of the excitation (absorption) process.

In our experimental setup, for each individual measurement on one molecule, we are only able to obtain the anisotropy (M) of the projection of the absorption ellipsoid onto the **x**–**y** plane. The full information on the absorption ellipsoid for a specific molecule is unknown until we can have some sort of three-dimensional measurement. However, we can rely on the distribution of measured anisotropy to get an estimation of the ensemble-averaged absorption ellipsoid anisotropy. We record M for a large number of single molecules and construct a histogram of M values (M_M as shown in Fig. 6.4.) The ensemble of values of the single-molecule property M reflects the distribution of chain orientations, chain conformations, and molecular weights. One can quickly conclude that the conformation of MEH-PPV does not fall

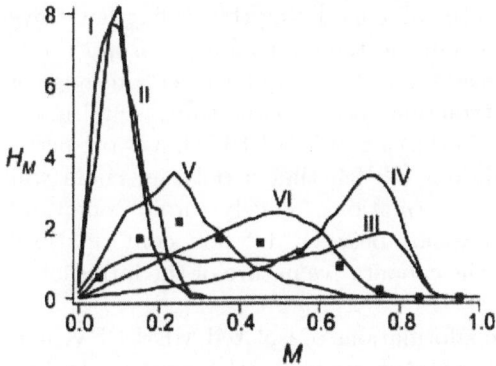

Fig. 6.4. The distribution of M from single-molecule polarization spectroscopy (squares) and from Monte Carlo simulations (curves I–VI)

into one of the two extreme cases. First, it is not a straight polymer chain where the individual transition dipoles fall on a straight line, and H_M is a delta function located at $M = 1$. Second, it is not a perfectly isotropic distribution of dipoles either (e.g. a random coil of infinite chain length), where the absorption ellipsoid is a sphere, and H_M is a delta function located at $M = 0$. We have experimentally verified the latter case by studying $\sim 10^4$ randomly oriented dye molecules imbedded in a 100-nm latex sphere. Those data demonstrate that there is an experimental error of ~ 0.1 in the M axis.

The important conclusion is that the measured single-molecule anisotropy is much bigger than expected if we assume the MEH-PPV molecules adopt a random coil conformation, which should give a much lower average M. This indicates a highly ordered collapsed structure of the MEH-PPV molecule when they were dissolved in the polycarbonate film. This result gives some insight into the extraordinarily fast exciton diffusion at the molecular level for MEH-PPV, demonstrated by the fluorescence intensity "flickering" [20]. The position of the fluorescence quencher along the polymer chain is apparently dictated by S_1 exciton migration. This process "funnels" excitons to exciton traps, which react with O_2 to form the quencher. This "energy funneling" is presumably a result of the ordered regions of the polymer chain containing parallel chains. These regions should favor rapid, directional energy transfer by resonance energy transfer due to aligned transitions dipole moments and short chromophore energy-transfer distances.

6.4 Determine the Conformation of the Polymer Chain by Computer Simulation

It is well established that a flexible polymer chain with little inter-segment attraction forms a self-avoiding random coil. For a real polymer chain that has moderate stiffness and/or inter-segment attraction, the common approach is to simply adjust the persistence length while assuming the scaling law stays unchanged as an ideal random coil: chain end-to-end distance is proportional to $N^{0.6}$ [21]. This approximation, however, is only valid for a certain range of chain stiffness and inter-segment attraction. When both stiffness and inter-segment attraction increase (such as in the case of MEH-PPV), it is predicted by theory and simulation conformations [22–24] that a polymer chain will adopt highly ordered and distinct conformations: namely toroid conformation and rod conformation. The transition between different conformations resembles a phase transition due to the cooperative nature of the intramolecular segment–segment interaction.

We endeavored to simulate the conformations of isolated MEH-PPV molecules using a "beads on a chain" type of model in the hope of finding the underlying physical reason for the big anisotropy observed in our single molecule polarization experiment. In our model, one bead represents a small segment of the polymer chain, i.e. 2.5 repeat units. The typical chain length in the simulation is 100 beads (250 repeat units of the polymer). The choice of the chain length is primarily limited only by the computation time. The contour length of such a chain is about 150 nm, which is no doubt much shorter than the real polymer chain in our experiments. The intra-chain interactions were modeled with either a square-well inter-segment potential or a Lennard–Jones like attractive potential. Both models gave similar results. It should be pointed out the attraction force we discussed here originates from the difference between interactions among polymer segments compared to polymer with solvent, as opposed to attraction force of two polymer segments in vacuum. In other words, the attraction force is solvent-dependent. The stiffness of the polymer was controlled by the parameter called the chain bending energy. The chain bending energy was assumed to take the form $b\alpha^2$, where α is the bending angle and b is the bending force constant. We estimated b from published force fields for solid polymer [25].

For a given set of input parameters, the simulation generates one conformation after 10^8 Monte Carlo steps (one step is one attempted move for each bead of the chain). From the simulated conformation we can calculate a radius of gyration (R_g) and laboratory frame anisotropy (or modulation depth) (M). For the anisotropy calculations, it was assumed that each local transition dipole was along the bond connecting the adjacent beads. Then we repeat the simulation 20 times for each different set of parameters. Using this approach we can get the ensemble averaged radius of gyration and histogram distribution of anisotropy H_M (Table 6.1).

A flexible chain ($b = 0$) with no attraction ($E_{cc} = 0$) resulted in the well-known self-avoiding random coil conformation. A typical example of a random coil conformation from the simulation is shown in Fig. 6.5 structure I. The random coil conformation has an H_M distribution (curve I in Fig. 6.4) that is considerably less anisotropic than the experimental results. The average

Table 6.1.

Type	E_{cc}/kT [a]	$b/kT{\cdot}rad^{-2}$	Number of Defects	$\langle M^2 \rangle^{1/2}$ [b]	$\langle R_g^2 \rangle^{1/2}$ [c] (nm)
I Random coil	0	0	N/A	0.11	13
II Molten globule	0.6	0	N/A	0.12	5.0
III Toroid	0.6	10	0	0.53	7.4
IV Rod [d]	0.6	10	0	0.67	10
V **Defect-coil**	0	10	15 [e]	0.28	21
VI **Defect-cylinder**	0.6	10	15 [e]	0.46	7.2

[a] E_{cc} is the depth of a Lennard–Jones attraction potential between bead pairs of the polymer chain. The bead attraction potential is:

$$\begin{cases} E_{LJ} = \infty & \text{if} \quad l \le 2.1\text{nm} \\ E_{LJ} = 4E_{cc}((\frac{2.1}{l})^{12} - (\frac{2.1}{l})^{6}) & \text{if} \quad 2.1nm < l \le 4.2\text{nm} \\ E_{LJ} = 0 & \text{if} \quad l > 4.2\text{nm} \end{cases}$$

[b] $\langle M^2 \rangle^{1/2}$ is the root mean square of the modulation depth of anisotropy.

[c] $\langle R_g^2 \rangle^{1/2}$ is the root mean square of radius of gyration.

[d] Rods composed of 2–6 straight chain regions are observed in the simulations. R_g strongly depends upon the number of straight chain regions in a rod and the distribution among types of rods is not converged in our limited number of simulations (20 trials). To avoid this source of error in Table 6.1, we restrict our analysis of R_g and M to rods with 4 straight chain regions.

[e] This number corresponds to 15 defects per polymer chain of 100 beads. Since the stiffness parameter (b) has been chosen such that each bead represents 2.5 repeat units, a polymer chain with 15 defects actually corresponds to 6% of the underlying 250 units.

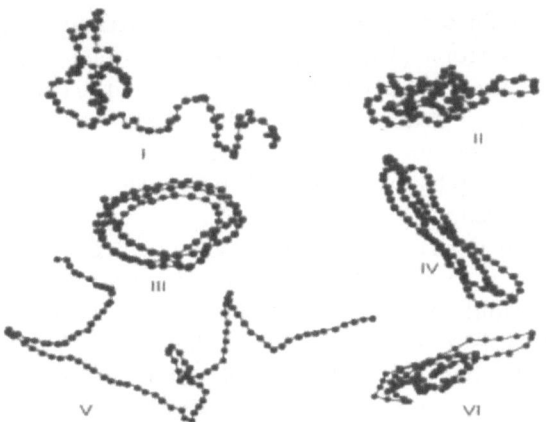

Fig. 6.5. Typical conformations generated by Monte Carlo simulations. They are denoted as: I random coil, II molten globule, III toroid, IV rod, V defect-coil, and VI defect-cylinder

anisotropy of 0.11 (Table 6.1) is consistent with theory for the random coil predicting that $\langle M^2 \rangle^{1/2} \propto \frac{1}{\sqrt{N}}$, where $N + 1$ is the number of beads [26]. A typical molten globule conformation from the simulations ($E_{cc} = 0.6$kT, $b = 0$) is shown by structure II in Fig. 6.5. The H_M curve for the molten globule is similar to the random coil, reflecting the "liquid like" disorder in the globule [27], but the radius of gyration is smaller for the molten globule (Table 6.1). Simulations with $E_{cc} = 0.6$ kT, $b = 10$ kT rad^{-2} reveal toroids and rods (Fig. 6.5). The simulated H_M curves (Fig. 6.4) are more similar to the experimental results than are the simulated H_M curves for the random coil and molten globule. The results for the toroid and rod are, however, peaked toward M values higher than experiment and tend to have a narrower distribution.

The simulations discussed so far, however, missed one important factor: chemical defects. The synthetic procedure for MEH-PPV chains results in chains with 1% or greater tetrahedral rather than conjugated links. Conjugated polymers also tend to be reactive, which allows for the further introduction of tetrahedral defects during photoexcitation. In fact, spectroscopic analysis of our MEH-PPV sample leads to a defect concentration estimate of ~5% [28]. Thus, we believe that it would be very unlikely that a real synthetic polymer molecule would form any of those ideal conformations such as a rod or a toroid, since the chemical defects would tend to break the "tension" needed for such a conformation.

We have introduced such chemical defects at random beads in the simulation by tetrahedral bond angles (unchanged force constant). The inclusion leads to two new types of conformations with different anisotropy characteristics. For the case of a stiff chain with defects but no inter-segment attraction, an extended conformation is found (V, Fig. 6.5) with approximately straight chain segments between defects, denoted here as a "defect-coil". The equilibrium structure of the same chain with strong inter-segment attraction is roughly cylindrical and more anisotropic (VI, Fig. 6.5), which we denote as a defect-cylinder. The defect-coil has a lower average anisotropy than the experimental behavior of MEH-PPV and is peaked at lower values. Defect-cylinder is arguably the most accurate representation of the realistic collapsed conformations of conjugated polymers such as MEH-PPV among the idealized structures that have been considered.

It is also plausible that both defect-coil and defect-cylinder coexist in our spin-coated sample. In fact, a combination of almost equal amount of defect-coil and defect-cylinder would give an M distribution that is in better agreement with our experiment. It is reasonable to assume that in liquid solution MEH-PPV molecules adopt a defect-coil conformation and the collapse to defect-cylinder conformation occurs during the spin-coating process when the solvent quality gets dramatically worse. Therefore, it is not surprising that some of the molecules were trapped in the defect-coil conformation even after the liquid solvent evaporated due to the kinetic nature of the spin-coating process. Interestingly, the discovery of highly ordered, cylindrical conformations reported here for MEH-PPV might be a critical factor resolving the puzzling properties of MEH-PPV and related polymers. These include the significant local anisotropy of thin films of pure materials [29]; and evidence that the structure of MEH-PPV in films can be controlled by the conformation in the solution used to spin-coat the film [12].

6.5   Fluorescence Spectra of Single MEH-PPV Molecules

Single MEH-PPV molecules were dispersed in polycarbonate host and illuminated with 488-nm laser light. The fluorescence spectra were taken from individual molecules using an imaging CCD. The averaged fluorescence spectrum (Fig. 6.6) of ~80 molecules of MEH-PPV reveals the typical band shape of a conjugated polymer. In comparison, the single-molecule fluorescence spectra of MEH-PPV (Fig. 6.6) are significantly narrower than the averaged spectrum in polycarbonate, but comparable in width to the ensemble spectrum in liquid toluene (Fig. 6.6). This suggests that in liquid solution at room temperature, the major contribution to the spectral broadening of MEH-PPV is from vibronic broadening, while in a solid state solution, the inhomogeneous broadening is much bigger. The ensemble distribution of the fluorescence spectral maxima (λ_{max}) of single molecules (Fig. 6.7) exhibits a double-peaked distribution with two maxima (560 and 580 nm). Similar double-peaked distributions (not shown) were observed in other polymer hosts, but with slightly different maxima (553 and 583 nm in polymethylmethacrylate and 555 and 575 nm in polystyrene). The presence of two distinct peaks in the spectral maxima distribution is highly unexpected because it implies that there are two different types of MEH-PPV trapped in the polymer host matrix. Con-

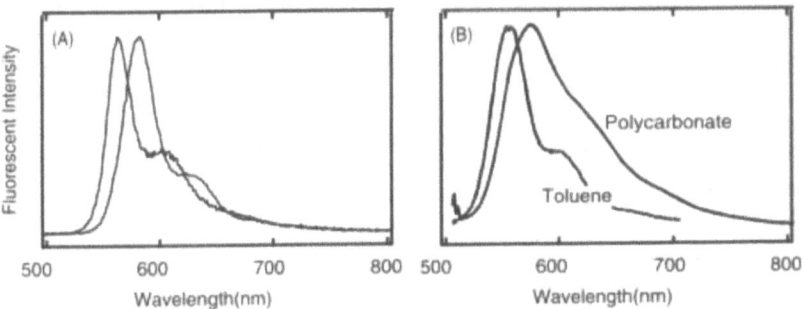

Fig. 6.6. a Fluorescence spectra of two single MEH-PPV molecules in polycarbonate host matrix. **b** Ensemble spectra of MEH-PPV in toluene liquid solution and in polycarbonate solid solution. The latter was constructed by adding up about 80 single-molecule spectra of MEH-PPV

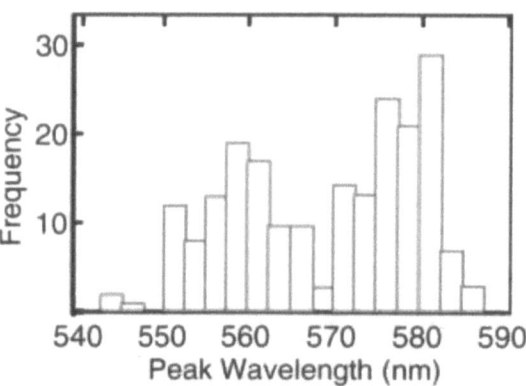

Fig. 6.7. Histogram of the peak wavelength (λ_{max}) of single-molecule fluorescent spectra of MEH-PPV in polycarbonate matrix

sidering how many conformational degrees of freedom a polymer molecule has, one would usually expect a broad, continuous distribution.

It is tempting to assign the two spectral maxima immediately to the two distinct conformations we got in the Monte Carlo simulation, namely defect-coil and defect-cylinder. The anisotropy data, however, do not give positive evidence supporting it. The simulation predicted that defect-cylinder, which is an ordered collapsed form, exhibits a more significant single-molecule excitation anisotropy ($\langle \mathbf{M}^2 \rangle^{1/2} = 0.46$) than the defect coil ($\langle \mathbf{M}^2 \rangle^{1/2} = 0.28$). We used methods described previously to measure $\langle \mathbf{M}^2 \rangle^{1/2}$ of the red and blue conformations of MEH-PPV in PMMA films. The resultant values (0.46 and 0.44, respectively) are too close to distinguish based on anisotropy. Another possibility is the existence of an exciton trap. Considering the likely defect-cylinder structure, a good candidate for the exciton traps in the red-shifted, 580-nm conformation of MEH-PPV would be chain–chain contacts that are sufficiently close to produce a significant lowering of localized transition energy through exciton interactions between nearby parallel-oriented chromophores, such as in molecular aggregates [30]. In the case of conjugated polymer, model compounds with nearby parallel chain contact produce a red-shift and broadening due to excimer-like interactions [31]. For the MEH-PPV 580-nm conformation, the absence of excimer broadening may simply result from a failure to obtain the precise spatial alignment of molecular orbitals that are necessary for an excimer interaction. A chain segment with an extended conjugation length could also be responsible for the 580-nm trap. This possibility seems unlikely, however, because the variation in spectral maxima versus conjugation length is relatively small for long conjugation lengths according to theory and results for model compounds [28]. To be consistent with the data, the conformational feature responsible for the 580-nm exciton trap would have to be relatively energetically unfavorable in the ground state of MEH-PPV, because the blue form of MEH-PPV appears to contain few

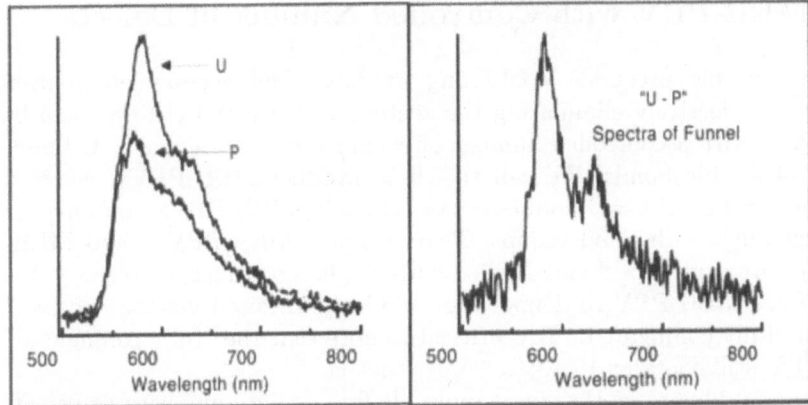

Fig. 6.8. a "Sorted" single-molecule fluorescent spectra of same MEH-PPV molecule in the unphotolyzed state (U) and photolyzed (P) state. **b** The difference spectra of U and P state

if any of these contacts. A careful examination of the spectra of individual MEH-PPV molecules reveals, however, the presence in many cases of both red- and blue-emitting spectral components. That may reflect incomplete energy transfer to the longest wavelength exciton traps, as described below.

There is further evidence showing that low-transition-energy exciton traps are involved in the energy funnels presented in Fig. 6.8. We took fluorescence spectra of single MEH-PPV molecules during the fluorescent intensity fluctuation process. Afterwards, the spectra were sorted for the "fully on" (unphotolyzed state U) and for the "intermediate-intensity level" (photolyzed state p). As stated earlier, the **P** state exists during the presence of a reversibly formed quencher. For the molecule studied in Fig. 6.8, the cumulative **U** spectrum is predominantly red emission with a small portion of blue-edge emission. The difference spectrum (**U–P**) is exclusively sharp, red-shifted emission, which suggests that energy transfer is extremely rapid and efficient within the collapsed region of the molecule. Thus, when a quencher is formed in this region, the emission is quenched by efficient funneling of S_1 excitons to the lowest energy chromophores. Indeed, the extraordinarily sharp and red-shifted emission in the **U–P** difference spectrum (Fig. 6.8) demonstrates that nearly 30% of the excited repeat units (~600 repeat units) all rapidly channel their S_1 excitons to a exciton trap that has a narrow red-shifted emission. The extraordinarily narrow emission spectrum suggests that very few chromophores, perhaps even one chromophore, is responsible for the **U–P** spectrum. The reproducibility of the **U** and **P** spectra during several quenching cycles demonstrates that while the quencher is repeatedly being formed and destroyed, the underlying conformationally induced "funnel" structure for exciton migration is permanent on these time scales.

6.6   MEH-PPV with Controlled Number of Defects

Using a suitable precursor containing methoxy and acetoxy elimination groups and selectively eliminating the acetoxy groups, "MEH-PPV" can be synthesized with a controlled number of chemical defects: single C–C bonds instead of double bonds. We call this final product MEH-PPVx, where x is the percentage of C=C double bonds. Namely, MEH-PPVx contains x% repeating units with vinyl groups. Two samples (MEH-PPV45 and MEH-PPV85) were synthesized and studied with single-molecule spectroscopy. We will still use MEH-PPV to denote the sample synthesized via the "normal" way and "fully conjugated." It is critical to note that the "fully conjugated" MEH-PPV still contains 1%–3% chemical defects.

Figure 6.9 illustrates the single-molecule fluorescence intensity transients $I(t)$ of MEH-PPV45 and MEH-PPV85 when the excitation light polarization was modulated between 0 and π. $I(t)$ of MEH-PPV is also shown in the plot for comparison. The fluorescent intensity dependence on the modulation angle $I(\theta)$ was obtained by averaging $I(t)$ before any intensity jump. $I(\theta)$ was then fitted to equation (6.1) to get the modulation depth M. The measurement was

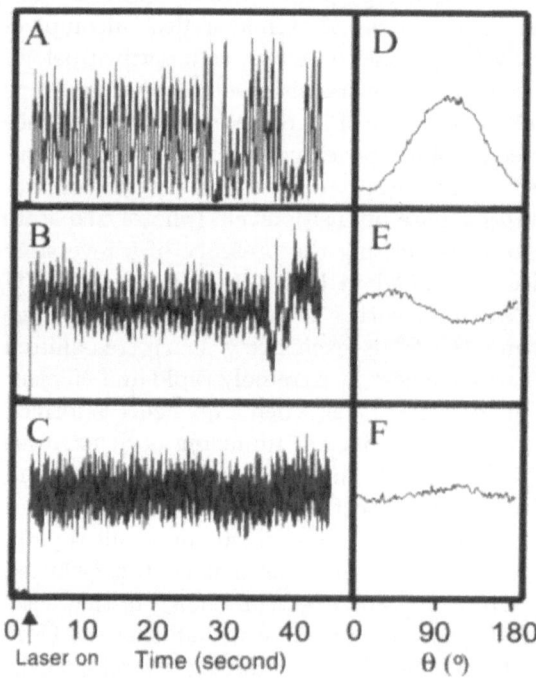

Fig. 6.9. Typical single-molecule fluorescence intensity transients $I(t)$ of MEH-PPV97 **a**, MEH-PPV85 **b**, and MEH-PPV45 **c** under linearly polarized excitation with 1-Hz angular rotation, and the corresponding $I(\theta)$ from averaging the intensity per cycle of polarization modulation

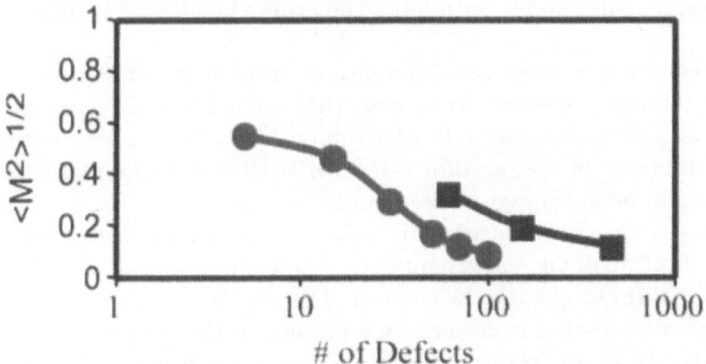

Fig. 6.10. The root mean square of modulation depth with a series of defect concentrations of simulated polymer chains (dots) and experimental study of MEH-PPVx single molecules (square)

performed on more than 100 single molecules for each sample. The histogram of M is shown in Fig. 6.10. The anisotropy value shows strong correlation to the defect concentration. Higher defect levels lead to lower anisotropy values. MEH-PPV single molecules have the highest anisotropy, indicating the defect-cylinder conformation predicted by the simulation. The anisotropy of MEH-PPV45 is the smallest (peak at <0.1) and that of MEH-PPV85 lies in between.

The experimental result on the anisotropy measurement agrees well with the simulation. Using the same simulation method discussed previously, but putting more defect sites in the chain, we can calculate the average modulation depth (M) for polymer chains with higher defect concentrations. The result is plotted in Fig. 6.10. The simulation data has the same trend as the experimental data. The difference between the simulation and experimental data may be due to the simulated chain (250 monomers) being much shorter than the average length of MEH-PPV for easier calculation. The chain length in simulation is uniform, while MEH-PPV length has a broad distribution. The correlation between defect level and anisotropy value can be understood as follows. MEH-PPV does not have a random roil conformation as predicted by theoretical analysis and Monte Carlo simulations. In fact, stiff straight-chain polymers with chain–chain attraction form highly ordered, liquid-crystal-like structures showing very high optical anisotropy. Defects are expected to introduce conformational disorder. The replacement of C=C by C–C single bonds not only breaks up conjugation but also makes the polymer chain into a more random-coil-like conformation, because the rotational barrier of the C–C single bond is small. The C–C single bond acts as a rotational joint of two chain-segments at room temperature. Therefore, at higher concentration of C–C bonds, the polymer chain become more and more random-coil like and no longer adopts the ordered collapsed conformation.

Figure 6.9 also demonstrates the difference of three polymer samples in terms of the "flickering" behavior. We found that while MEH-PPV usually shows large-amplitude fluorescence intensity jumps, MEH-PPV85 shows considerably fewer intensity jumps, and for ~100 MEH-PPV45 single molecules we scanned through, none showed obvious intensity jumps.

As discussed before, the intensity "flickering" is due to the fast exciton migration to the bottom of an "energy funnel," where a fluorescent quencher-defect could easily form to quench the exciton. The absence of "flickering" in MEH-PPV45 indicates that it is either much harder for the quencher-defect to form in MEH-PPV45, or fast exciton migration is not feasible in MEH-PPV45. Considering the results from anisotropy measurements, it is plausible that MEH-PPV45, being much more disordered than MEH-PPV, does not have the right conformation to have an efficient exciton migration pathway. In other words, a collapsed highly-ordered conformation with spatially close and parallel chain segments is critical for fast exciton migration in MEH-PPV. This result demonstrates the interesting link between polymer conformation and photophysical properties.

Acknowledgments. This work was supported by grants from the NSF (PFB), the Robert A. Welch Foundation (PJR, PFB), and the Texas Advanced Research Program (PJR). Further support was provided by the Institute for Theoretical Chemistry and by the Laboratory for Spectroscopic Imaging, University of Texas. We also would like to thank G. Padmanaban and S. Ramakrishnan for providing the defect MEH-PPV samples, and thank Arun Yethiraj for helpful discussions.

References

1. J. H. Burroughes, D. D. C. Bradley, A. R. Brown, R. N. Marks, K. Mackay, R. H. Friend, P. L. Burn, and A. B. Holmes, Nature **347**, 539 (1990)
2. Marsella M. J. and Swager T. M., J. Am. Chem. Soc. **115**, 12214 (1993)
3. J. R. Sheats, Y. Chang, D. B. Roitman, and A. Stocking, Acc. Chem. Res. **32**, 193 (1999)
4. Y. Cao, I. D. Parker, G. Yu, C. Zhang, and A. J. Heeger, Nature **397**, 414 (1999)
5. M. A. Diazgarcia, F. Hide, B. J. Schwartz, M. D. McGehee, M. R. Andersson, and A. J. Heeger, Appl. Phys. Lett. **70**, 3191 (1997)
6. t C. G. Granqvis, A. Azens, A. Hjelm, L. Kullman, G. A. Niklasson, D. Ronnow, M. S. Mattsson, M. Veszelei, and G. Vaivars, Solar Energy **63**, 199 (1998)
7. t D. A. Vandenbou, W. T. Yip, D. H. Hu, D. K. Fu, T. M. Swager, and P. F. Barbara, Science **277**, 1074 (1997)
8. A. Kohler, D. A. Dossantos, D. Beljonne, Z. Shuai, J. L. Bredas, A. B. Holmes, A. Kraus, K. Mullen, and R. H. Friend: Nature **392**, 903 (1998)

9. J. Yu, D. H. Hu, and P. F. Barbara, Science **289**, 1327 (2000)

10. D. H. Hu, J. Yu, K. Wong, B. Bagchi, P. J. Rossky, and P. F. Barbara, Nature **405**, 1030 (2000)

11. R. Jakubiak, C. J. Collison, W. C. Wan, L. J. Rothberg and B. R. Hsieh, J. Phys. Chem. **103**, 2394 (1999)

12. T. Q. Nguyen, V. Doan, and B. J. Schwartz, J. Chem. Phys. **110**, 4068 (1999)

13. H. Sirringhaus, P. J. Brown, R. H. Friend, M. M. Nielsen, K. Bechgaard, B. M. W. Langeveld-Voss, A. J. H. Spiering, R. A. J. Janssen, E. W. Meijer, P. Herwig, and D. M. de Leeuw, Nature **401**, 685 (1999)

14. H. S. Woo, O. Lhost, S. C. Graham, C. B. D. D., R. H. Friend, C. Quattrcchi, J. L. Bredas, R. Schenk, and Mullen K., Synth. Met. **59**, 13 (1993)

15. S. Mukamel, S. Tretiak, T. Wagersreiter, and V. Chernyak, Science **277**, 781 (1997)

16. T. J. Savenije, J. M. Warman, ands A. Goossen, Chem. Phys. Lett. **287**, 148 (1998)

17. T. W. Hagler, K. Pakbaz, and A. J. Heeger, Phys. Rev. B **49**, 10968 (1994)

18. T. Ha, J. Glass, T. Enderle, D. S. Chemla, and S. Weiss, Phys. Rev. Lett. **80**, 2093 (1998)

19. E. W. Thulstrup and J. Michl, in: *Elementary Polarization Spectroscopy* (VCH, 1989)

20. D. Hu, J. Yu, and P. Barbara, J. Am. Chem. Soc. **121**, 6936 (1999)

21. P. d. Gennes, in: *Scaling Concepts in Polymer Physics*, (Cornell University Press, Ithaca, 1979)

22. H. Noguchi and K. Yoshikawa, J. Chem. Phys. **109**, (1998)

23. Y. A. Kuznetsov and E. G. Timoshenko, J. Chem. Phys. **111**, 3744 (1999)

24. V. A. Ivanov, W. Paul, and K. Binder, J. Chem. Phys. **109**, 5659 (1998)

25. I. Orion, J. P. Buisson, and S. Lefrant, Phys. Rev. B **57**, 7050 (1998)

26. T. P. Lodge and G. H. Fredrickson, Macromolecules **25**, 5643 (1992)

27. Y. Q. Zhou, M. Karplus, J. M. Wichert, and C. K. Hall, J. Chem. Phys. **107**, 10691 (1997)

28. G. Padmanaban and S. Ramakrishnan, J. Am. Chem. Soc. **122**, 2244 (2000)

29. J. W. Blatchford, T. L. Gustafson, A. J. Epstein, D. A. Vandenbout, J. Kerimo, D. A. Higgins, P. F. Barbara, D. K. Fu, T. M. Swager, and A. G. Macdiarmid, Phys. Rev. B **54**, R3683 (1996)

30. V. Czikkley, H. D. Försterling, and H. Kuhn, Chem. Phys. Lett. **6**, 11 (1970)

31. G. Bazan, W. J. Oldham, R. J. Lachicotte, S. Tretiak, V. Chernyak, and S. Mukamel, J. Am. Chem. Soc. **120**, 9188 (1998)

7 Confining and Probing Single Molecules in Synthetic Liposomes

C. F. Wilson, D. T. Chiu, R. N. Zare, A. Strömberg, A. Karlsson, and O. Orwar

As organisms, we are amazingly complex living laboratories. As we move, breathe, think, and eat, seemingly endless chemical reactions and interactions occur inside us. The test tubes, beakers, and flasks used to separate and selectively mix the myriad of reactants involved are cells, vesicles, and organelles. Taking the analogy further, whereas chemists typically mix chemicals milliliters or more in volume, biological systems carry out their biochemistry in containers that are femtoliters or less in volume. As researchers we assume, with good reason, that the material surfaces of our laboratory test tubes do not substantially affect the kinetics we measure. This assumption might not hold were we to shrink our containers to the femtoliter scale. At such a small scale, collision rates between reactants and their container walls become significant [1], and the inner surface, particularly in biological containers, is chemically complex. The bilayers of cells and organelles are composed of a variety of lipids. These varieties assemble into domains [2] in a process partly controlled by the transmembrane proteins in them [3]. Cellular and organellar control of chemical reactions may thus come, in part, from alterations in the composition and arrangement of the molecular species making up the bilayer membrane [4]. How do systematic alterations to the bilayer composition of a liposome alter the kinetics of reactions within the liposome interior? We may find that the potential physiological significance of lipid domains within bilayers to the kinetics of in-plane reactions [5] (i.e., for proteins and other molecules moving within the bilayer) has applications to molecules within liposomes that interact with the inner bilayer surface.

A method for investigating these questions has been elusive. The challenge is to make, manipulate, and mix the contents of biomimetic containers for biochemical studies in the laboratory. Our approach is to use synthetic liposomes (vesicles) to carry out chemistry on the femtoliter scale, with the specific aim of studying single-molecule kinetics in a biorelevant nanoenvironment.

In the past we have carried out work in the area of solution-phase single-molecule detection [6,7] and manipulation [8,9] as well as single liposome [10] and organelle [11] manipulation and analysis. We developed a protocol in which liposomes are formed at room temperature in two minutes in order to minimize the denaturing and degradation of biological molecules [12].

We have used these liposomes as chemical containers and reaction vessels for both large numbers of molecules and for single enzymes [1,13]. To improve the protocol used in these experiments, we developed a micropipette capable of holding for over an hour lipid vesicles one to ten micrometers in diameter with little or no change in the suction applied at the micropipette tip and with minimal liposome deformation at the docking point. One-micron-diameter vesicles match well the probe volume of many single-molecule detection (SMD) schemes. The micropipette allows the liposome to be mechanically held in the probe region of a laser beam so that studies can be made of the molecules confined within the liposome. The use of liposomes also allows for the addition of reagents to their interiors to initiate reactions by electroporation or by fusion, for example.

The power of single-molecule studies comes from their ability to detect physical processes that would otherwise be hidden within the ensemble averaging of bulk measurements. The individual steps and chemical intermediates of physical processes can be detected, and the static (between molecules) and dynamic (for a single molecule) heterogeneities contributing to ensemble measurements can be discerned [14]. Exciting evidence for conformational changes of single molecules has been provided by localizing molecules in agarose gel [15], in capillaries [16], in microwells [17], and on glass slides [18,19] or by observing them directly in solution [20,21].

Lu et al. [15], for example, demonstrated that individual cholesterol oxidase molecules turn over substrate molecules with a time-varying rate (on the order of seconds). Over longer periods of time, and for ensembles of cholesterol oxidase molecules, standard kinetics are followed.

Where molecules do not act differently from one another, the same statistics used to describe the ensemble can be used to describe the individual molecular processes therein. Differences might become quite significant where molecules have complex internal structures (such as those leading to conformational changes) or exist within a complex microenvironment (such as in a cell) [22]. To address this latter scenario, we have carried out Monte Carlo simulations of single-molecule Brownian dynamics. These calculations estimate the collision frequencies between individual molecules within a liposome and between those molecules and the inner liposome wall. In addition, we have performed experiments in individual vesicles in which we monitor the turnover rate of substrate molecules by an enzyme [1,13]. In this chapter we review these results, discuss their implications, and report on our related work aimed at exploiting liposomes as tools to look at single molecules and low concentrations of molecules.

Computer simulations were done to estimate the collision frequencies between a substrate molecule (S, with radius r_S), an enzyme molecule (E, with radius r_E), and the inner wall of the vesicle containing them [1]. The substrate was fluorescein diphosphate and the enzyme alkaline phosphatase to correlate with our experimental work. For simplicity, each molecule was

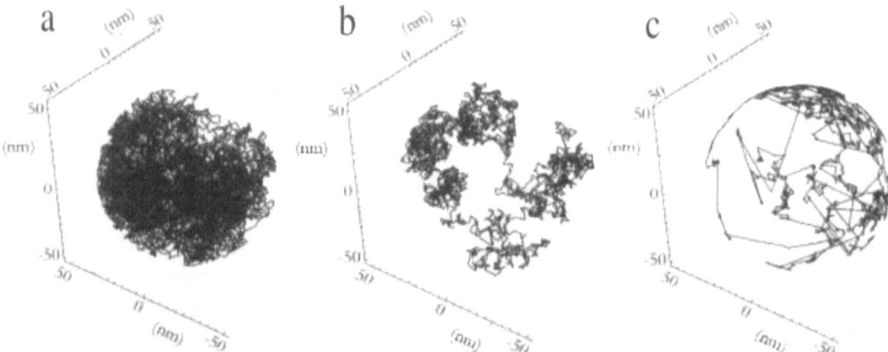

Fig. 7.1. Trajectories of (**a**) a single substrate and (**b**) a single enzyme inside a 104-nm diameter sphere modeled using a Brownian dynamics Monte Carlo simulation. The substrate was followed for 10^4 steps with 10 ns between each step, and the enzyme for 2.5×10^3 steps with 40 ns between each step. (**c**) A trace showing collisions between a single substrate and the spherical wall (40 x 10^6 steps with 1.5 ps between steps). Reprinted from Chemical Physics, vol. 247, Chiu et al., "Manipulating the biochemical nanoenvironment around single molecules contained within vesicles," pp. 133–139, 1999, with permission from Elsevier Science

modeled as a sphere and the vesicle as having hard walls. A collision occurs between S and E when they are within the distance $r_{SE} = r_S + r_E$. The radii were estimated from the Stokes–Einstein equation by using diffusion coefficients of molecules of similar masses. For the enzyme and substrate we took respectively $D_E = 7 \times 10^{-11}$ m^2/s and $D_S = 4.4 \times 10^{-10}$ m^2/s. The Stokes–Einstein equation does not account for charge effects, and it is most applicable to solute molecules that are large compared to the solvent molecules. Consequently, this approximation probably introduces inaccuracies for r_S [23]. The effective charge and size of S will be reduced and increased, respectively, by S's solvation. Both these effects act to improve the accuracy of the calculation. For these reasons, as well as for simplicity, we use the Stokes–Einstein approximation for r_S.

Each molecule (S and E) within the vesicle was allowed to move a random distance in each step. The random displacement lengths were uniformly distributed between maximum displacements in both directions for each dimension. The time step, Δt, needed for the simulation is chosen so that the diffusion coefficient calculated from it ($D = \langle x_n^2 \rangle / 2n\Delta t$) reproduces the estimated values for D_S and D_E. The maximum displacements used for S and E were kept roughly the same, making each time step for S one-fourth as long as that for E. This difference in time steps reflects the fact that D_S is larger than D_E and allows for a simulation in which the solvent affects the diffusive path more for S than for E.

Figures 7.1a and b show traces of the diffusive paths followed by S and E, respectively, within a 104-nm diameter sphere in 100 μs. Figure 7.1c is

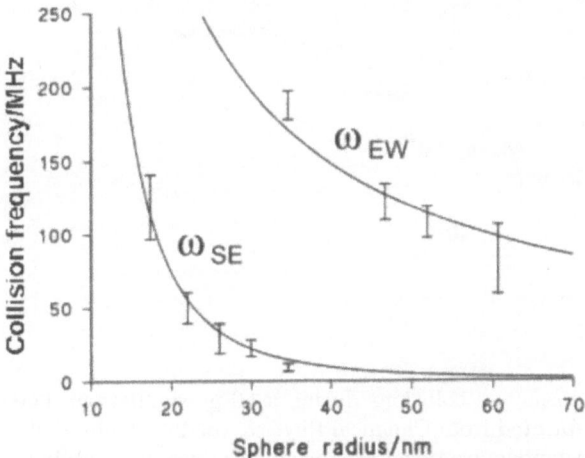

Fig. 7.2. A plot showing collision frequency vs. vesicle radius r. ω_{SE} is the substrate–enzyme collision frequency (40×10^6 steps with 1.5 ps between steps), and ω_{EW} is the enzyme–wall collision frequency (10×10^6 steps with 6 ps between steps). To demonstrate the volume and radius dependence of collision rates respectively ω_{SE} and ω_{EW} vs. r were fit to the function $y = kx^{-r}$. For ω_{SE} $r = -3$ (correlation coefficient = 0.997) and for ω_{EW} $r = -1$ (correlation coefficient = 0.999). Reprinted from Chemical Physics, vol. 247, Chiu et al., "Manipulating the biochemical nanoenvironment around single molecules contained within vesicles," pp. 133–139, 1999, with permission from Elsevier Science

a trace of the collisions between S and the spherical wall over 60 µs [1]. Figure 7.2 plots the collision frequency between S and E (ω_{SE}) and between E and inner sphere (vesicle) wall (ω_{EW}) versus vesicle size [1]. In a 170-nm diameter sphere, ω_{SE} is 0.3 MHz, ω_{EW} is above 50 MHz, and ω_{SW} (the frequency at which S collides with the wall) is 200 MHz. Thus, the single-molecule collision rates for E and S with their container wall are respectively, in a 170-nm sphere, over 150 and almost 1000 times as frequent as the rates of their collisions with each other. Our simulation assumed a perfectly reflective boundary, not taking into account any chemical interactions (from charge or hydrophobic effects, for example). Including such effects could strongly bias where the molecules reside. While some groups of molecules may be repelled from the liposome wall, which would raise their effective concentration in the center of the liposome, others may be biased to spend significant periods near the wall each time they collide with it. If both molecules involved in a reaction spend sufficient time near the bilayer wall, most of their reactions might occur there. Molecules diffusing in two dimensions are more likely to collide than those diffusing in three [24].

Figure 7.3 shows the number of collisions occurring between E and S in a 60-nm sphere over a 60-µs simulation time [1]. The collisions occur in clusters because once a collision occurs, the two molecules remain in the

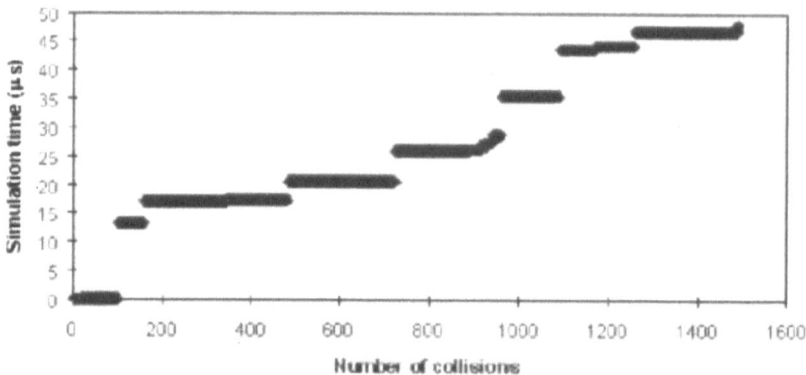

Fig. 7.3. Number of substrate–enzyme collisions during a 60-μs simulation. The vesicle diameter was 60 nm. Reprinted from Chemical Physics, vol. 247, Chiu et al., "Manipulating the biochemical nanoenvironment around single molecules contained within vesicles," pp. 133–139, 1999, with permission from Elsevier Science

same region for some time. This localization makes subsequent collisions more likely. There appears to be no correlation in time between collision clusters. Based on this behavior alone, reactions requiring multiple collisions could be expressed by non-Poissonian rate distributions if they occur within the same collision cluster. Poissonian rate distributions assume random events uncorrelated in time.

To investigate relative reaction rates over time within liposomes, we loaded into liposomes fluorescein diphosphate (from Molecular Probes, Leiden, Netherlands) and alkaline phosphatase (from Sigma). Extravesicular analytes were removed using a size-exclusion column. The liposomes were manipulated using an optical trap, and molecules inside the liposome were detected using laser-induced fluorescence (LIF). The optical trapping and LIF experimental setup have been described elsewhere [1,13]. Briefly, the near-IR light from a MOPA diode laser was sent through a microscope objective to trap the liposome containing the molecules to be studied. An argon ion laser (488-nm line) was then used to excite the enzyme–substrate reaction product (fluorescein). The fluorescence photons were sent through a pinhole via the microscope objective and collected on a single-photon-avalanche diode detector. The starting concentration of fluorescein diphosphate (at the time the liposomes were made) was 0.75 mM. The enzyme concentration was sufficiently low that only one to a few enzymes would be expected to reside inside each liposome. This number varies from vesicle to vesicle and depends strongly on vesicle size and encapsulation efficiency.

Figure 7.4 shows the fluorescence intensity versus time of the enzyme–substrate reaction inside a 3-μm (left) and a 1-μm (right) diameter liposome [1]. Every 60 seconds the reaction product formed was simultaneously probed to measure the fluorescence intensity and photobleached to reset the reaction

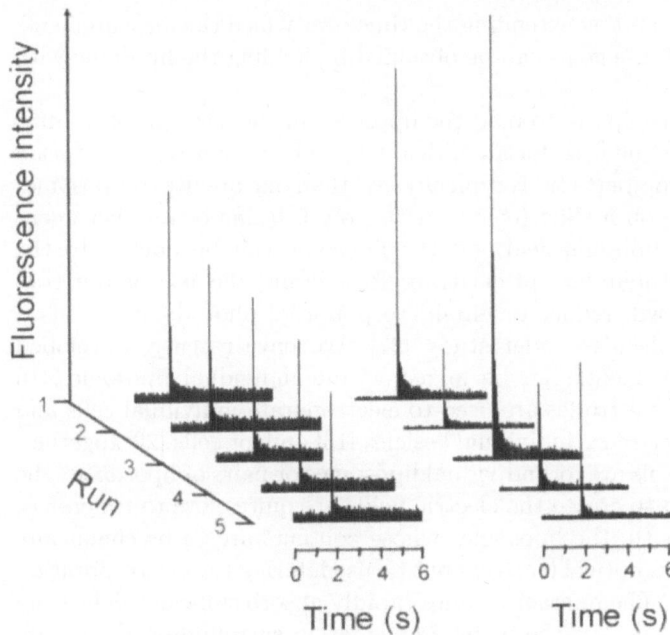

Fig. 7.4. Fluorescence intensity vs. time as a measure of alkaline phosphatase catalytic activity inside an optically trapped vesicle. The substrate, fluorescein diphosphate, was co-encapsulated with the enzyme within the vesicle. At 60-s intervals, the amount of fluorescent product accumulated was simultaneously probed and bleached at 488 nm. The bleaching resets the reaction clock for each run. The vesicles used had radii of 1.5 µm (*left panel*) and 500 nm (*right panel*). Reprinted from Chemical Physics, vol. 247, Chiu et al., "Manipulating the biochemical nanoenvironment around single molecules contained within vesicles," pp. 133–139, 1999, with permission from Elsevier Science

clock. In this way, the relative catalytic activity of alkaline phosphatase was measured every minute over several minutes. In the 3-µm vesicle this activity was nearly homogeneous in time, in stark contrast to the nonhomogeneous product formation in the 1-µm vesicle. While the reasons for the difference are unknown, in the case of alkaline phosphatase the cause of the anomalous substrate turnover rates is unlikely to be conformational changes on the order of minutes [25] (the time scale of each of our experiments), unless bilayer-wall effects cause such changes [26]. In another 3-µm vesicle the turnover rate decreased exponentially with time over a period of 9 minutes [13]. This exponential decrease would be expected for a system in which the substrate concentration has been reduced to the point where diffusion no longer replaces substrate molecules as fast as the enzyme can turn them over.

The experiments just described can be improved by reducing the use of laser light, by sequentially delivering reactants to the liposome while it is held in the probe volume, and by extending the time over which the measurements are carried out. All these goals can be obtained by holding the liposome with a micropipette.

The use of a micropipette to hold the liposome in the laser probe volume reduces or eliminates the need for an optical trap. The time it takes to dock a liposome to the micropipette tip is typically less than one minute. If liposomes are allowed to settle on a slide (coated with poly-L-lysine or another agent to reduce potential liposome leakage) the liposome can be docked to the micropipette tip without an optical trap. Restricting the use of the trap to a minute or less will reduce or eliminate potential photodamage caused to the biological molecules under study [27]. We have recently developed methods by which reagents can be mixed within individual liposomes. In this procedure microelectrodes are used to electroporate individual cells and organelles [28] or electrofuse individual vesicles [13] and/or cells [29] together. Because these protocols are for individual liposomes or pairs of liposomes, the microelectrodes used to create the electric fields are quite close to (microns), or even in contact with, the liposomes whose contents are to be chemically altered. Consequently, optical traps cannot be used during the electroporation or electrofusion step. The microelectrodes rapidly absorb sufficient light from the trap to form bubbles in the water immediately surrounding them [30]. Any liposomes in the vicinity move away from the microelectrodes quite rapidly. To overcome this problem, we have been working with liposomes that adhere to the surface of glass slides or that are held on the ends of micropipettes. In either case the liposomes are placed, or docked, by the use of optical traps, but those traps are then turned off before positioning the microelectrode tips. In our work with single enzymes, described above, we held onto the liposomes containing the enzyme and substrate molecules with an optical trap, because placing liposomes on slides can in some instances result in liposome leakage, and holding them at the tips of micropipettes has proven to be difficult. Micropipettes capable of providing robust support for a micron-sized liposome would not only facilitate a reduction or elimination in the use of an optical trap but also allow for the introduction of reagents or substrate molecules into the liposome by electroporation or fusion after the liposome is placed into the LIF probe volume. The amount of reactants that is required as a whole would be much reduced, lessening or eliminating the need for a photobleaching event to start the reaction clock in kinetics experiments (again reducing potential photodamage).

We report here the development of micropipettes with tips characterized by a narrow inside diameter (i.d.) and a large liposome-seating surface. The small-tip i.d. (less than one micron) prevents the vesicle from going inside the micropipette. The large liposome-seating surface provides mechanical stability during electroporation, electrofusion, and long-duration (over an hour) experiments for one-to-ten micron vesicles. The seating surface surrounding the tip opening is funnel-shaped so that a liposome can be docked there with-

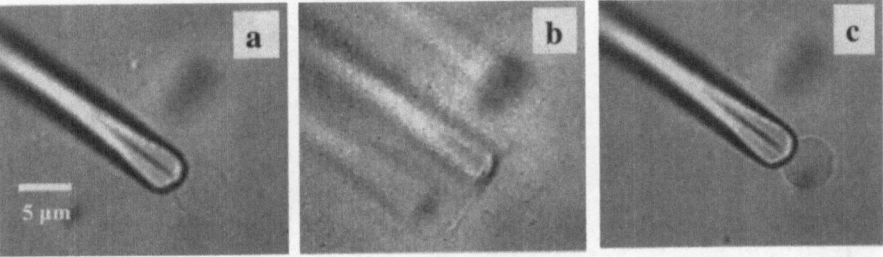

Fig. 7.5. Flick test. Image **a** shows a liposome docked onto a pipette whose inner-tip surface is funnel-shaped. Image **b** showns the liposome and pipette 0.1 seconds after the flick-test impulse is delivered to the manipulator holding the pipette. Several flick tests were performed in rapid succession. The liposome and pipette oscillate at greater than 30 Hz (the video collection rate) with an initial back-and-forth displacement of 20 microns. The oscillation is essentially damped out in 2 s. Image **c** is 3 s after the fourth flick test. No visible instability of the liposome on its seating surface occurs. Application of a small increase in pressure inside the pipette followed by a milder version of such a test allows the liposome to be undocked (see text)

out significant deformation of its surface curvature. Suction and pressure – to dock or undock a liposome, respectively – are provided to the micropipette tip by a system of syringes (10 μL for fine control and 100 μL for coarse control, part numbers 80065 and 81320, respectively, Hamilton, Reno, NV). The syringe plungers are translated by rotating 127-thread-per-inch adjustment screws (part no. AJS127-0.5, Newport Corp., Irvine, CA). The ends of the adjustment screws push against the plunger heads. When a screw is retracted to generate suction, a spring around the associated syringe's plunger barrel pushes the plunger against the screw. Ultrafine pressure control is obtained by the installation of a piezoelectric stack (part no. AE0505D08, Thorlabs, Inc., Newton, NJ) between the 10-μL syringe plunger and its associated control screw. The system uses water as the hydraulic medium, and a 5-mL syringe is installed to act as a reservoir for filling the fine and coarse control elements.

Figure 7.5a shows a 5-μm vesicle docked on a holding micropipette with a 4-μm outer diameter (o.d.) and a 0.5-μm i.d. The micropipette appears to be round where the liposome is seated owing to the extended, round upper lip of the funnel opening. If the line of the vesicle boundary is followed, the vesicle is seen to be seated in the funnel with a small deformation or bud extending into the funnel itself. Each funnel shape corresponds to a particular vesicle size. In this case the micropipette would more naturally hold a smaller vesicle, perhaps 3 μm in size. The extent to which a vesicle deforms into the seating surface is precisely controlled with the fine-pressure syringe (use of the piezo is not necessary for this). To test the mechanical stability of the docked vesicle on the micropipette tip, a "flick test" is performed.

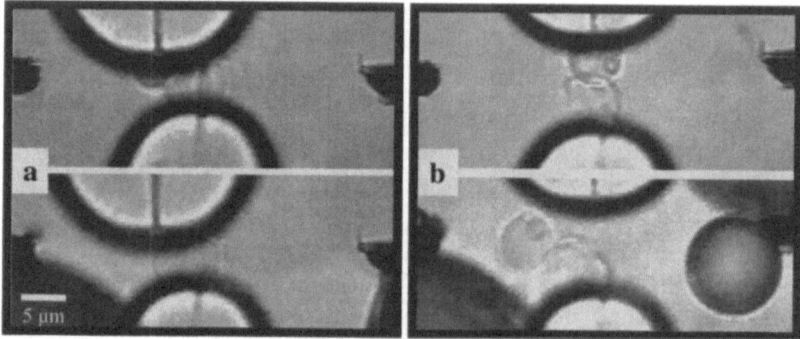

Fig. 7.6. A pair of vesicles is oriented with their lipid bilayers at the point of contact (**a**, top) perpendicular and (**b**, top) parallel to applied electric field lines. The black tips of the microelectrodes used to generate the field are visible on the left and right of each panel. After an electric-field pulse, the perpendicular orientation results in fusion (**a**, bottom) whereas the parallel orientation does not (**b**, bottom)

This test, shown in Fig. 7.5b, consists of literally flicking with one's finger the micromanipulator that is holding the micropipette. In the trial shown, the initial vibration of the micropipette moves the vesicle back and forth over a distance of 20 µm at a frequency of greater than 30 Hz (the video collection rate). The motion is effectively damped out in 2 seconds. Several flick tests were done in rapid succession. Figure 7.5c shows the vesicle after the last test in the series, approximately 10 s after Fig. 7.5a and 9 s after Fig. 7.5b. Using a different vesicle, we did three series of flick tests over 1/2 hour (initially, after 1/4 hour and after 1/2 hour) with similar results. No alterations in suction pressure were needed, and no change in vesicle size was seen. Immediately following the final series of flick tests, the pressure inside the micropipette was increased using the fine-pressure control. A subsequent flick test (of milder nature than shown in Fig. 7.5, consisting of two flicks) jarred loose and then released the vesicle from its seat. It slowly drifted through the field of view and could easily have been optically trapped for subsequent manipulation. We anticipate that for work involving unilamellar liposomes, which are very difficult to optically trap, this micropipette design will offer a stable mechanical method of manipulation.

Figure 7.6 demonstrates vesicle-fusion attempts using microelectrodes (tips seen on the far left and right of each image) and holding micropipettes (tips seen on the top and bottom of each image). These images were taken before we developed a technique to reduce the micropipette tip o.d. (compare them with the more recently designed micropipettes in Fig. 7.5). In Figs. 7.6a (top) and b (top), vesicles are held together so that the plane of their bilayers where they touch are nearly perpendicular or parallel, respectively, to the field lines generated by the microelectrodes. When an electric-field pulse is applied, the electrolysis of water in the region of the microelectrode tips produces bubbles. These bubbles are seen as large dark spheres extending off the field of view in Figs. 7.6a (bottom) and b (bottom). If fusion does not occur, one or both of the vesicles can be jarred from their micropipette seats as a result of the rapid flow patterns produced by the bubbles. The perpendicular configuration (again referring to the vesicle bilayers where they contact relative to the applied electric-field lines) used in Fig.7.6a and our other work [13,29] results in fusion. Multiple attempts to fuse two vesicles in the parallel configuration, one of which is shown in Fig. 7.6b, were not successful. The trials in Figs. 7.6a and b were done within ten minutes of each other using vesicles from the same drop of solution placed on the microscope stage and used the same micropipettes, microelectrodes, voltage, and pulse-length settings. Thus, the only effective difference between the trials was the configuration of the vesicles with respect to each other and the microelectrodes. The mechanism of electrofusion is not completely understood but may be directly related to electroporation [31–34]. Electroporation occurs in a cell's bilayer membrane predominantly where applied electric-field lines are perpendicular to the cell's surface [35]. The fusogenicity of cells in the perpendicular as compared to the parallel configuration was demonstrated by Tessié and Blangero in 1984 [36]. The use of micropipettes that hold liposomes in a robust way has facilitated our observing the vectoral character of electrofusion on liposomes directly in real time.

Micropipettes with fine-pressure control have been used previously in other applications. For example, micropipettes have been employed to measure the elastic properties of a wide range of bilayer compartments [37], the strength of adhesion between cells [38], and the strengths of interactions between biomolecules bound to liposome surfaces [39]. Sensitive micropipette systems have also been used as tools to build nanoscale biomembrane conduits [40] and to investigate how mechanical stress affects cellular cytoskeletal rearrangements [41], liposome electroporation [42], liposome surface topology [43], and the redistribution of membrane components in cells [44]. Submicrometer-i.d. micropipettes were used to investigate the properties of the membrane cortex of cells as well as to help validate the existence of such a cortex [45]. These various applications differ from those described in our work in that they involve bilayer deformation, often to measure physical properties. Our goal, in contrast, has been to hold liposomes on a micropipette tip with a large funnel-shaped seating surface and a small i.d. to minimize liposomal deformation into the micropipette. Further improvements to the micropipette from those shown here have recently been applied to sensitive (single-digit pN force control) but robust manipulationsof single cells and liposomes as small as 1µm in diameter [46]. Using light scattering, vibration rates of holding micropipettes during flick tests has been shown to average ∼125 Hz, corresponding to holding forces in excess of 2000 pN for some liposomes [46].

The flick test, undocking, and fusion-orientation experiments were done to test the integrity of the recently developed micropipette system. What lies ahead are many experimental opportunities afforded by the ability to dock, hold, move in three dimensions, electroporate, electrofuse, and undock lipsomes. These opportunities include not only single-molecule studies but also efforts in single organelle and cell manipulation (mechanical, chemical, and genetic). We believe that the use of liposomes as single-molecule containers will provide a strong complement to other techniques as answers are sought to the ever-expanding biochemical complexity that lies before us. The microenvironments surrounding single biomolecules are integral to that expanding complexity. Analysis taking different aspects of *in vivo* conditions into account will provide an improved perspective on the living laboratories of which those single molecules are a part.

References

1. D. T. Chiu, C. F. Wilson, A. Karlsson, A. Danielsson, A. Lundqvist, A. Strömberg, F. Ryttsén, M. Davidson, S. Nordholm, O. Orwar, R. N. Zare, Chem. Phys. **247**, 133 (1999)

2. See, for example, E. J. Shimshick, H. M. McConnel, Biochem. **12**(12), 2351 (1973) entitled "Lateral Phase Separation in Phospholipid Membranes"; J. Korlach, P. Schwille, W. W. Webb, G. W. Feigenson, PNAS-USA **96**, 8461 (1999) demonstrating the exact superposition of like phase domains in the outer and inner leaflet of three-component synthetic vesicles; J. Hwang, L. A. Gheber, L. Margolis, M. Edidin, Biophys. J. **74**, 2184 (1998) entitled "Domains in Cell Plasma Membranes Investigated by Near-Field Scanning Optical Microscopy", A. Rietveld, K. Simons: Biochim. Biophys. Acta. **1376**(3), 467 (1998) entitled "The Differential Miscibility of Lipids as the Basis for the Formation of Functional Rafts." For a discussion of how lipid domains change over a cell cycle for the bacterium *Micrococcus luteus*, see M. Welby, Y. Poquet, J. F. Tocanne, FEBS Lett. **384** (2), 107 (1996)

3. See for example J. Tocanne, L. Cézanne, A. Lopez, B. Piknova, V. Schram, J. Tournier, M. Welby, Chem. Phys. Lipids **73**, 139 (1994); M. B. Sankaram, D. Marsh, L. M. Gierasch, T. E. Thompson: Biophys. J. **66**(6), 1959 (1994); S. Morein, E. Standberg, J. A. Killian, S. Persson, G. Arvidson, R. E. Koeppe II, G. Lindblom, Biophys. J. **73**(6), 3078 (1997); V. Schram, T. E. Thompson, Biophys. J. **72**(5), 2217 (1997)

4. Cells alter their bilayer-lipid composition in response to temperature changes to achieve a particular phase state in their bilayer [E. F. Terroine, C. Hatterer, P. Roehrig, Bull. Soc. Chim. Biol. **12**, 682 (1930); E. R. L. Gaughran: J. Bacteriol. **53**, 506 (1947); A. G. Marr, J. L. Ingraham, J. Bacteriol. **84**, 1260 (1962)]. The phase state achieved corresponds to a very narrow temperature band and results in an anomalously high diffusion coefficient for a probe molecule in bilayers formed from total lipid extracts of *E. coli* at the specific temperatures at which different *E. coli* groups were grown [A. J. Jin, M. Edidin, R. Nossal, N. L. Gersheld, Biochem. **38**, 13275 (1999)]. Corresponding critical states in synthetic bilayers have been shown to provide for unusual mechanical properties [N. L. Gershfeld, L. Ginsberg, J. Membr. Biol. **156**(3), 279 (1997)] and an anomalous heat capacity [N. L. Gershfeld, C. P. Mudd, K. Tajima, R. L. Berger,

Biophys. J. **65**, 1174 (1993)]. Thus, the bilayer membranes of cells have special structural properties very sensitive to small changes in temperature and their protein and lipid composition. The existence of domains within lipid bilayers in living systems is an indication of tremendous phase complexity under cellular control.

5. T. E. Thompson, M. B. Sankaram, R. L. Biltonen, D. Marsh, and W. L. C. Vaz, Mol. Membr. Biol. **12**, 157 (1995)

6. S. Nie, D. T. Chiu, and R. N. Zare, Science **266**, 1018 (1994)

7. S. Nie, D. T. Chiu, and R. N. Zare, Anal. Chem. **67**, 2849 (1995)

8. D. T. Chiu and R. N. Zare, J. Am. Chem. Soc. **118**, 6512 (1996)

9. D. T. Chiu and R. N. Zare, Chem. Eur. J. **3**(3), 335 (1997)

10. D. T., Chiu, A. Hsiao, A. Gaggar, R. A. Garza-López, O. Orwar, and R. N. Zare, Anal. Chem. **69**, 1801 (1997)

11. D. T. Chiu, S. J. Lillard, R. H. Scheller, R. N. Zare, S. E. Rodriguez-Cruz, E. R. Williams, O. Orwar, M. Sandberg, and J. A. Lundqvist, Science **279**, 1190 (1998)

12. A. Moscho, O. Orwar, D. T. Chiu, B. P. Modi, and R. N. Zare, PNAS-USA **93**, 11443 (1996); This paper claimed that most of the vesicles made were unilamellar. Subsequent work (O. Orwar and coworkers, unpublished), has shown that although some of the liposomes produced by this technique are unilamellar, most are multilamellar.

13. D. T. Chiu, C. F. Wilson, F. Ryttsén, A. Strömberg, C. Farre, A. Karlsson, S. Nordholm, A. Gaggar, B. P. Modi, A. Moscho, R. A. Garza-Lopéz, O. Orwar, and R. N. Zare, Science **283**, 1892 (1999)

14. X. S. Xie and H. P. Lu, J. Biol. Chem. **274**, 15967 (1999)

15. H. P. Lu, L. Xun, and X. S. Xie, Science **282**, 1877 (1998)

16. Q. Xue and E. S. Yeung, Nature **373**, 681 (1995)

17. W. Tan and E. S. Yeung, Anal. Chem. **69**, 4242 (1997)

18. T. Ha, A. Y. Ting, J. Liang, W. B. Caldwell, A. A. Deniz, D. S. Chemla, P. G. Schultz, and S. Weiss, PNAS-USA **96**, 893 (1999)

19. L. Edman, Z. Földes-Papp, S. Wennmalm, and R. Rigler, Chem. Phys. **247**, 11 (1999)

20. L. Edman, Ü. Mets, and R. Rigler, PNAS-USA **93**, 6710 (1996)

21. M. Borsch, P. Turina, C. Eggeling, J. R. Fries, C. A. Seidel, A. Labahn, and P. Graber, FEBS Lett. **437**, 251 (1998)

22. C. Bai, C. Wang, X. S. Xie, and P. G. Wolynes, PNAS-USA **96**, 11075 (1999)

23. D. Ermak, J. Chem. Phys. **62**, 4189 (1975)

24. For multicomponent bilayers, reactant molecules might associate differently with different lipid domains. In that case the two-dimensional diffusion of a molecule interacting with the bilayer wall might undergo complex variations. Restricted, time-dependent diffusion has been observed for proteins in cell membranes [see T. J. Feder, I. Brust-Mascher, J. P. Slattery, B. Baird, and W. W. Webb, Biophys. J. **70**, 2767 (1996)] as well as for molecules that associate with transmembrane proteins [R. Simson, B. Yang, S. E. Moore, P. Doherty, F. S. Walsh, and K. A. Jacobson, Biophys. J. **74**, 297 (1998); P. R. Smith, I. E. G. Morrison, K. M. Wilson, N. Fernández, and R. J. Cherry, Boiphys. J. **76**, 3331 (1999)]. Diffusion coefficients for green fluorescent protein (GFP) molecules and derivitized GFP molecules within *E. coli* cells cannot be explained by a single effective cytoplasmic viscosity [M. B. Elowitz, M. G. Surette, P. Wolf, J.

B. Stock, and S. Leibler, J. Bacteriol. **181**, 197 (1999)]. Cytoplasmic diffusion likely varies due to a combination of bilayer wall, cytoskeletal, and other effects.

25. D. B. Craig, E. A. Arriaga, J. C. Y. Wong, H. Lu, and N. J. Dovichi, J. Am. Chem. Soc. **118**, 5245 (1996)

26. *In vivo* alkaline phosphatase activity can be influenced by the phase state of nearby lipid bilayers. See V. M. Bresler, S. N. Valter, M. A. Jerebtsova, V. V. Isayev-Ivanov, E. N. Kazbekov, A. R. Kleiner, Y. N. Orlov, I. A. Ostapenko, A. T. Suchodolova, and V. N. Fomichev, Biochim. Biophys. Acta **982**, 288 (1989); A. S. Molina, A. Paladini, and M. S. Gimenez, Horm. Metab. Res. **29**(4), 159 (1997).

27. Biological molecules transparent to the wavelength used for optical trapping do not undergo photodamaging processes while in an optical trap. Complex systems, especially within living systems, are susceptible to damage over the entire trapping range normally utilized (790-1064 nm). See K. C. Neuman, E. H. Chadd, G. F. Liou, K. Bergman, and S. M. Block, Biophys. J. **77**, 2856 (1999) and the references therein.

28. J. A. Lundqvist, F. Sahlin, M. A. I. Åberg, A. Strömberg, P. S. Eriksson, and O. Orwar, PNAS-USA **95**, 10356 (1998)

29. A. Strömberg, F. Ryttsén, D. T. Chiu, M. Davidson, P. S. Eriksson, C. F. Wilson, O. Orwar, and R. N. Zare, PNAS-USA **97**, 7 (2000)

30. The spatial coincidence of microthermocouples and the focus of an optical trap has been shown to produce temperature jumps greater than $500°C$, vaporizing the surrounding media. A method has been developed by which beads of the butyl ester of stearic acid are used to measure the local heating rate due to an optical trap at its focus. See S. C. Kuo, "A simple assay for local meating by optical tweezers," in: *Laser Tweezers in Cell Biology* (*Methods in Cell Biology*, vol. 55) M. P. Sheetz (ed.), pp. 43–45 (Academic Press, Orlando, FL, 1998).

31. D. S. Dimitrov: "Electroporation and electrofusion of membranes," in: *Structure and Dynamics of Membranes (Handbook of Biological Physics*, vol. 1B, series ed. A. J. Hoff), R. Lipowski and E. Sackmann (eds.) pp. 851–901 (Elsevier, Amsterdam, 1995)

32. U. Zimmermann: "Electrofusion and electropermeabilization in genetic engineering," in: *Membrane Fusion*, J. Wilschut and D. Hoekstra (eds.) pp. 665–695 (Marcel Dekker, New York, 1991)

33. J. Teissié and M. P. Rols, "Interfacial membrane alteration associated with electropermeabilization and electrofusion" and A. E. Sowers, "Mechanisms of electroporation and electrofusion," respectively, in: *Guide to Electroporation and Electrofusion*, pp. 139–153 and 119–138 (Academic Press, San Diego, 1992)

34. E. Neumann, "The relaxation hysteresis of membrane electroporation," and G. B. Melikyan and L. V. Chernomordik, "Electrofusion of lipid bilayers," in: *Electroporation and Electrofusion in Cell Biology*, pp. 61–82 and 181–192, respectively (Plenum Press, New York, 1989). The latter develops a cogent hypothesis stating that more complex intermediate structures may exist between the formation of pores and the fusion event. See also L. V. Chernomordik, G. B. Melikyan, and Y. A. Chizmadzhev, Biochim. Biophys. Acta **906**, 309 (1987). Other work has shown that a critical level of permeabilization must exist for fusion to occur when cells are placed in contact. See J. Teissié and C. Ramos, Biophys. J. **74**, 1889 (1998).

35. Consider the simplified case of an insulating spherical membrane containing a conducting aqueous medium. Let this spherical membrane be placed between two parallel plates at a known potential difference. Application of Gauss's law shows that the transmembrane potential has a cos θ dependence. Here θ is the angle between the normals to the bilayer surface at (1) the location of interest and (2) the point where the electric field is perpendicular to the surface. Opposing charges build up on opposite sides within the sphere. Pore formation occurs when a critical breakdown potential for the bilayer is exceeded. The spatial distribution of electroporation has been demonstrated by tracking fluorescent dyes entering [W. Mehrle, U. Zimmermann, and R. Hampp, FEBS Lett. **185**, 89 (1985)] and exiting [D. S. Dimitrov and A. E. Sowers: Biochim. Biophys. Acta **1022**, 381 (1990)] electroporated cells. Optical images showing the spatial distribution of the transmembrane potential during electroporation have been produced in cells by using dyes whose fluorescence depends on the local electric-field strength. See D. Gross, L. M. Loew, and W. W. Webb, Biophys. J. **50**, 339 (1986), and M. Hibino, H. Itoh, and K. Kinosita Jr., Biophys. J. **64**, 1789 (1993).

36. J. Tessié and C. Blangero, Biochim. Biophys. Acta **775**, 446 (1984)

37. Micropipette aspiration to measure the mechanical properties of bilayer membranes was first used to study the elastic properties of sea urchin eggs; J. M. Mitchinson and M. M. Swann, J. Exp. Biol. **31**, 443 and 461 (1954). Since that time this method, or modified forms of it, has been used to study the elastic properties for the membranes of many systems. These systems include, for example, red blood cells [R. P. Rand and A. C. Burton, Biophys. J. **4**, 115 (1964); E. A. Evans, Biophys. J. **13**, 941 (1973)], liposomes [R. Kwok and E. Evans, Biophys. J. **35**, 637 (1981); E. Evans and D. Needham, Faraday Discuss. Chem. Soc. **81**, 267 (1986)], and compartments made of polymer bilayers ("polymersomes") [B. M. Discher, Y. Won, D. S. Ege, J. C. Lee, F. S. Bates, D. E. Discher, and D. A. Hammer, Science **284**, 1143 (1999)].

38. E. Evans: "Physical actions in biological adhesion," in: *Structure and Dynamics of Membranes* (Handbook of Biological Physics, vol. 1B, series ed. A. J. Hoff) R. Lipowsky and E. Sackmann (eds.), pp. 723–754 (Elsevier, Amsterdam, 1995)

39. F. Pincet, W. Rawicz, E. Perez, L. Lebeau, C. Mioskowski, and E. Evans, Phys. Rev. Lett. **79**, 1949 (1997)

40. E. Evans, H. Bowman, A. Leung, D. Needham, and D. Tirrell, Science **273**, 933 (1996)

41. D. V. Zhelev and R. M. Hochmuth, Biophys. J. **68**, 2004 (1995)

42. D. Needham and R. M. Hockmuth, Biophys. J. **55**, 1001 (1989)

43. E. Evans and W. Rawicz, Phys. Rev. Lett. **64**, 2094 (1990)

44. D. E. Discher and N. Mohandas, Biophys. J. **71**, 1680 (1996)

45. D. V. Zhelev, D. Needham, and R. M. Hochmuth, Biophys. J. **67**, 696 (1994)

46. C. F. Wilson, G. J. Simpson, D. T. Chiu, A. Strömberg, O. Orwar, N. Rodriguez, and R. N. Zare, Anal. Chem. **73**, 787 (2001)

8 Single Molecule Detection Using Near Infrared Surface-Enhanced Raman Scattering

K. Kneipp*, H. Kneipp, I. Itzkan, R. R. Dasari, and M. S. Feld

Detecting single molecules and simultaneously identifying their chemical structures represents the ultimate limit in chemical analysis, and is of great practical interest in many fields. For instance, further progress in human DNA sequence studies will mainly depend on developing methods for selective and rapid detection of the four single DNA bases [1]. Monitoring molecules and molecular interactions at the single molecule level in cells or biological membranes and identifying single DNA fragments would be of great interest in the fields of biology, medicine, and pharmacology. Up to now, optical trace detection with single-molecule sensitivity has been mainly based on laser-induced fluorescence [2–5]. Effective fluorescence cross sections can reach about 10^{-16} cm^2 per molecule for high-quantum yield fluorophores. The fluorescence method provides ultrahigh sensitivity but, particularly at room temperature, the amount of molecular information which can be obtained is limited.

In the Raman effect, incident light is inelastically scattered from a sample and gets shifted in frequency by the energy of its characteristic molecular vibrations and thus, Raman spectroscopy provides very high information content on the chemical structure of a molecule. However, Raman scattering is a very weak effect, resulting in cross sections anywhere between 10^{-30} cm^2 and 10^{-25} cm^2 per molecule, the larger values occurring under favorable resonance Raman conditions. Such small Raman cross sections require a large number of molecules to achieve adequate conversion rates from excitation laser photons to Raman photons, thereby making single-molecule Raman spectroscopy almost impossible.

This situation is dramatically altered for so-called surface-enhanced Raman scattering (SERS). The exciting phenomenon of a strongly increased Raman signal from molecules attached to rough metal surfaces or metal colloids was discovered in 1977 by Van Duyne, Jeanmaire, Albrecht, and Creighton [6,7] (for reviews see: [8,9]).

In 1996, unexpectedly large SERS-enhancement factors of about 10^{14} were inferred for molecules attached to silver colloidal clusters using a straightforward method based on steady-state population redistribution due to the

* Corresponding author: kneipp@usa.net

pumping of molecules to the first excited vibrational state via a very strong
Raman process [10]. The corresponding effective Raman cross sections, which
are on the order of 10^{-16} cm^2 per molecule are comparable to effective fluorescence cross sections of common laser dyes and, as we know from fluorescence
experiments, sufficient for single-molecule detection [11,12] (for review see
[13,14]).

In this article, we discuss Raman spectroscopy at the single molecule
level based on surface-enhanced Raman scattering using nonresonant near
infrared excitation. The target molecules are attached to colloidal silver and
gold clusters. In addition to the Stokes Raman signal, which depends linearly
on excitation laser intensity, "pumped" anti-Stokes Raman scattering and
surface enhanced hyper Raman scattering are considered as nonlinear or two-
photon single-molecule Raman probes, where the Raman scattering signal
depends quadratically on the excitation laser intensity [15]. This article also
discusses prospects and limitations for Raman single-molecule detection.

8.1   Physical Background: Surface-Enhanced Raman Scattering (SERS) at Extremely High Enhancement Level

Figure 8.1a presents a schematic of surface-enhanced Raman scattering. Single molecules are attached to a metallic nanostructure which is a section of

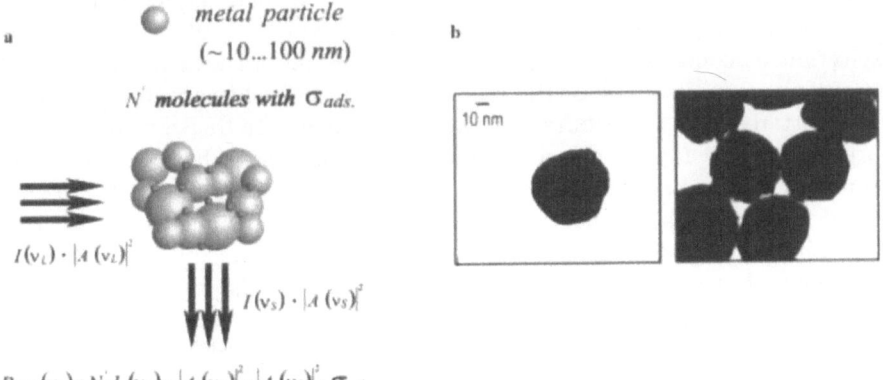

Fig. 8.1. a Schematic of surface-enhanced Raman scattering. Single molecules
(small dots) are attached to a metallic nanostructure, which is a section of a cluster
formed by aggregation of metal colloids. The formula describes the estimate of the
SERS Stokes signal power P$_{SERS}$ of N$'$ adsorbed molecules. (For more explanation
of the formula see text) **b** Electron micrographic views of colloidal gold clusters.
Reprinted with permission from [13] (copyright 1999 American Chemical Society)
(a), and [24] (copyright 1998 Society Applied Spectroscopy) (b)

a cluster formed by aggregation of metal colloids; for comparison, see the electron micrographic views of colloidal gold clusters in Fig. 8.1b.

It is generally agreed that more than one effect contributes to the observed large enhancement of many orders of magnitude for molecules attached to silver or gold structures tens of nanometers in size. The enhancement mechanisms are roughly divided into so-called "electromagnetic field enhancement" and "chemical first layer" effects [8,9,16]. The latter effects include enhancement mechanism(s) of the Raman signal, related to specific interactions between molecule and metal, resulting in an "electronic" enhancement [17] of the Raman cross section σ_{ads} of the adsorbed molecule compared to the Raman cross section of a "free" molecule in a "normal" Raman experiment. Possible electronic SERS mechanisms involve a resonance Raman effect due to a new metal–molecule charge transfer electronic transition [8,18] or a dynamic charge transfer between metal and molecule, which can be described by the following four steps [16,17]: a) photon annihilation, excitation of an electron into a hot electron state; b) transfer of the hot electron into the LUMO of the molecule; c) transfer of the hot electron from the LUMO (with changed normal coordinates of some internal molecular vibrations) back to the metal; d) return of the electron to its initial state and Stokes photon creation. "Atomic scale roughness" seems to play an important role by providing pathways for the hot electrons to the molecule. The magnitude of chemical enhancement has been estimated to reach not more than factors of 10 to 1000 [8].

The electromagnetic or field enhancement factors $A(\nu)$ arise from enhanced local optical fields at the site of the molecule near the metal particle, due to excitation of electromagnetic resonances which result from collective excitation of conduction electrons in small metallic structures, also called surface plasmon resonances. Because the excitation field as well as the Raman scattered field contribute to this enhancement, the SERS signal is proportional to the fourth power of the field enhancement factor. Maximum values for electromagnetic enhancement are on the order of 10^6 to 10^7 on the surfaces of isolated silver and gold spheres tens of nanometers in size. The enhancement decreases as the twelfth power of the distance between molecule and metal [19,20]. Electromagnetic fields can also be enhanced near sharp features and large curvature regions, which may exist on silver and gold nanostructures (for more information on electromagnetic SERS enhancement see [8,9,21]).

However, in many experiments, SERS-active substrates consist of a collection of nanoparticles, such as colloidal clusters formed by aggregation of colloidal particles or metal island films [22–25]. In experiments performed on cluster structures, "classical" electromagnetic field enhancement, which is based on the electromagnetic response of isolated individual particles, is not valid and estimates of electromagnetic SERS enhancement must be based on the properties of electromagnetic fields in fractal small-particle composites

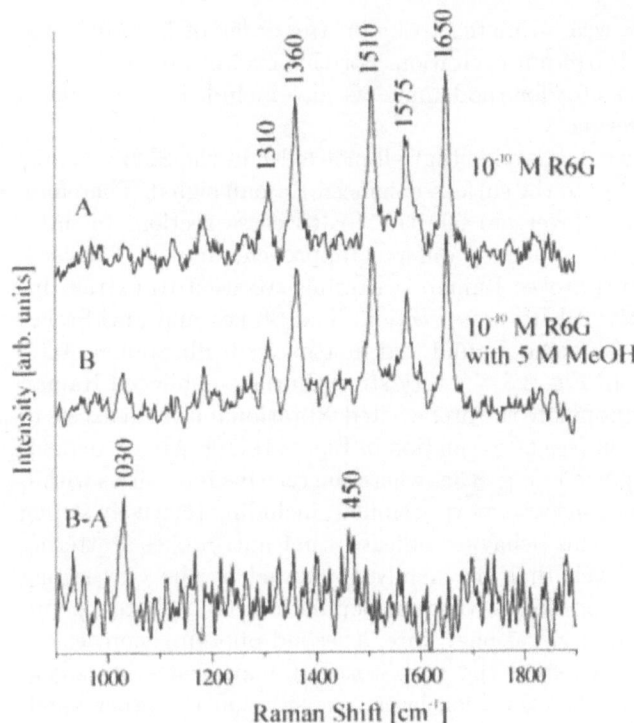

Fig. 8.2. SERRS spectrum from 8×10^{-11} M rhodamine 6G in silver colloidal solution (top), with addition of 5M methanol (middle). Spectra were collected at 514.5-nm resonant excitation, laser intensity was about 10^3 W/cm^2. No fluorescence was obtained at such low concentration because all dye molecules can find a place on the colloidal silver particles where the fluorescence is quenched. The bottom curve depicts the exact subtraction of the top curve from the middle one and shows only the methanol lines. The methanol Raman signal is not enhanced on colloidal silver and shows a Raman cross section on the order of 10^{-30} cm^2 per molecule. Reprinted with permission from [13] (copyright 1999 American Chemical Society)

[26]. Optical excitation in such structures tends to be spatially localized in so-called "hot" areas, where extremely large electromagnetic fields are theoretically predicted [27–30], and also experimentally verified [31–33]. The local field intensities very close to the cluster surface, are predicted to be very heterogeneous so that the variation in field intensities can exceed 10^5, implying local electromagnetic SERS enhancement factors in excess of 10^{10}.

In general, experimental estimates of enhancement factors are based on the comparison of SERS intensity with fluorescence intensity or with normal Raman intensity. In Fig. 8.2 the enhancement factor of rhodamine 6G in silver colloidal solution was determined by comparing the intensity of the nonenhanced methanol Raman line to the enhanced Raman lines of rhodamine 6G, taking into account the different concentrations of both compounds. The total enhancement factor was estimated to be on the order of 5×10^{11} [34]. Spectra were taken with 514-nm excitation. For this excitation wavelength, the total enhancement factor for rhodamine 6G also includes effects due to resonance Raman scattering.

In these estimates, we did assume that all molecules in the SERS sample contribute in a similar way to the surface-enhanced Raman signal. Therefore, we estimated a minimum (average) effective SERS cross section. In order to avoid this problem, we applied a different approach, in which surface-enhanced Stokes *and* anti-Stokes Raman scattering are used to extract information on the effective SERS cross section. The Stokes and anti-Stokes Raman processes start from the ground and first excited vibrational state, respectively (see sketch in Fig. 8.3). A very strong surface-enhanced Raman process can measurably populate the first excited vibrational level in excess of the Boltzmann population (see also equation in Fig. 8.3) [10]. This is demonstrated by the spectra shown in Fig. 8.3a, where the relative intensities within the Stokes and anti-Stokes spectra are very similar, including relatively strong higher frequency modes. This behavior indicates that anti-Stokes scattering arises from vibrational levels that are mainly populated by the very strong SERS Stokes process in addition to thermal population. One photon populates ("pumps") the excited vibrational state, a second photon generates the anti-Stokes scattering. Therefore, the anti-Stokes Raman scattering signal depends quadratically on the excitation laser intensity. On the other hand, the Stokes process, which starts from the vibrational ground state, is a linear process. Figure 8.3b shows the expected linear and quadratic dependence on excitation intensity of Stokes and anti-Stokes data measured from crystal violet on colloidal gold clusters. The formulas for anti-Stokes and Stokes signal power in Fig. 8.3b can be derived for the stationary case and far from saturation. The "vibrational population pumping," which is reflected in the quadratic dependence of the anti-Stokes signal on laser excitation intensity together with the deviation of the anti-Stokes to Stokes signal ratio from that expected from a Boltzmann population, allows a rough estimate of the size of Raman cross sections effective in "pumping." According to the equations shown in Fig. 8.3b, fits to the experimentally observed data show that the product of the cross section and vibrational lifetime, $\sigma^{SERS} \cdot \tau_1$, must be of the order of 10^{-27} cm^2 s. Assuming vibrational lifetimes on the order of 10 ps, the surface-enhanced Raman cross section is then estimated to be at least $\sim 10^{-16}$ cm^2.

It should be noted that the very large SERS cross sections were derived from experiments performed on several molecules subject to near-infrared excitation. The 830-nm laser light is clearly far from any electronic transitions in the target molecules, particularly in the case of adenine, thereby ruling out any contribution to the enhancement due to resonance Raman scattering. The extremely large enhancement factors are related to the existence of colloidal silver and gold clusters. Giant local fields in the "hot" zones of

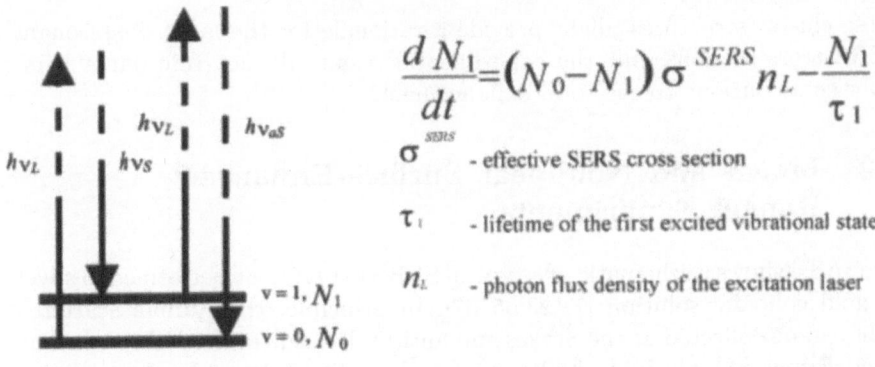

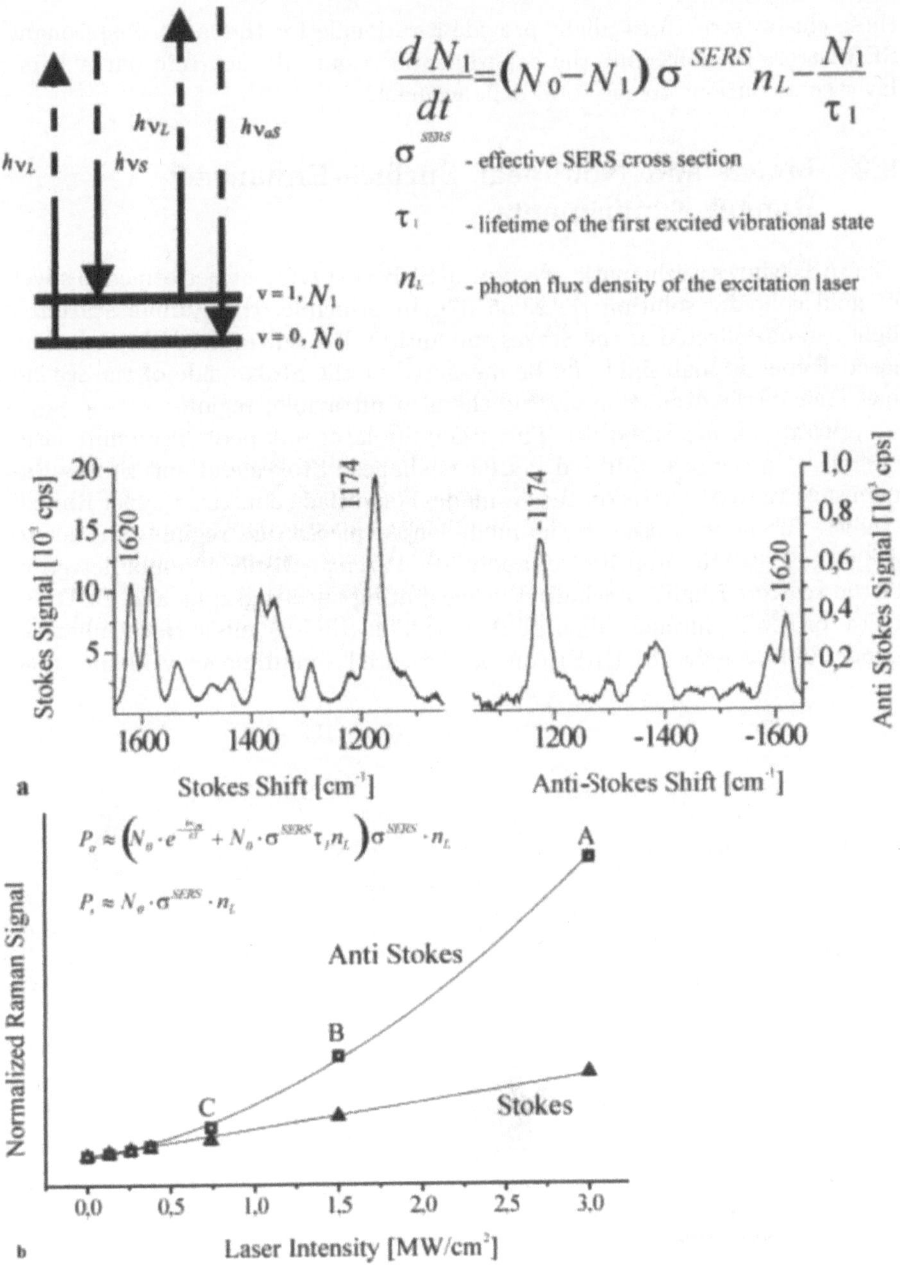

$$\frac{dN_1}{dt} = (N_0 - N_1)\,\sigma^{SERS}\,n_L - \frac{N_1}{\tau_1}$$

σ^{SERS} - effective SERS cross section

τ_1 - lifetime of the first excited vibrational state

n_L - photon flux density of the excitation laser

$$P_a \approx \left(N_0 \cdot e^{-\frac{h\nu}{kT}} + N_0 \cdot \sigma^{SERS}\tau_1 n_L\right)\sigma^{SERS} \cdot n_L$$

$$P_s \approx N_0 \cdot \sigma^{SERS} \cdot n_L$$

Fig. 8.3. a Pumping of the first vibrational level due to the strong Raman process (see text) and surface-enhanced Stokes and anti-Stokes spectra of crystal violet on gold colloidal clusters. **b** Surface-enhanced Stokes (Δ) and anti-Stokes (o) Raman scattering signal of the 1174 cm^{-1} line of crystal violet on colloidal gold clusters plotted vs. 830-nm cw excitation intensity. The lines are quadratic and linear fits to the experimental data; for point A–C see Fig. 8.7b

these cluster structures might provide a rationale for the large nonresonant SERS cross sections, but the experimental results do not rule out a "first layer contribution" to the total enhancement.

8.2   Linear and Nonlinear Surface-Enhanced Raman Experiments

Figure 8.4 shows a schematic of a typical SERS experiment performed in silver or gold colloidal solution [11,23,35–37]. In principle, the Raman scattered light can be collected at the Stokes and anti-Stokes side of the NIR excitation laser. Hyper Raman light can be measured at the Stokes side of the second harmonic of the excitation laser in the near ultraviolet region.

Spectra are measured using an argon-ion laser pumped Ti:sapphire laser operating in the near infrared. Surface-enhanced Stokes and anti-Stokes Raman spectra were excited in the cw mode. For surface-enhanced hyper Raman studies, the laser is used in the mode-locked picosecond regime in order to achieve excitation intensities of about 10^7 W/cm^2. Dispersion and detection of the scattered light is achieved using grating spectrographs and CCD detector or PMT. Surface-enhanced Raman light (SERS) and surface-enhanced hyper Raman light (SEHRS) can be measured simultaneously in the same

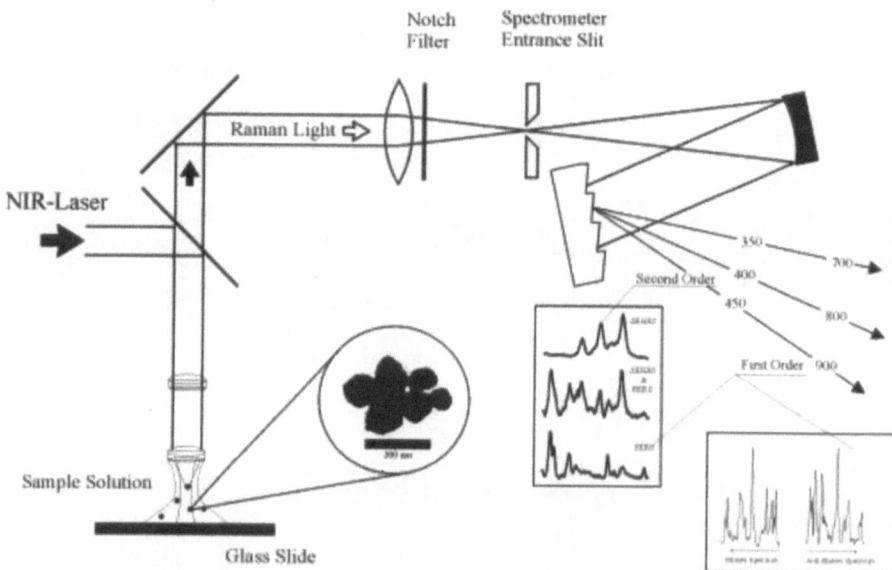

Fig. 8.4. Schematic experimental setup for linear and nonlinear single-molecule Raman experiments. Stokes and anti-Stokes light is collected in the first diffraction order of the spectrograph, Stokes Raman and hyper Raman light can be measured simultaneously using the first and second order of diffraction. The insert shows an electron micrograph of a SERS-active colloidal silver cluster

spectrum using the first and second diffraction order of the spectrograph. This allows a direct determination of the ratio between SEHRS and SERS signal intensities. Nonlinear Raman experiments exploiting pumped anti-Stokes scattering and hyper Raman scattering will be discussed in Sect. 8.2.2. In Sect. 8.2.1 we focus on the Raman measurements performed at the low-energy Stokes side of the near-infrared excitation laser [11,23,36].

8.2.1   Single-Molecule Stokes SERS Experiments and Data Analysis

In single-molecule experiments, the analyte is provided as solution at very low concentration (smaller 10^{-11} M) which is added to the solution of small colloidal silver or gold clusters. Concentration ratios of silver clusters to target molecules of at least 10 make it unlikely that more than one analyte molecule will be attached to the same colloidal cluster. A microscope objective is used for laser excitation and collection of the Raman scattered light. Sizes of the probed volumes are on the order of femtoliters to picoliters. Therefore, analyte concentrations on the order of 10^{-12} to 10^{-14} M result in an average number of one or fewer target molecules in the focus volume.

Brownian motion of the silver clusters loaded with a single analyte molecule into and out of the probed volume results in strong statistical changes in the height of Raman signals measured from such a sample in time sequence. This is demonstrated in Fig. 8.5a which shows typical unprocessed SERS spectra measured in time sequence from a sample with an average of 0.6 crystal violet molecules in the probed 30 pl volume. Fig. 8.5b displays the peak heights of the 1174 cm^{-1} line for the 100 SERS spectra, the background level of the colloidal solution with no analyte present, and 100 measurements of the 1030 cm^{-1} Raman line of 10^{14} methanol molecules in colloidal silver

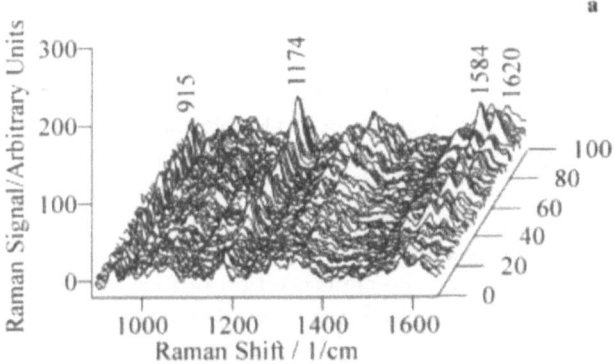

Fig. 8.5a. 100 SERS spectra from crystal violet single molecules. Each spectrum is acquired in 1 second. Reprinted with permission from [11] (copyright 1997 American Physical Society)

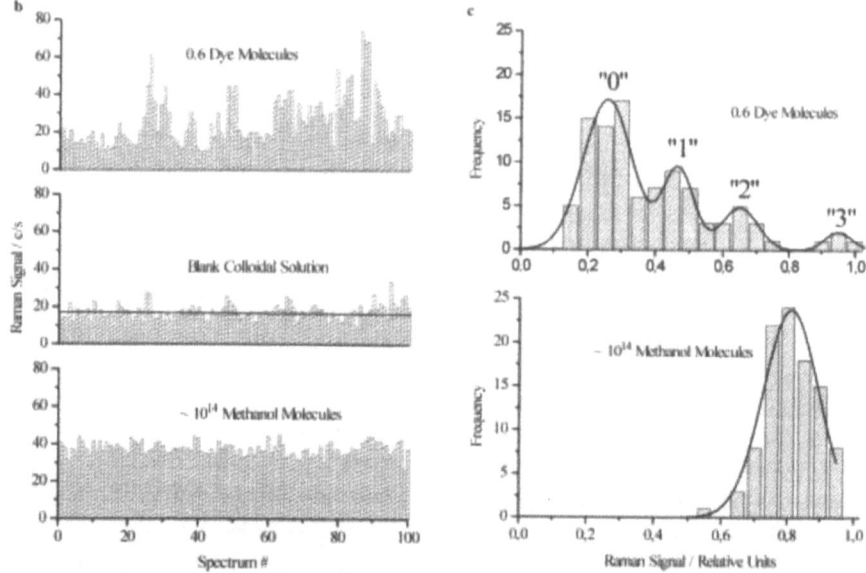

Fig. 8.5b and c. b Peak heights of the 1174 cm^{-1} line for the 100 SERS spectra shown in Fig. a (top). Signals measured at 1174 cm^{-1} from a sample without crystal violet (middle), peak heights of the 1030 cm^{-1} Raman line measured from 3M ($\sim$10^{14} molecules in 30 pL) methanol in silver colloidal solution (bottom). The single-molecule events in the top trace appear at about 38 counts/sec, which corresponds to the signal level of $\sim$10^{14} methanol molecules in the bottom trace. **c** Statistical analysis of 100 SERS measurements for an average of 0.6 crystal violet molecules in the probed volume using 20 bins whose widths are 5% of the maximum of the observed signals (x axis). The y axis displays the frequency of the appearance of the appropriate signal levels of the bin (top). Statistical analysis of 100 "normal" Raman measurements at 1030 cm^{-1} of 1014 methanol molecules. The solid line is a Gaussian fit to the data (bottom). Reprinted with permission from [11] (copyright 1997 American Physical Society)

solution. The normal Raman signal of the 10^{14} methanol molecules appears at the same level as the SERS signal of a single crystal violet molecule, confirming an enhancement factor on the order of 10^{14} for nonresonant Raman scattering. As expected, the methanol Raman signals collected in time sequence displays a Gaussian distribution (Fig. 8.5c bottom). In contrast, the statistical distribution of the "0.6 molecules SERS signal" exhibits four relative maxima which are reasonably fit by the superposition of four Gaussian curves whose areas are roughly consistent with a Poisson distribution for an average number of 0.5 molecules, in good agreement with the number inferred from concentration and probed volume. The Poisson statistics reflects the probability of finding zero, one, two, or three molecules in the scattering volume during the actual measurement. The change in the statistical dis-

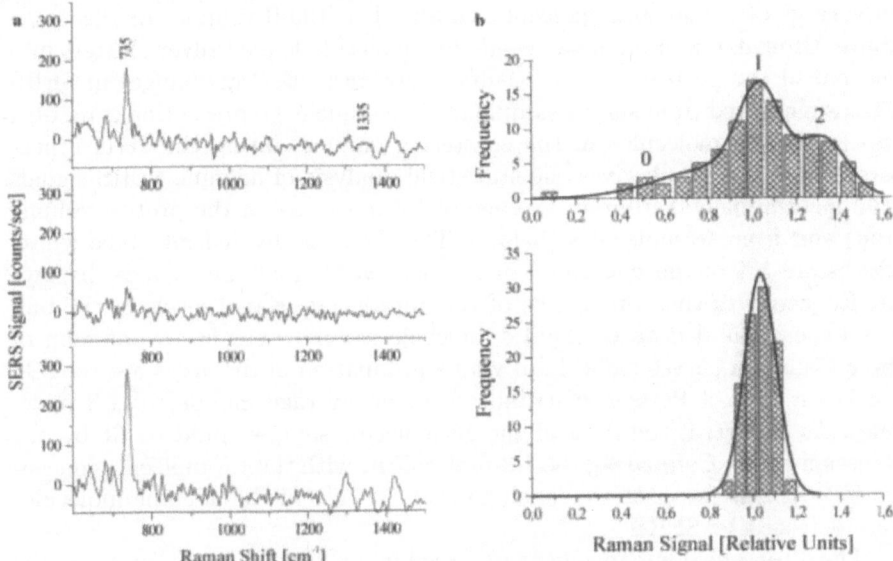

Fig. 8.6. a Typical SERS Stokes spectra representing approximately 1 (top), 0 (middle), or 2 (below) adenine molecules in the probed volume (collection time 1 s, 80 mW NIR excitation). **b** Statistical analysis (see Fig. 8.5c) of 100 SERS measurements at an average of 1.8 adenine molecules (top) and for 18 adenine molecules (below) in the probed volume. Reprinted with permission from [23] (copyright 1998 American Physical Society)

tribution of the Raman signal from Gaussian to Poisson when the average number of dye molecules in the scattering volume is one or fewer is evidence for single-molecule detection by SERS. The relatively "well-quantized" signals for one, two, or three molecules suggest relatively uniform enhancement despite the nonuniform shape and size (ca 10–50 nm) of the silver particles forming the clusters. This might be explained by the "cluster-based enhancement," which has been found to be independent of the individual particle in the cluster and also of the size of the cluster after exceeding a critical cluster size [23,24,38]. In order to generate a relatively "good" statistical distribution in Fig. 8.5c, we choose relatively large probed volumes to generate a balance between dwell time of the analyte molecule in the probed volume and collection time of one spectrum, which allows the direct measurement of relatively well separated zero, one, and two molecule events in the Poisson statistical distribution.

Recently, extremely large effective surface-enhanced Raman cross sections on the order of 10^{-16} cm^2 per molecule have also been derived for adenine and adenosine monophosphate (AMP) on colloidal silver clusters [23]. Such cross sections are sufficient for single-molecule detection. Figure 8.6a shows selected typical spectra collected in 1 second from a sample which contains an average of 1.8 adenine molecules in a probed 100-fl volume. As discussed above, Brownian motion of single adenine molecule-loaded silver clusters into and out of the probed volume results in strong statistical changes in SERS signals measured from such a sample in time sequence representing zero, one, two (or three) molecules in the scattering volume during the actual measurement. Figure 8.6b gives the statistical analysis of adenine SERS-signals (100 measurements) from an average of 1.8 molecules in the probed volume (top) and from 18 molecules (below). The x axis is divided into bins whose widths are 5% of the maximum of the observed signal. The y axis displays the frequency of the appearance of the appropriate signal levels in the bin. The experimental data of the 1.8-molecule sample were fit by the sum of three Gaussian curves (solid line) whose graduation of the areas are roughly consistent with a Poisson distribution for an average number of 1.3 molecules. As expected, the data of the 18-molecule sample could be fit by one Gaussian curve. Comparing the 1.3-molecule fit with the 1.8-molecule concentration/volume estimate we conclude that 70–75% of the adenine molecules were detected by SERS.

The strong field enhancement provided by colloidal clusters might be the key for understanding the enormous nonresonant surface-enhanced Raman cross sections exploited in the single-molecule adenine experiments. Due to this very general "physical" origin of Raman enhancement, it should be possible to achieve SERS cross sections on the same order of magnitude for other bases when they are attached to colloidal silver or gold clusters. The nucleotide bases show well-distinguished surface-enhanced Raman spectra [39,40]. Thus, after cleaving single native nucleotides from a DNA or RNA strand into a medium containing colloidal silver or gold clusters, for instance into a flowing stream of colloidal solution, or onto a moving surface with silver or gold cluster structure, NIR-SERS provides a method for detecting and identifying a single DNA base, which does not require any labeling because it is based on the intrinsic surface-enhanced Raman scattering of the base.

8.2.2   Nonlinear Raman Probe

To use Raman scattering at the single molecule level, the effective Raman cross section has to be surface-enhanced about 14 orders of magnitude. We obtained such extremely high enhancement factors at near infrared excitation for molecules adsorbed on colloidal silver or gold clusters. Strong fields in the so-called "hot areas" on the clusters can provide a rationale for such an extreme SERS effect. In this electromagnetic field enhancement model, nonlinear Raman effects are surface enhanced to a greater extent than "normal" Raman scattering since they depend nonlinearly on the strong fields in the hot areas.

"Pumped" anti-Stokes Raman scattering and surface enhanced hyper Raman scattering are selected as effects where the Raman scattering signal depends quadratically on the excitation laser intensity [15].

Surface-enhanced anti-Stokes Raman scattering. Figure 8.7a shows anti-Stokes and Stokes SERS spectra of 10^{-9} M crystal violet in solution of colloidal gold clusters measured in one second collection time using a confocal microscope Raman setup. Based on a concentration volume estimate, an average of about one molecule contributes to the spectra collected from about 1 fl probed volume. The relatively strong band at 1620 cm^{-1} in the anti-Stokes spectrum is an indication that the anti-Stokes signal arises from pumped vibrational levels and not from thermally populated ones. At the applied intensity of about 10^{24} laser photons/cm^2 s, surface-enhanced anti-Stokes signals appear at about ten times lower signal level than Stokes signals, which allows single-molecule detection at a signal-to-noise level of about 2:1. Methanol anti-Stokes signals are not detected under these experimental conditions.

To illustrate the nonlinear dependence on laser intensity of the anti-Stokes signal, Fig. 8.7b shows anti-Stokes spectra at different excitation intensities together with the Rayleigh background, which depends linearly on the excitation intensity. Due to the quadratic dependence of the anti-Stokes signal, the signal-to-background ratio is improved for increasing excitation laser intensities.

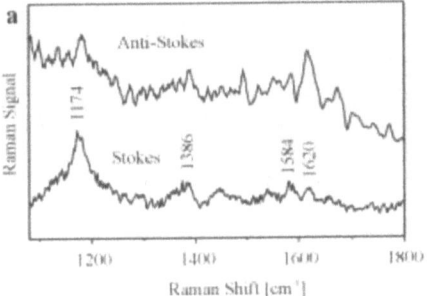
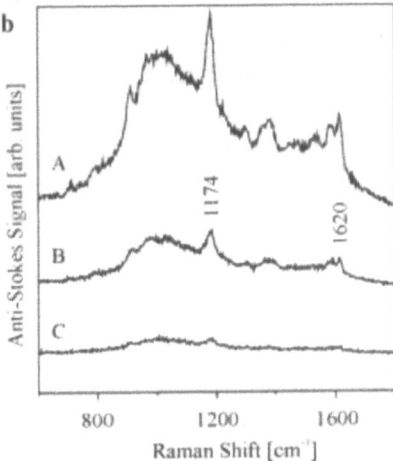

Fig. 8.7. a Typical Stokes and anti-Stokes spectra of crystal violet on gold colloidal clusters measured for an average of 1 molecule in the probed volume. **b** Selected anti-Stokes spectra (3 MW/cm^2 (A), 1.4 MW/cm^2 (B), and 0.7 MW/cm^2 (C)). The Rayleigh background is suppressed by a notch filter up to about 900 cm^{-1}. Reprinted with permission from [15] (copyright 1999 Elsevier)

Surface-enhanced hyper Raman scattering (SEHRS). Strong surface enhancement factors can overcome the inherently weak nature of hyper Raman scattering and surface-enhanced hyper Raman spectra and surface-enhanced Raman spectra can appear at very similar signal levels [35]. This is demonstrated in the middle spectrum in Fig. 8.8, which displays surface-enhanced hyper Raman and Raman signals of crystal violet on colloidal silver clusters measured in the same spectrum (see also Fig. 8.4).

The experimental ratios between SERS and SEHRS intensities can be combined with the corresponding estimated "bulk" intensity ratio between

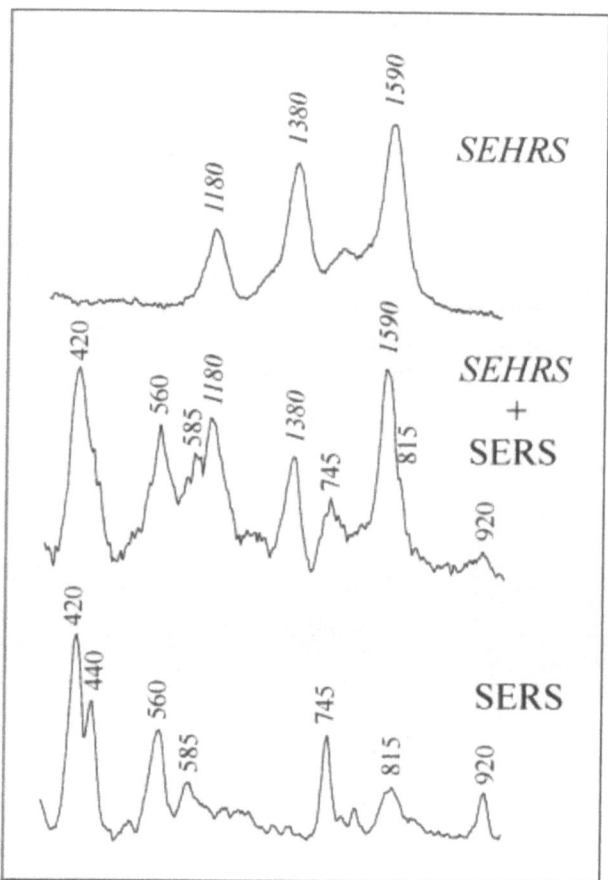

Fig. 8.8. SEHRS and Raman SERS signals of crystal violet on colloidal silver clusters measured in the same spectrum using 10^7 W/cm^2 NIR excitation (middle trace, see also Fig. 8.4). In the upper and the lower trace, SEHRS and SERS spectra are differentiated by placing an NIR absorbing filter in front of the spectrograph or by switching off the mode locker, respectively. Reprinted with permission from [35] (copyright 1995 Elsevier)

Raman scattering and hyper Raman scattering for the applied 10^7 W/cm^2 excitation intensity to infer a ratio between surface-enhancement factors of hyper Raman scattering and Raman scattering of about 10^6. Combining this ratio with NIR–SERS enhancement factors on colloidal silver clusters on the order of 10^{14}, total surface enhancement factors of hyper Raman scattering on crystal violet adsorbed on colloidal silver clusters can be inferred to be on the order of 10^{20} [15].

The appearance of surface-enhanced hyper Raman and Raman scattering at the same signal intensity suggests that surface-enhanced hyper Raman scattering is a spectroscopic technique that can be applied at the single-molecule level. The hyper Raman spectrum can provide a special "fingerprint" of a molecule by showing modes that are forbidden in normal Raman scattering.

8.3   Prospects and Limitations of Single-Molecule Raman Detection

The ability to perform Raman spectroscopy at effective cross sections comparable to or even better than those of fluorescence opens up exciting opportunities for detecting and identifying single molecules. The frequency shifts, line widths, and relative intensities of the Raman lines characterize the Raman spectrum of a molecule. Due to nonuniform enhancement level of the different vibrations, and interaction between the molecule and the metal surface, a SERS spectrum of a molecule can exhibit deviations from the normal Raman spectrum. Nevertheless, in most cases the SERS spectrum provides a clear "fingerprint" of the molecule. For experiments performed at room temperature and in solutions, this is superior to the broad and nonspecific fluorescence spectrum obtained under similar conditions. Furthermore, SERS provides the possibility of detecting and identifying nonfluorescent molecules such as nucleotides and amino acids at the single-molecule level without the need for labeling.

Single molecule Raman spectroscopy provides the opportunity to retrieve individual properties of single molecules, which are hidden within the inhomogeneous linewidth or averaged out in an ensemble measurement. The Raman spectrum of a single molecule can also provide a very sensitive probe of the environment of the molecule.

In single-molecule SERS detection, the target molecule has to be attached to a SERS-active substrate, such as a silver or gold colloidal cluster. Attachment of the single target molecule to such a much bigger particle can be an advantage for single-molecule detection. Increased diffusion times compared to that of the free molecule minimize the possibility of the target molecule escaping too rapidly from the focal volume of the excitation beam.

In our experiments, performed with nonresonant near infrared excitation, SERS enhancement factors sufficient for single-molecule detection are clearly related to the formation of a silver or gold colloidal cluster [10,11,23,24,36]. Isolated small spherical silver and gold particles generated maximum enhancement factors of 10^6–10^7, in agreement with "classical" electromagnetic theories. Very large field enhancement [30,33,38] provides the rationale for performing single-molecule SERS experiments on colloidal clusters with NIR excitation. The uniform enhancement factor, independent of cluster size, explains the relatively well-quantized SERS signals for one, two, and three molecules observed in our experiments, despite the nonuniform sizes and shapes of the SERS active clusters and of the individual particles forming the clusters. Of course, strong field enhancement does not rule out, and may even support, an additional "chemical" enhancement, which might be essential to the SERS effect.

It should be noted that cross sections of about 10^{-16} cm^2 describe conversion into Stokes photons derived from a single Raman line. The total cross section for generating Stokes light over a frequency range comparable to fluorescence, typically covered by 5 to 10 Raman lines, is on the order of 10^{-15} cm^2 per molecule, which is about an order of magnitude higher than the best known effective fluorescence cross sections. This total cross section for generating Stokes-shifted light is an interesting parameter if one wishes to detect a known molecule without identifying its structure.

The maximum number of photons emitted by a molecule in fluorescence or Raman emission under saturation conditions is inversely proportional to the lifetime of the excited states involved in the optical process. Due to the shorter vibrational relaxation times compared to electronic relaxation times, a molecule can undergo more Raman excitation–emission cycles per unit time than for fluorescence [42]. Therefore, the number of Raman photons emitted by a molecule under saturation conditions can be higher than the number of fluorescence photons by a factor of 10^2 to 10^3. This allows shorter integration and/or higher counting rates for detecting single molecules. Of course, due to the shorter vibrational lifetimes, the saturation intensities for a SERS process will be characteristically higher than for fluorescence. However, the use of higher excitation intensities should not be a problem in experiments that employ colloidal clusters and nonresonant near infrared photons, thus avoiding photodecomposition of the probed molecule.

Single molecule spectra can be measured with good signal-to-noise ratios ($\sim$10:1) in a 1 second collection time, using an excitation power of 100 mW focused to an area of about 3×10^{-7} cm^2. Assuming a SERS cross section on the order of 10^{-17}–10^{-16} cm^2 and a vibrational lifetime on the order of 10 picoseconds, saturation of SERS is achieved at excitation intensities of 10^8–10^9 W/cm^2. Applying the same Raman system (same focusing geometry of the excitation laser beam, same signal collection and detection efficiency), under saturation conditions the collection time for single-molecule spectra could be reduced by a factor of 1000, and it should be possible to measure single-molecule SERS spectra in milliseconds, i.e. at kHz detection rates.

To the best of our knowledge, as of now at least five different molecules have been reported to be detected by Raman scattering at the single-molecule level [11,12,23,36,41]. Several experimental findings suggest a very strong electromagnetic origin of the extremely large SERS enhancement, at least on colloidal clusters and with nonresonant near infrared excitation. Therefore, effective SERS cross sections sufficient for single-molecule detection should be available for a wide range of molecules. However, there are molecules, methanol, for example, that do not show any SERS enhancement. A better understanding of the SERS effect is an important prerequisite for further development of SERS as a tool for single-molecule detection. In particular, it will be important to better understand the nature of the "chemical" contribution.

Pumped anti-Stokes Raman scattering and surface-enhanced hyper Raman scattering appear at comparable signal levels as surface-enhanced Stokes Raman scattering. Both effects provide potential tools for a nonlinear or two-photon Raman probe of single molecules.

References

1. R. Keller et al., Los Alamos Annual Report (1990)
2. R. Keller, W. P. Ambrose, P. M. Goodwin, J. H. Jett, J. C. Martin, and M. Wu, Appl. Spectr. **50**, 12A (1996)
3. R. Rigler, J. Widengren, and Ü. Mets, *Fluorescence Spectroscopy*, O. S. Wolfbeis (ed.), p. 13 (Springer-Verlag, Berlin, 1992)
4. S. Nie and R. N. Zare, Ann. Rev. Biophys. Biomol. Struct. **26**, 567 (1997)
5. X. S. Xie and J. K. Trautman, Ann. Rev. Phys. Chem. **49**, 441 (1998)
6. D. L. Jeanmaire and R. P. V. Duyne, J. Electroanal. Chem. **84**, 1 (1977)
7. M. G. Albrecht and J. A. Creighton, J. Am. Chem. Soc. **99**, 5215 (1977)
8. A. Otto, in *Light Scattering in Solids IV. Electronic Scattering, Spin Effects, SERS and Morphic Effects*, M. Cardona and G. Guntherodt (eds.), vol. 1984, p. 289 (Springer-Verlag, Berlin, 1984)
9. M. Moskovits, Rev. Mod. Phys. **57**, 783 (1985)
10. K. Kneipp, Y. Wang, H. Kneipp, I. Itzkan, R. R. Dasari, and M. S. Feld, Phys. Rev. Lett. **76**, 2444 (1996)
11. K. Kneipp, Y. Wang, H. Kneipp, L. T. Perelman, I. Itzkan, R. R. Dasari, and M. S. Feld, Phys. Rev. Lett. **78**, 1667 (1997)
12. S. Nie and S. R. Emory, Science **275**, 1102 (1997)
13. K. Kneipp, H. Kneipp, I. Itzkan, R. R. Dasari, and M. S. Feld, Chem. Rev. **99**, 2957 (1999)
14. K. Kneipp, H. Kneipp, I. Itzkan, R. R. Dasari, and M. S. Feld, Current Science **77**, 915 (1999)
15. K. Kneipp, H. Kneipp, I. Itzkan, R. R. Dasari, and M. S. Feld, Chem. Phys. **247**, 155 (1999)
16. A. Otto, I. Mrozek, H. Grabhorn, and W. J. Akemann, J. Phys. Chem. Condens. Matter **4**, 1143 (1992)

17. A. Otto, in *Int. Conference on Raman Spectroscopy XVI*, Kapstadt, South Africa, 1998
18. A. Campion, Chem. Soc. Rev. **4**, 241 (1998)
19. M. Kerker, O. Siiman, L. A. Bumm, and D. S. Wang, Applied Optics **19**, 3253 (1980)
20. D. S. Wang and M. Kerker, Phys. Rev. B **24**, 1777 (1981)
21. E. J. Zeman and G. C. Schatz, J. Phys. Chem. **91**, 634 (1987)
22. D. A. Weitz and M. Oliveria, Phys. Rev. Lett. **52**, 1433 (1984)
23. K. Kneipp, H. Kneipp, V. B. Kartha, R. Manoharan, G. Deinum, I. Itzkan, R. R. Dasari, and M. S. Feld, Phys. Rev. E **57**, R6281 (1998)
24. K. Kneipp, H. Kneipp, R. Manoharan, E. B. Hanlon, I. Itzkan, R. R. Dasari, and M. S. Feld, Appl. Spectrosc. **52**, 1493 (1998)
25. Y. Yamaguchi, M. K. Weldon, and M. D. Morris, Appl. Spectrosc. **53**, 127 (1999)
26. V. M. Shalaev, *Nonlinear Optics of Random Media* (Springer-Verlag, Berlin, Heidelberg, 2000)
27. E. Y. Poliakov, V. M. Shalaev, V. A. Markel, and R. Botet, Opt. Lett. **21**, 1628 (1996)
28. E. Y. Poliakov, V. A. Markel, V. M. Shalaev, and R. Botet, Phys. Rev. B **57**, 14901 (1998)
29. V. M. Shalaev and A. K. Sarychev, Phys. Rev. B **57**, 13265 (1998)
30. P. Gadenne, E. Brouers, V. M. Shalaev, and A. K. Sarychev, J. Opt. Soc. Am. B **15**, 68 (1998)
31. D. P. Tsai, J. Kovacs, Z. Wang, M. Moskovits, V. M. Shalaev, J. S. Suh, and R. Botet, Phys. Rev. Lett. **72**, 4149 (1994)
32. P. Zhang, T. L. Haslett, C. Douketis, and M. Moskovits, Phys. Rev. B **57**, 15513 (1998)
33. V. A. Markel, V. M. Shalaev, P. Zhang, W. Huynh, L. Tay, T. L. Haslett, and M. Moskovits, Phys. Rev. B **59**, 10903 (1999)
34. K. Kneipp, Y. Wang, R. R. Dasari, and M. S. Feld, Appl. Spectr. **49**, 780 (1995)
35. H. Kneipp, K. Kneipp, and F. Seifert, Chem. Phys. Lett. **212**, 374 (1993)
36. K. Kneipp, H. Kneipp, G. Deinum, I. Itzkan, R. R. Dasari, and M. S. Feld, Appl. Spectr. **52**, 175 (1998)
37. K. Kneipp, H. Kneipp, R. Manoharan, I. Itzkan, R. R. Dasari, and M. S. Feld, Bioimaging **6**, 104 (1998)
38. M. I. Stockman, V. M. Shalaev, M. Moskovits, R. Botet, and T. F. George, Phys. Rev. B **46**, 2821 (1992)
39. R. Sheng, F. Nii, and T. M. Cotton, Anal. Chem. **63**, 437 (1991)
40. J. Thornton and R. K. Force, Appl. Spectrosc. **45**, 1522 (1991)
41. H. Xu, E. J. Bjerneld, M. Kåll, and L. Borjesson, Phys. Ref. Lett.**83**, 4357 (1999)
42. K. Kneipp, Exp. Tech. Phys. **36**, 161 (1988)

9 Single-Molecule Fluorescence –
Each Photon Counts

C. G. Hübner, V. Krylov, A. Renn, P. Nyffeler, and U. P. Wild

The first experiments on single molecules were driven by the challenge of reaching the ultimate limit in chemical analysis. Following the first absorption experiments [1], fluorescence excitation detection facilitated single-molecule spectroscopy with an almost unbelievable signal-to-noise ratio. Fluorescence excitation spectroscopy allowed for the investigation of properties of single molecules and for the comparison of the time average of these properties with the ensemble average. In all of these measurements, the detected photons were integrated to obtain an intensity as a function of one or several experimental parameters in a way similar to ensemble fluorescence spectroscopy. In other words, each photon was just a "click" that increased the number of counts in a certain channel by one.

The questions we asked ourselves recently are: What can we learn from a single photon? Is it possible to gain information from all simultaneously measured properties of a single photon? We found that a possible application of single-molecule fluorescence detection is closely related to answering these (for the time being) abstract questions, namely, identification of single molecules from a fixed set of different species.

9.1 Properties of Single Molecules

If one goes from ensemble spectroscopy to single-molecule (SM) spectroscopy, two principal requirements have to be fulfilled: the signal has to be distinguishable both from that of other molecules and from the background noise. Different strategies are followed in order to meet the requirements. The spatial separation of the molecules is determined by the ratio between the detection volume and the concentration of the molecules. Thus, single molecules can in principle be distinguished in a solution with a very low concentration. But, since the number of solute molecules contributing to the background noise is dependent on the detection volume, it is desirable to reduce this as much as possible. With optical far-field microscopy the detection volume is limited by diffraction, while optical near-field techniques enable one to overcome this barrier [2]. The requirement for distinguishing the single-molecule signal from the background is that the signal exceed the statistical fluctuations of the background. This means the molecule has to "survive" for a time that is sufficient to collect enough photons.

These requirements were met for the first time in a solid host doped with a low concentration of fluorophores placed in a liquid helium cryostat. The interaction of the fluorophore with the surrounding matrix differs slightly from molecule to molecule, thus shifting their resonances to slightly different frequencies. If a small volume of the sample is illuminated by a narrow-band laser just one molecule in this volume is excited. This spectral sensitivity facilitates the discrimination of single-molecule fluorescence [3].

At very low temperatures in a liquid helium cryostat, the photodestruction rate is much lower than under ambient conditions. Recently, from single terylene molecules doped into naphtalene, more then 500 million photons were detected [4] at 1.4 K. The huge number of detectable photons allows one to perform spectroscopy on a single molecule, i.e. to determine its properties, e.g. orientation [5], fluorescence lifetime [6], Stark effect [7], and pressure effect [8]. Since fluorescence lifetime will play an outstanding role in the SM identification experiments presented in the next section, its determination for single molecules will be reviewed in the following.

9.1.1   Fluorescence Lifetime

The time distribution of the photons emitted by single molecules shows fascinating properties. Photon bunching has been observed by Bernard et al. [9] on a μs time scale. At high excitation rate, the fluorescence photons are emitted in bunches which are separated by dark periods whenever the molecule is shelved through intersystem crossing in its triplet state. In the nanosecond time region, photon antibunching has been reported by Basché et al. [10] by measuring the number distribution of the time interval of consecutive photon pairs. Just after emitting a photon, the single molecule will definitely occupy its electronic ground state, and the probability of immediately emitting a further photon is zero.

Fluorescence lifetimes of single pentacene molecules in the O_1 site (16882.7 cm^{-1}) of a p-terphenyl crystal at a temperature of 1.8 K have been measured by time-correlated single-photon counting (TCSPC) [6].

In a typical TCSPC apparatus the sample is excited by a short pulse from a synchronously pumped dye laser. The time lag between the laser pulse and the detection of the photon is measured by a time-to-amplitude converter (TAC). The output pulses of the TAC, with an amplitude proportional to the time lag, are usually analyzed by a multichannel analyzer (MCA) that generates a histogram of the pulse amplitudes in a number of channels. Pulse durations on the order of tens of picoseconds can easily be obtained, and the frequency of the exciting light can be changed by rotating a Lyot filter in the dye laser cavity.

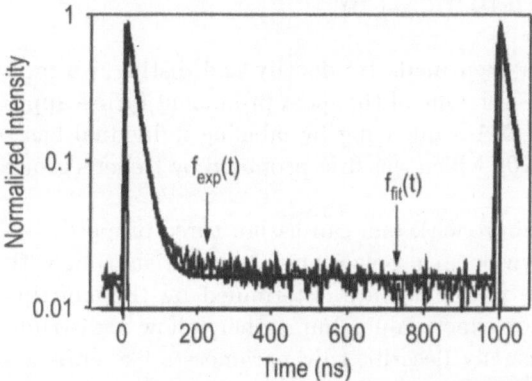

Fig. 9.1. Fluorescence decay curve for a single pentacene molecule measured at 1.8 K. The thick line represents the fitted curve. Adapted from [6]

For the excitation of extremely narrow molecular absorption lines at low temperatures, we decided to use a single-mode dye laser as the basic excitation source, and to employ an acousto-optic modulator to produce Fourier transform limited excitation pulses of a suitable duration, giving the desired spectral width. The shortest pulses from the AOM had an almost Gaussian time profile with a FWHM of 9 ns.

In Fig. 9.1, the curve fit to the measured fluorescence decay of a single pentacene molecule is shown. The plot contains the excitation and emission pulse averaged over the repetition interval as well as the fitted curve (thick line). The total recording time was determined by the fluorescence emission rate of the single molecule and the detection efficiency of the apparatus. Recording times up to 1200 s were necessary in order to get a reliable statistic (100,000 counts) for the fluorescence decay curve. In our experiment, the excitation pulse was of the same order of magnitude (9 ns) as the fluorescence decay (24 ns), and it was therefore necessary to deconvolve the data. For this reason, and in order to get better precision, more counts had to be be accumulated than described by Köllner *et al.* [11].

Molecules in the center and on the red side of the inhomogeneously broadened O_1 site were selected. The lifetimes obtained were between 23.9(1) ns and 24.5(1) ns. Within the error limits of our experiment no variation in the fluorescence lifetimes of the different molecules could be observed. All these lifetimes agreed with the lifetimes measured on higher concentrated samples. The experiment has shown that the fluorescence lifetimes of single molecules can be measured at low temperatures with high accuracy. Under ambient conditions, however, it is difficult to collect such a high number of photons.

9.2   Single-Molecule Identification

In recent years much effort has been made to identify and distinguish individual molecules in solution [12–19]. One of the most prominent future applications of such a technique is DNA sequencing by labeling individual bases with different fluorescent dyes [20], which was first proposed by Keller's group [21].

In principle (in a classical approach), one can assign three properties to every detected single photon: wavelength, polarization, and arrival time with respect to excitation. All these properties are determined by the emitting molecule. However, they are not exact values but rather follow a distribution. These distributions are usually described by parameters like emission maximum or fluorescence lifetime. Single-molecule identification thus seems to be an easy task: determine the respective parameters of single-molecule fluorescence and compare them to the known values. In order to determine such parameters with high accuracy it is necessary to integrate many photons, which is strongly hindered by fast photodestruction.

On the basis of this insight we tried to find a way to analyze the maximum information that we get from a single photon. In our method, each photon emitted by the molecule is analyzed with regard to a probability distribution obtained from reference experiments. The probability distributions which serve as a fingerprint are established from measurements of the photon properties mentioned above for pure bulk samples of the different species. For each photon, a certain probability for having either of the molecules as the emission source can be computed. By combining these probabilities photon by photon, eventually the molecule can be identified. An equivalent algorithm was recently proposed by Enderlein and Sauer [26]

9.2.1   Single-Photon Analysis

The normalized photon histogram from molecules of sort m_i ($1 \leq i \leq N$), where i denotes the different species, with respect to one of the above-mentioned photon properties, provides the relative occurrence of photons in the respective bin. This relative occurrence is equal to the probability of detecting a photon in bin b_k ($0 \leq k \leq K$):

$$P(b_k \mid m_i) = \frac{c_i(b_k)}{\sum\limits_{b_n=0}^{K} c_i(b_n)}, \tag{9.1}$$

where $c_i(b_n)$ is the number of counts and K the number of bins. Under the condition that a detected photon was emitted by a molecule of the fixed set of species, the probability that the molecule m_i of sort i is the source of a count in bin b_k is

$$P(m_i|b_k) = \frac{P(m_i)P(b_k|m_i)}{\sum\limits_{q=1}^{N} P(m_q)P(b_k|m_q)}. \tag{9.2}$$

If the occurrences of the different molecules in the sample are comparable, Eq. 9.2 can be approximated by

$$P(m_i|b_k) = \frac{P(b_k|m_i)}{\sum\limits_{q=1}^{N} P(b_k|m_q)}. \tag{9.3}$$

The probabilities $P(m_i|b_k)$ from Eq. 9.3 establish a lookup table for the different types of molecules with an entry for each bin b_k and each species m_i. These entries are the probabilities, which can be addressed for each detected photon.

Let us consider as an example two dyes A and B that have fluorescence lifetimes of 2 and 5 ns, and emission maxima at 600 and 550 nm, respectively. A graphical representation of the lookup tables for this case are shown in the left column of Fig. 9.2. Here, the stacked probabilities according to Eq. 9.3 for the properties of arrival time after excitation (top row) and wavelength (bottom row) are plotted. For the sake of simplicity, single-exponential decays and Gaussian emission spectra were assumed. The ratio of the probabilities for both species for a certain value of the property corresponds to the ratio of the normalized photon histograms (also plotted in Fig. 9.2) at the respective point. Thus, at the crossing point of the normalized histograms, both probabilities have the same value of 0.5. Since only photons detected within the limits of both observables are considered, the normalization was performed within the temporal and spectral windows.

If L photons are detected from one molecule in the respective bin b_k^j, where the index j denotes the photon number, the probability that these photons are emitted by the molecule m_i is given by

$$P_L(m_i) = \frac{\prod\limits_{j=1}^{L} P(m_i|b_k^j)}{\sum\limits_{q=1}^{N} \prod\limits_{j=1}^{L} P(m_q|b_k^j)}. \tag{9.4}$$

In the right column of Fig. 9.2 the probabilities for a sequence of photons of the model species B (5 ns fluorescence lifetime, 550 nm emission maximum) are shown for the properties of arrival time (top row) and wavelength (bottom row). The photon arrival times and wavelengths were simulated with the help of the Monte Carlo method. In the case shown, after a few detected photons, the probability that species B was the emitter is higher than 90% for both

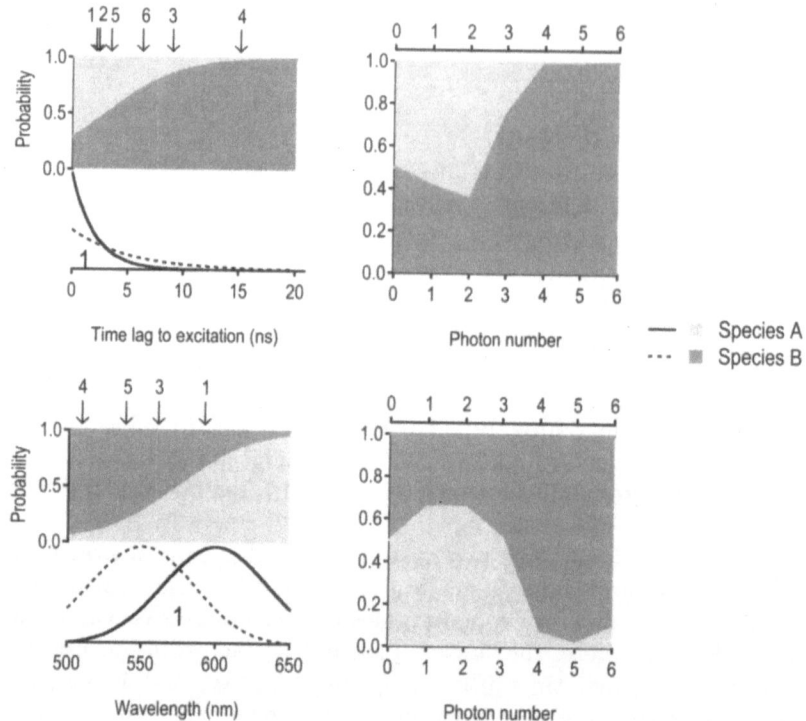

Fig. 9.2. Left column: Graphical representations of the lookup tables of two model species A and B together with the normalized photon histograms. Right column: Graphical representations of the assignment probabilities of a simulated photon sequence of species B. The two properties arrival time after excitation (top row) and spectrum (bottom row) are used. The bins where the photons of the sequence were detected are indicated on top of the lookup representations

observables. The values of the arrival time and the wavelength of the detected photons are marked by arrows on top of the lookup tables in the left column of Fig. 9.2. Photon 2 was outside the spectral window. Therefore, there is no change in the probability with regard to the property of wavelength (bottom right panel) between photon 1 and photon 2.

If the photon properties are independent, the probabilities obtained from their respective lookup tables can be multiplied. However, if they are not independent, two(or more)-dimensional photon histograms from bulk samples have to be recorded. Also, a multidimensional lookup table has to be constructed. In this case, the bins have more than one index: $b_{k,l,m...}$.

The course of the analysis can be summarized as follows. First, the observables are measured in reference samples containing only one species of molecule. The resulting photon histograms with respect to the observables are normalized to obtain the probability distribution $P(b|m_i)$. If only photons within a range of values of the observables are to be considered, *e.g.*, within a photon arrival time window, the normalization has to be performed within the limits of this window. A lookup table is generated with the aid of Eq. 9.3, which assigns to each bin of the observable x a certain probability for each type of molecule (see Fig. 9.2). This lookup table is multidimensional if more than one observable is measured.

In order to identify the molecules, for each photon the probabilities $P(m_i|b)$ can be read from the lookup table, and the probabilities $P_L(m_i)$ according to Eq. 9.4 are calculated (see Fig. 9.2). The analysis can be stopped as soon as the probability $P_L(m_i)$ has reached a preset accuracy level.

The assumptions we made to enable this technique of single-photon analysis – discrimination of SM fluorescence against fluorescence of other molecules, negligible number of background photons – make special demands on the instrumentation used for the experiments.

9.2.2   Instrumentation

In recent years, the most widely applied technique for single-molecule fluorescence spectroscopy at room temperature has been based on confocal excitation and fluorescence detection with a microscope objective. The excitation laser is focused onto a pinhole that is imaged to a diffraction-limited spot by a microscope objective. Fluorescence emerging from the molecule is collected by the same objective and spatially filtered by a second pinhole placed in front of the single-photon sensitive detectors. In fact, the excitation and detection pinholes are imaged by the microscope objective to the same diffraction-limited spot, thus defining the confocal volume. This way a detection volume of ≈ 100 aL can be realized, which guarantees a low background light level. A dichroic mirror usually separates the excitation and the fluorescence light. Remaining laser light is blocked by holographic notch and/or color filters.

To identify single molecules with confocal fluorescence detection, it has to be ensured that every molecule passes the detection volume. This has been done by constraining the flow of a solution of the analyte to a diameter smaller than the excitation spot either in a microcapillary [18] or by hydrodynamic focusing in a sheath flow [22]. In our experiment, molecules embedded in a thin polymer film were moved through the confocal volume by (raster) scanning the sample.

Pulsed lasers are preferably used as excitation sources. The advantages of the use of pulsed lasers are twofold. First, with a pulsed laser it is possible to measure the time interval between excitation and emission of the molecule. Second, scattered laser light that still passes the notch filter can be completely removed by rejecting photons that arrive in a time window at the temporal position of the laser pulse. Thus, the background can be reduced to a very low level.

TCSPC on single molecules makes special demands on the photon counting electronics. In ensemble lifetime measurements, the fluorescence count

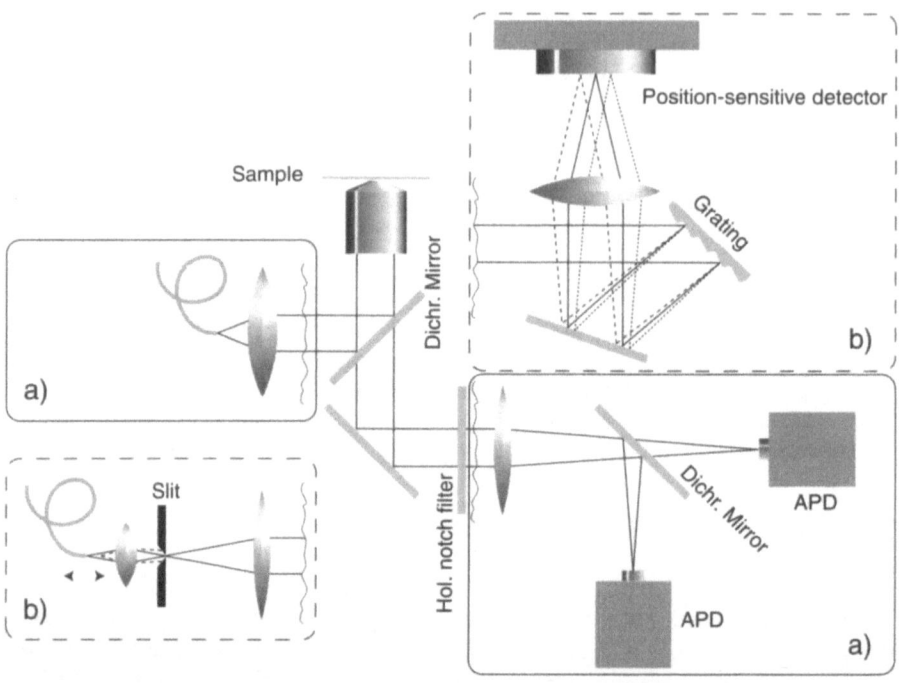

Fig. 9.3. The optical setups used for the experiments. **a** Scanning confocal optical microscope. **b** Slit illumination setup with spectrally and spatially resolved detector (for details see text)

rate has to be at least two orders of magnitude lower than the excitation rate in order to accomplish Poissonian statistics. In contrast, from a single molecule only one photon is emitted per excitation (provided that the laser pulses are short compared to the lifetime). The transit time of the molecule through the excitation volume in solution is on the order of a millisecond. A short dead time of the counting electronics is mandatory to detect enough photons during this short time period. Since an ensemble measurement is *per se* averaging, all photons can be integrated in a multichannel analyzer, whereas the need to distinguish single molecules makes it necessary to store each photon separately in SMD. The complete counting electronics that meets these requirements is available now on a single PC board [23].

In our experiment, the arrival time after pulsed excitation and the emission wavelength of the fluorescence photons were observed. An active mode-locked Nd:YAG laser (Coherent Antares, Palo Alto, CA), frequency-doubled to 532 nm, was used as the excitation light source. The width of the pulses was ≈ 150 ps with a repetition rate of 76 MHz.

The two optical arrangements we used are sketched in Fig. 9.3.

Confocal Setup. In the confocal setup, an optical fiber served as the excitation pinhole that was imaged by a lens (f=60 mm) and by an oil immersion objective (Numerical Aperture 1.3, Leica, Germany) to a diffraction-limited spot onto the sample. The dichroic mirror that separates the fluorescence – collected by the same microscope objective – had a 50% dichroic ratio at 550 nm. An additional holographic notch filter at 532 nm suppressed scattered excitation light. Fluorescence light was focused onto the detectors by a second lens (f=150 mm). The end of the fiber as well as the detectors are placed exactly in the focus of the respective lenses, thus creating an image at infinity. This way, the distance between the objective and the other components can be varied which makes the setup more flexible.

For fast and efficient detectors, two avalanche photodiodes (APD) (SPQ 141, EG&G, Canada) with dark count rates of 100 counts per second (cps) were used. Two-channel spectral resolution was achieved by a second dichroic mirror (50% transmission at 575 nm) separating the light detected by the APDs. Both APD signals were fed through a routing device into a PC-card based single-photon counting device (SPC 402, Becker&Hickl, Germany). A trigger from the mode-locker driver served as the synchronization signal. Having dead times of 100 ns and 150 ns, respectively, the APDs and the SPC card were capable of detecting photons with a rate of 8 MHz. Considering an overall detection efficiency of the setup of $\eta \simeq 0.065$ this number well matched the excitation repetition rate.

For each photon, the arrival time τ with respect to the excitation pulse (256 channels, 100 ps width), the time t with respect to the start of the experiment (50 ns resolution), and the information on which detector counted the photon were stored. The timing resolution was $\approx$ 400 ps FWHM. A home-built piezo-driven x-y scanning stage was used for raster scanning the sample, thus mimicking the motion of the molecules in a flowing solution. Scanning the sample led to a fluorescence burst each time a molecule traversed the excitation volume.

Slit Excitation Setup. In a second setup, an additional lens and an adjustable slit were placed in the excitation path. The fiber could be shifted between the focus and twice the focal length of the first lens in order to change the illuminated spot on the slit. The slit was placed exactly at the focus of both lenses and was therefore imaged onto the sample plane. This way a stripe that can be reduced to a thickness of $\approx$ 500 nm was illuminated on the sample. The length of the stripe could be varied between a confocal spot and 50 µm by shifting the fiber.

In this setup, we used a photon-counting image acquisition system (PIAS) (Hamamatsu, Japan). The detector is a multichannel plate (MCP) photomultiplier with a rectangular anode that has a finite resistance. The anode has contacts on each of the four corners. Depending on the position of the incidence of a photon on the photocathode, the avalanche of the multiplied

Table 9.1. Bulk emission maximum and fluorescence lifetime of the different dyes

Dye	Emission maximum (nm)	Fluorescence lifetime (ns)
Dibenzanthranthene (DBATT)	586	10.15
Sulforhodamine B (SRB)	584	4.15
Rhodamine 6G (R6G)	555	3.75
DiI	572	3.05
Pyridine (Pyr1)1	692	2.27
Nile red (Nred)	618	4.25

photoelectron leads to pulses of a certain amplitude on each of the four contacts of the anode. An arithmetic unit computes the x-y position from these pulses. An incident photon leads to a drop in MCP voltage, thus providing the arrival time information. In summary, the PIAS is a spatially and temporally resolved single photon detector. The drawbacks of the PIAS are its lower maximum count rate of 10^4 s^{-1} compared to the APD, and its factor of ten lower quantum efficiency.

A 400-mm photo objective focused to infinity in front of the detector created an image on the photocathode. The image from the microscope objective was reflected by a reflection grating (400 lines/mm, 500 nm blaze) and a second mirror onto the detector. Thus, the y coordinate of the position-sensitive PMT is the spatial image of the illuminated stripe on the sample, while the x coordinate provides spectral resolution for each point of the excitation stripe. A Picomotor micrometer-replacement actuator (MRA) (New Focus, USA) with a step size of 30 nm moved the sample perpendicular to the excitation stripe. Therefore, spectrally and temporally resolved fluorescence bursts can be detected in parallel along the excitation stripe. With the help of a flippable mirror in front of the grating, a spatial image can be taken by a highly sensitive CCD (Sensicam, PCO, Germany), which facilitates focusing.

The dyes rhodamine 6G (R6G), sulforhodamine B (SRB), dibenzanthranthene (DBATT), 1,1'-dioctadecyl-3,3,3',3' tetramethylindocarbocyanine perchlorate (DiI), pyridine 1 (Pyr1), and nile red (Nred) were chosen for the experiments because of their distinct emission spectra and fluorescence lifetimes (cf. Table 9.1). Thin films with a thickness of $\approx$ 50 nm were spin cast on microscope coverslips from a solution of polymethylmethacrylate (PMMA) in toluene with dye concentrations of order 10^{-9} M.

Bulk samples were prepared in the same way as the single-molecule samples, but with micromolar dye concentrations. In this case several hundred molecules were in the excitation volume, which allowed for recording of the arrival-time histograms and spectra without a burst analysis.

9.2.3   SM Identification – Lifetime and 2 Spectral Channels

The arrival-time histograms in both spectral channels for the four dyes are shown in Fig. 9.4a on a logarithmic scale. The slopes for the dyes DiI and DBATT are different in the two channels. We can conclude that the wavelength and arrival time are not independent. The increase of the fluorescence lifetime in the red wing of the spectrum can be explained by solvent relaxation [25]. Therefore, a two-dimensional lookup table has to be constructed that has just two bins in the spectral dimension. A graphical representation of this lookup table is shown in Fig. 9.4b (cf. Fig. 9.2). The stacked probabilities to have a given molecule as the source of a photon are plotted versus the photon arrival time for both spectral channels. Due to the pulse repetition interval of 13 ns, almost two complete fluorescence decays occur within the 25-ns range of the TAC. Therefore, for each of the two decays a separate time window is set. The borders of the time window were adjusted in such a way that laser light scattered by the matrix surrounding the molecule is excluded. However, in order to optimize the number of analyzed photons, the time windows were kept as large as possible.

Figure 9.5 visualizes the identification experiment for four different single molecules, in each case one of the dyes R6G, DBATT, SRB, and DiI, embedded in a PMMA film. Photons from a burst emitted by the transit of a molecule through the confocal volume are analyzed by the method described above. The time lag between 50 subsequently detected photons served as the

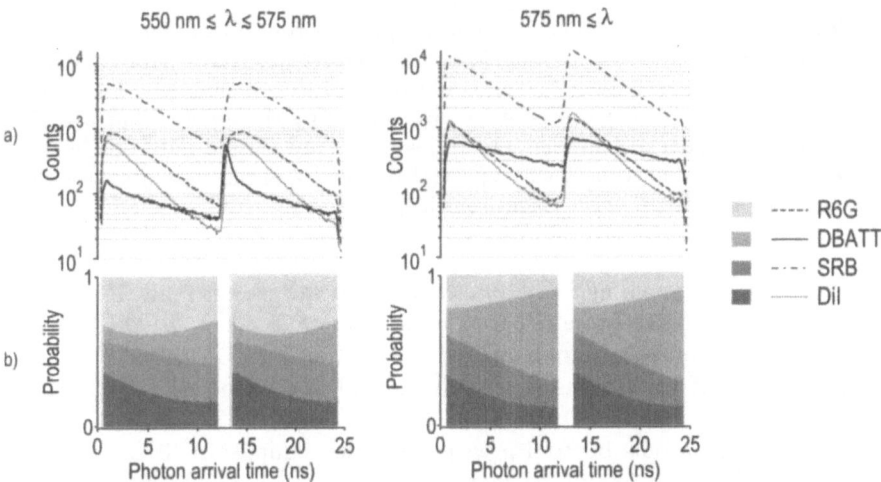

Fig. 9.4. a Photon arrival-time histograms in the two spectrally distinguished channels obtained from bulk measurements of the four dyes R6G, SrB, DBATT, and DiI. **b** Graphical representation of the lookup table computed from the histograms (**a**) according to Eq. 9.3 in two time windows within the TAC range. For each photon arrival time in the respective channel the probabilities for the four species can be read from this table. Data taken from [24]

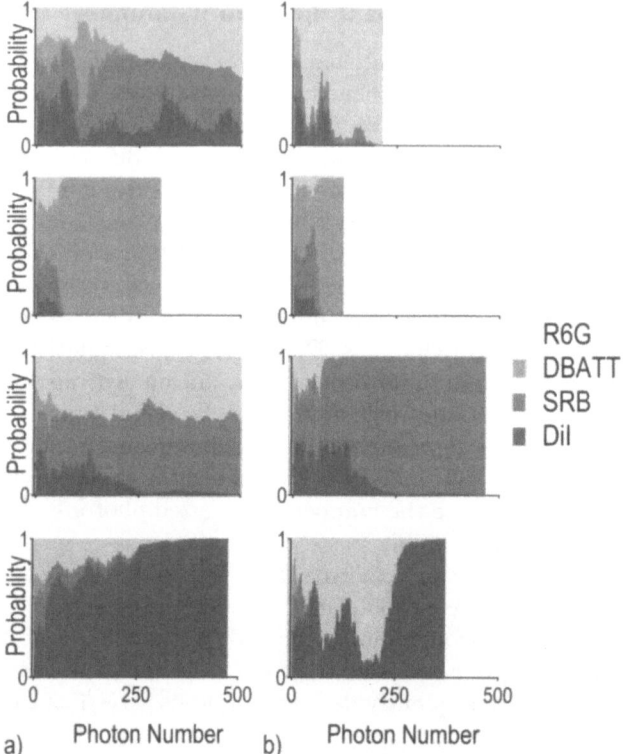

Fig. 9.5. Analysis of single-molecule identification experiments for four molecules of the respective dyes. Stacked probabilities to have either dye versus photon number, a) without spectral resolution, b) spectrally resolved in two channels. The graphs show the probabilities that the photons are emitted by the different species after detection of a number of photons. The probabilities are plotted until one of them reaches the value 0.999, which, with spectral resolution, is the case after less than 500 photons for all dyes. Data taken from [24]

criterion for distinguishing SM fluorescence from the background. If the time lag is below 2 ms (corresponding to a count rate of 25,000 s^{-1}), the analysis is started [24]. The background count rate is on the order of 1000 s^{-1}. During the transit time of the molecules of $\approx$ 15 ms, 500–3000 photons were detected with excitation pulses of $\approx$ 0.1 mJ/cm^{-2}.

From the top to the bottom row, pure dilute samples of R6G, DBATT, SRB, and DiI were investigated in this experiment. The stacked probabilities according to Eq. 9.4 are plotted versus the photon number within the photon sequence for four selected molecules. Each panel shows how the probability of each of the four molecules develops with the photon number of the fluorescence burst. As soon as the confidence level of 99.9% is reached, the analysis is stopped. The fluorescence photons of each burst are analyzed without (a) and with (b) spectral information taken into account. In the case of spectrally resolved time-correlated single-photon counting (b), the four molecules are identified with less then 500 photons. In contrast, if spectral resolution is discarded (a), no decision can be made regarding the dyes R6G and SRB within the transit time of the molecule in this example, since the fluorescence lifetimes of these dyes are similar. The analysis of several fluorescence bursts of the different dyes showed similar results, in agreement with Monte Carlo simulations obeying the experimental conditions [24]. Figure 9.5 demonstrates the stochastic nature of the identification experiment. For small numbers of photons the probabilities show large fluctuations which decrease as the number of detected photons increases.

9.2.4   SM Identification – Lifetime and 128 Spectral Channels

Since the properties of arrival time and wavelength are not independent, two-dimensional photon histograms (128 by 128 bins) were recorded with the PIAS setup. Again four dyes were chosen, in this case DBATT, DiI, Pyr1, and Nred. From the normalized histograms, a two-dimensional and two one-dimensional lookup tables were created. The one-dimensional lookup tables, one for the property arrival time and one for the spectrum, were created in order to compare the two-dimensional to the one-dimensional identification.

The traces of several single DiI molecules passing the excitation stripe in the PIAS setup are shown in Fig. 9.6. In the image in the top panel, the vertical coordinate corresponds to the y coordinate of the position-sensitive detector, whereas the horizontal coordinate is the scan time, which corresponds

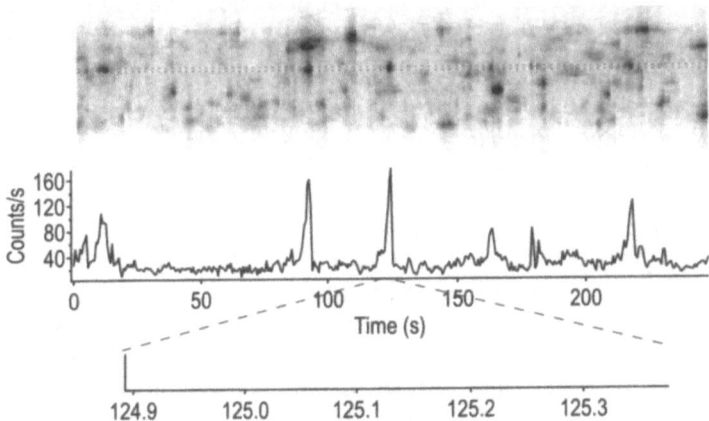

Fig. 9.6. Trace of single DiI molecules passing the illumination stripe. Top panel: Intensity as a function of y position and time. Middle panel: Intensity (counts per second) along the line marked in the top panel. Bottom panel: Single photon arrivals in the specified spatiotemporal window

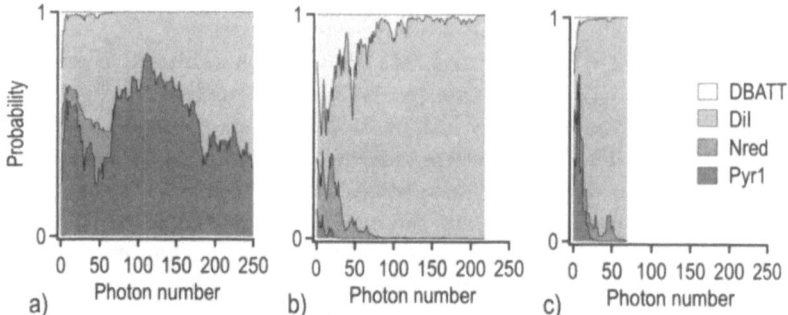

Fig. 9.7. Assignment probabilities for a sequence of photons from a DiI fluorescence burst analyzing a) arrival time, b) wavelength, and c) both properties in a two-dimensional lookup table

to the x coordinate on the sample. In fact, this image is a projection of the four-dimensional (sample-x-y, spectrum, arrival time) photon histogram into the sample plane. Photon bursts are observable as dark black features in the image. Blinking and one-step photobleaching, both visible in the image, are proof that the observed features are single molecules. Due to the spectral and temporal resolution of the detection setup, the wavelength and the arrival time with respect to the excitation are known for each photon of the image in Fig. 9.6. As in the case of the confocal setup, single-molecule fluorescence is selected with the help of a count rate criterion. For this purpose, the image in Fig. 9.6 is divided into small stripes (1 μm on the sample) along the horizontal direction. Because of the lower maximum count rate of the PIAS compared to the APDs and of the parallel detection, the sample had to be scanned at a lower velocity resulting in a transit time of the molecules on the order of 1 s. With excitation pulses of ≈ 0.01 mJ/cm^{-2}, a number of photons comparable to the confocal setup was detected in one burst.

In Fig. 9.7 the assignment probabilities for a photon sequence of a DiI fluorescence burst are shown. The photons are analyzed with respect to arrival time after excitation (a), wavelength (b), and with respect to both properties in a two-dimensional lookup table (c). Because of the similarity of the fluorescence lifetimes of DiI and Pyr1 they cannot be distinguished by analyzing exclusively the arrival time information. If the wavelength information is exploited, the DiI molecule is identified with 99.9% probability with less than 250 photons. This number is further decreased to less than 70 if all spectral and temporal information is matched with the two-dimensional lookup table. Single molecules of the other dyes DBATT, Pyr1, and Nred are identified with comparable numbers of photons. The higher background in the PIAS experiments impedes even more efficient identification.

In some cases, the identification can be finished with fewer photons. In rare cases, the one-dimensional identification outmatches the two-dimensional identification, demonstrating the stochastic nature of the photon properties

9.3   Conclusion and Outlook

The detailed experiments show that it is possible to learn from each detected single photon. Spectrally-resolved time-correlated single-photon counting was demonstrated to be suitable for identifying single molecules that are similar in fluorescence lifetime, but different in emission spectra, and vice versa. With this method of analysis, the information associated with each photon can be transparently exploited for the identification of the emitting single molecule.

With a position- and time-resolved detector it is possible to measure the spectrum and the fluorescence lifetime of different single molecules in parallel in one experiment. Besides single-molecule identification, another possible application of this detector is spectrally and time-resolved imaging of single molecules. With one scan of the slit almost all information obtainable from a single molecule sample could be measured.

Thus, the improvement of the experimental methods presented opens up a realm of new and fascinating experiments and empowers the application of single-molecule spectroscopy to other fields of science, most notably to life sciences.

Acknowledgment. The Swiss National Science Foundation and the Swiss Federal Institute of Technology are acknowledged for financial support.

References

1. W. Moerner, L. Kador, Phys. Rev. Lett. **62**(21), 2535 (1989)
2. J. Trautman, J. Macklin, L. Brus, and E. Betzig, Nature **369**, 40 (1994)
3. M. Orrit and J. Bernard, Phys. Rev. Lett. **65**(21), 2716 (1990)
4. T. Nonn, T. Plakhotnik, and U. P. Wild, to be published (2000)
5. F. Güttler, J. Sepiol, T. Plakhotnik, A. Mitterdorfer, A. Renn, and U. Wild, J. Lumin. **56**, 29 (1993)
6. M. Pirotta, F. Güttler, H. Gygax, A. Renn, und U. Wild, Chem. Phys. Lett. **208**, 379 (1993)
7. U. P. Wild, F. Güttler, M. Pirotta, and A. Renn, Chem. Phys. Lett. **193**, 451 (1992)
8. M. Croci, H. J. Müschenborn, F. Güttler, A. Renn, and U. Wild, Chem. Phys. Lett. **212**, 71 (1993)
9. J. Bernard, L. Fleury, H. Talon, and M. Orrit, J. Chem. Phys. **98**(2), 850 (1993)
10. T. Basché, W. Moerner, M. Orrit, and H. Talon, Phys. Rev. Lett. **69**(10), 1516 (1992)
11. M. Kollner and J. Wolfrum, Chem. Phys. Lett. **200**, 1 (1992)

12. S. A. Soper, L. M. Davis, and E. B. Shera, J. Opt. Soc. Am. B **9**(10), 1761 (1992)
13. S. Nie and R. Zare, Anal. Chem. **70**, 431 (1995)
14. L. Brand, C. Eggeling, C. Zander, K. H. Drexhage, and C. A. M. Seidel, J. Phys. Chem. A **101**(24), 4313 (1997)
15. J. R. Fries, L. Brand, C. Eggeling, M. Kollner, and C. A. M. Seidel, J. Phys. Chem. A **102**(33), 6601 (1998)
16. M. Kollner, A. Fischer, J. ArdenJacob, K. H. Drexhage, R. Muller, S. Seeger, and J. Wolfrum, Chem. Phys. Lett. **250**(3-4), 355 (1996)
17. C. Zander, M. Sauer, K. H. Drexhage, D. S. Ko, A. Schulz, J. Wolfrum, L. Brand, C. Eggeling, and C. A. M. Seidel, Appl. Phys. B **63**(5), 517 (1996)
18. C. Zander, K. H. Drexhage, K. T. Han, J. Wolfrum, and M. Sauer, Chem. Phys. Lett. **286**(5-6), 457 (1998)
19. A. VanOrden, N. P. Machara, P. M. Goodwin, and R. A. Keller, Anal. Chem. **70**(7), 1444 (1998)
20. K. Dorre, S. Brakmann, M. Brinkmeier, H. Kyung Tae, K. Riebeseel, P. Schwille, J. Stephan, T. Wetzel, M. Lapczyna, M. Stuke, R. Bader, M. Hinz, H. Seliger, J. Holm, M. Eigen, and R. Rigler, Bioimaging **5**(3), 139 (1997)
21. H. R. Jett, R. A. Keller, J. C. Martin, B. L. Marrone, R. K. Moyzis, R. L. Ratliff, N. K. Seitzinger, E. B. Shera, and C. C. Stewart, J. Biomol. Struct. Dynamics **7**, 301 (1989)
22. P. M. Goodwin, W. P. Ambrose, and R. A. Keller, Acc. Chem. Res. **29**(12), 607 (1996)
23. W. Becker, H. Hickl, C. Zander, K. H. Drexhage, M. Sauer, S. Siebert, and J. Wolfrum, Rev. Sci. Instr. **70**(3), 1835 (1999)
24. M. Prummer, C. G. Hübner, B. Sick, B. Hecht, A. Renn, and U. P. Wild, An. Chem. **72**(3), 433 (2000)
25. E. Görlach, H. Gygax, P. Lubini, and U. Wild, Chem. Phys. **194**, 185 (1995)
26. Enderlein and Sauer, J. Phys. Chem. A **105**(1), 48 (2001)

10 Fluorescence Correlation Spectroscopy in Single-Molecule Analysis: Enzymatic Catalysis at the Single Molecule Level

R. Rigler, L. Edman, Z. Földes-Papp, and S. Wennmalm

Fluorescence correlation spectroscopy (FCS) was introduced in the early seventies for the analysis of thermodynamic fluctuations of chemical systems [15,18,28] in an attempt to complement chemical relaxation spectroscopy as introduced by Eigen and de Meyer for the analysis of ultrafast kinetics. Chemical relaxation refers to the adjustment of chemical reactions into a new equilibrium state after an instantaneous change of intensive parameters such as temperature, pressure, or electric field [16]. Chemical fluctuations depend on the spontaneous change in number density of chemical systems due to processes involving transitions into the excited state [15], Brownian motion [15,28] as well as chemical kinetics [18,29].

Fluctuation amplitudes are linearly related to the inverse number of molecules or particles defining the system. In the early stages of FCS correlation amplitudes of about 10^{-4} to 10^{-6} could be obtained due to limitations in the signal-to-background ratios (S/B 1:1000) attainable at that time.

A substantial change occurred with the discovery that using confocal volume elements the signal-to-background ratio could be increased to 1000:1 and above [30,37]. It also provided the possibility of single-molecule detection and spectroscopy for chemical and biochemical systems. With the sensitivity in recording molecular fluctuations increased by more than six orders of magnitude FCS finally became the powerful routine [17,39,40] at the envisaged start.

10.1 Single-Molecule Detection in Solution and Correlation Functions

The first successful experiments in detecting single molecules were carried out by W. E. Moerner and M. Orrit (see review [33]). They used the inhomogenous spectral broadening of the emission of organic molecules at cryotemperatures to detect the emission of a single terylene molecule in pentacene matrices. The absence of photodynamic processes at cryotemperatures and high absorption cross sections (10^{-20} cm^2) were important ingredients in the sucessful detection by absorption (Moerner) and fluorescence (Orrit) spectroscopy. With the introduction of confocal volume elements and the suppression of the solvent background consisting of Raman scattering, the observation of dye molecules such as Rhodamin in solution and at room temperature became possible [30,36,37].

An alternative way for single-molecule detection was used by Richard Keller and collaborators in their attempt to sequence DNA at the level of single molecules [3,41]. Background suppression is possible by using the differences in the lifetime of the excited state of dye and solvent. The separation of coherent and incoherent emission processes is achieved by time gating [41]. This technology was used to record emission spectra of nucleic acids with short excited-state lifetimes and low quantum yields [38]. Compared to lifetime gating, using pulsed-excitation confocal single-molecule detection with continuous laser excitation provides superior sensitivity, since the excitation frequency is only limited by the lifetime of the excited state and not by the pulse frequency of the excitation source.

Our experiments were carried out using avalanche photodiodes (APD) as detectors, counting rates for single Rhodamin G molecules of 100,000 cps were achieved when 50 mW of the Ar-laser line at 514.5 nm excited a Gaussian confocal volume element of radii $\omega_{xy} = 0.5$ μm and $\omega_z = 2$ μm [37,39,40]. At this excitation level about 10% of the molecules are in the triplet state, which has a lifetime of around 1 μs [52] as compared to the singlet state, with a lifetime around 2 ns. The burst size spectrum of the emission of single Rh 6G molecules was analyzed by a multichannel scaler in bin sizes compatible with the characteristic diffusion time of Rh 6G of 40 μs (Fig. 10.1).

The question whether the emission bursts are due to single molecules can be answered by calculating the autocorrelaation function for diffusion $G(t)$:

$$G(\tau) = 1 + 1/N\{1 + \tau/\tau_D)^{-1}\{1 + \tau/\tau_D(\omega_z^2/\omega_x^2)]^{-1/2}\} \qquad (10.1)$$

with $G(\tau) = \langle I(t)I(t+\tau)\rangle/\langle I\rangle^2$

and $\tau_D = \omega_x/4D$,

N = number of molecules in the VE .

The value of N after correction for the background is found to be 0.005, indicating that on average 5 Rh 6G molecules can be found in 1000 volume elements; thus N equals the average probability P of finding a single Rh 6G molecule in the VE. Hence the probability for detecting 2 molecules occurring simultaneously in the VE as well its relative contribution can be calculated [11,39].

Given a time-averaged Poissonian distribition of Rh molecules in the VE we calculate the probability of finding one (p_1) or two (p_2) molecules in the VE from the average detection probability P:

$$p(n, P) = \frac{P^n e^{-P}}{n!} . \qquad (10.2)$$

Then $p_1 = 5 \times 10^{-3}$, $p_2 = 12.5 \times 10^{-6}$, and $p_2/p_1 = 2.5 \times 10^{-3}$.

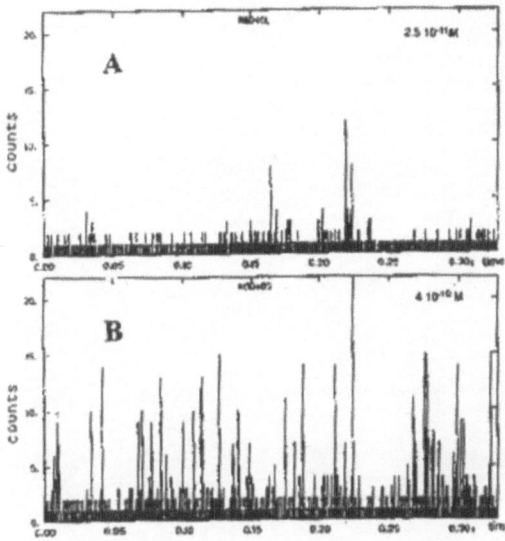

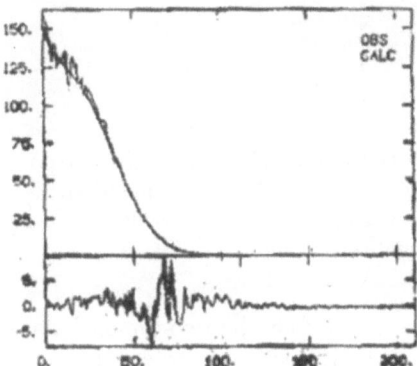

Fig. 10.1. Single molecule recordings of Rhodamine G. **a** 2.5×10^{-11} M. **b** 4×10^{-10} M, bin size 40 µs. **c** autocorrelation function of trace A (from [37])

This means that more than 1000 single-molecule events have to be registered before a double-molecule event is detected given the concentrations used (Fig. 10.1).

10.2   Confocal Single-Molecule Imaging

A direct consequence of single-molecule detection by confocal VE was the idea to use the confocal principle to image single molecules [7] (Fig. 10.2a) as an alternative to near-field imaging (e.g., [43]). Combining confocal SM

A.

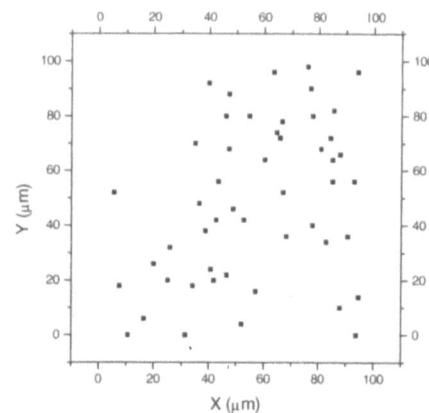

B.

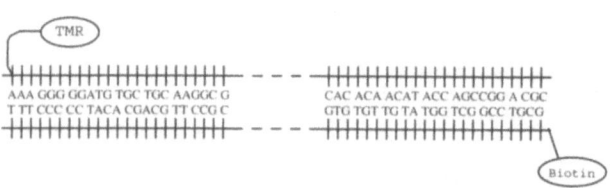

Fig. 10.2. a Confocal image of single 217 bp DNA molecules attached to strep-
tavinized cover slip by a biotin residue linked to one 5'end and a tramethylrho-
damin (TMR) residue linked to the other 5'end. Approximately 50 molecules per
100×100 µm^2 area. **b** the 217 bü DNA molecule (from [46]

detection with scanning over the area of interest reveals the two-dimensional
position of single molecules adsorbed or attached specifically to surfaces. A
typical image obtained from single DNA fragments mounted on a strepatvi-
dinized glass surface by a biotin group linked to the 5'-end on one strand and
visualized by a fluorescent tag (tetramethyl-rhodamin, TMR) attached by a
linker to the 5'-end of the other strand is demonstrated in Fig. 10.2b. This
started the analysis of conformational kinetics of single biomolecules in our
laboratory [11,46] and examplified the importance of FCS in the analysis of
single-molecule events.

10.3   Conformatial Transitions
in Single DNA Molecules

From the determination of the excited states lifetime of 217-bp long DNA
fragments tagged by TMR it became evident that the dye–DNA complex

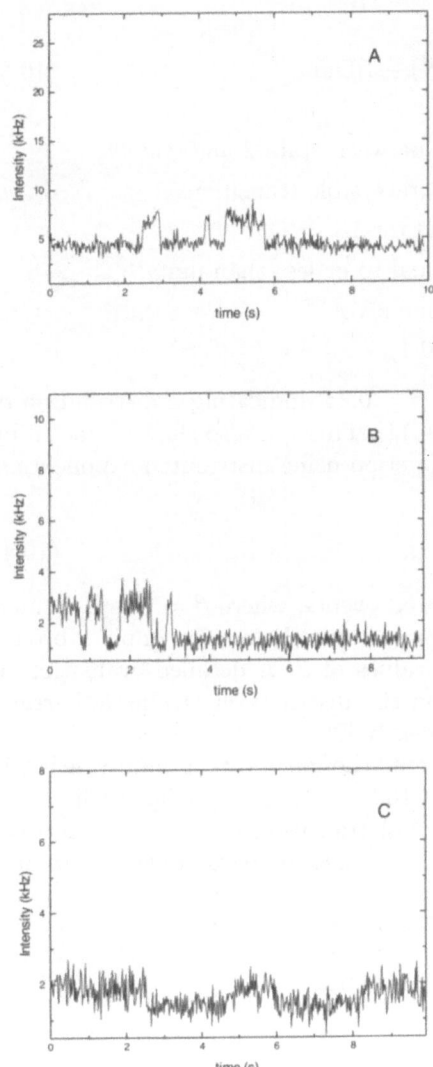

Fig. 10.3. Three cases (a,b,c) of conformational fluctuations in single 217 bp DNA molecules attached to a cover slip as sensed by the TMR probe (from [37])

must exist in two different states which exchange slower than the characteristic time of the experiment, i.e., the diffusion time of the DNA fragment through the volume element (5–30 ms) [10,11]. It also became clear from FCS measurements (Fig. 10.3a) that the correlation function contained in addition to the diffusion term a component, which was best described by a transition with stretched exponential behavior $\exp(-kt)^{\beta}$. The correlation function could be described by [11,50]

$$G(\tau) = 1 + 1/N[A(\exp(-kt)^{\beta}) + 1]\text{Diff} , \qquad (10.3)$$

with $A = K(1 - Q)^2/(1 + KQ)^2 ,$

$K =$ equilibrium constant between state 2 and state 1,

$k =$ relaxation rate of the reversible transitions
    between state 1 and state 2,

$\beta =$ stretch parameter, equal to or less than unity,
    but always greater than zero,

Diff $=$ diffusion term (Eq. 10.1).

We found parameters $k = 20$ s^{-1} and $\beta = 0.44$, indicating a distribution of rates over several orders of magnitudes [11]. This was also shown later on by a direct evaluation of the correlation function using distributed exponentials [46]:

$$G(\tau) = 1 + 1/N[\Sigma_i A_i \exp{-(k_i\tau)}]\text{Diff} . \qquad (10.4)$$

β refers to a short notation of distributed events, where $\beta = 1$ represents a Dirac distribution (single exponential) and values below one imply a broadening of the distribution to the lower values of β. A detailed evaluation of the impact of the stretch parameter on the distribution of kinetic barriers (activation energies) has been given recently [9].

The relaxation rate of 20 s^{-1} represents the mean rate around which k is distributed at higher and lower values (100 s^{-1} to 1 s^{-1}) (Fig. 10.3b). The explanation is a matrix of conformational transition in which the vertical transitions equilibrate on a time scale that is fast compared to the horizontal

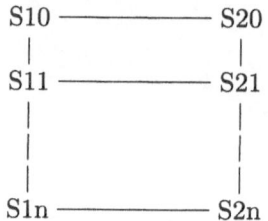

transitions. A likely scenario is a multitude of interactions of the linker-bound TMR with Guanosin bases in its vicinity preceding further steps which may involve insertion into the large groove, as well as intercalation.

Similar experiments have been carried out by Hochstrasser et al. [24] on single tRNA molecules, confirming the existence of different conformations, as found in the case of a DNA double helix [10,11]. A detailed theoretical analysis of the probability density of conformational states using both data sets [10,11,24] has been performed by Geva and Skinner [20] and Berezhkovskii et al. [5].

10.4   Single-Molecule Traces

The technique developed in this laboratory is a "stochastic experiment" where a surface covered with single molecules is scanned point by point until a single-molecule trace and its correlation function are observed. At that moment the scan is stopped, and traces and their correlation functions are recorded [46].

A typical trace obtained from the tagged DNA fragments demonstrates the spontaneous occurence of the transition between the two different fluorescence states, the intensity of which is proportional to the lifetime of their excited states [11,46] (Fig. 10.3). While in the ensemble transition rates are averages, in the single-molecule case the transition is a spontaneous, unpredictable event, which however follows a distribition of exponential character [46,47]. From the time intervals between the transition times of S1 to S2 ($\tau 1$) and vice versa ($\tau 2$), the relaxation rates $k = 1/\tau 1 + 1/\tau 2$ can be determined. Equivalent to this evaluation, where the transition times must be determined individually, is recording the correlation function for the same trace (Fig. 10.4). The relaxation rate can then be determined directly from the correlation trace.

10.5   Homogeneous and Heterogeneous Behavior

The erogodic principle states the equivalence of ensemble average and time average. Hence the behavior of a single molecule observed over a sufficiently long time should show all molecular facets as demonstrated by the molecular ensemble incorporating a vast number of individual molecules. From a kinetic point of view the distribution of transition rates observed in the ensemble of DNA fragments should be observable for a single molecule if the system is ergodic and homogeneous. In reality, however, the relaxation rates of individual DNA fragments constitute subfractions of the distribution of the ensemble (Fig. 10.5) and a heterogeneous behavior. The most likely explanation is the fact that during the time during which single molecules can be observed (ca. 10 s), not all transitions can be performed which are possible in an ensemble with a long time history. A different description is a scenario with a high-dimensional landscape of activation barriers of different height which limit transitions to certain areas, given a limited time span for transitions to take place. A special reason for the heterogeneous behavior is in this case the chemical lifetime of the fluorescent molecule, which has been determined in great detail [47].

10.6   Time Resolution of Single-Molecule Behavior

The availability of avalanche photodiodes with substantially increased quantum efficiency was an important reason for successful single-molecule detection and significant reduction of the collection times in FCS. Their time

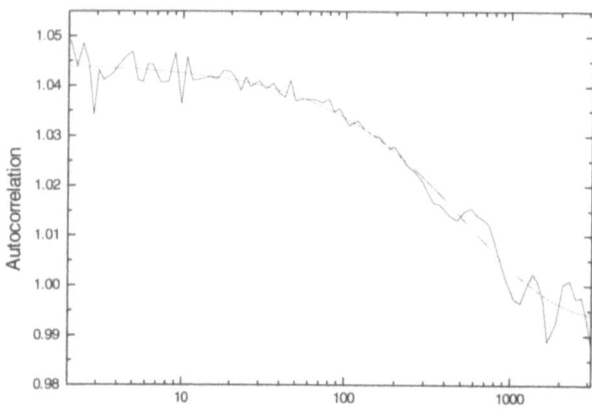

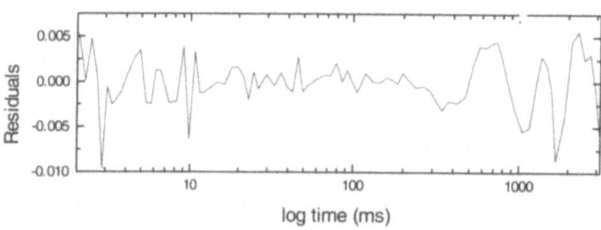

log time (ms)

Fig. 10.4. Autocorrelatin function of trace A of Fig. 3. Below mean squares difference between experimental and parametrized autocorrelation function (from [37])

resolution is limited by their recovery time, which in older versions is around
200 ns and in modern versions around 40 ns. Their linearity in response to im-
pinging photons is strongly dependent on the recovery time which constitutes
a time interval during which the APD is not active. Compared to our first
single-molecule measurements of Rh 6G, where 100,000 cps could be recorded
with a 200-ns dead time APD, a factor 2 or more can be gained using short
dead-time versions. The important feature of APDs as detectors is their high
quantum yield in combination with a fast response time. Since the recovery
process is a stochastic phenomenon, the time limitation can be surmounted
by using two APDs and cross-correlating the signal from the same source.
Time resolution down to a few ns could be obtained in this way, and was
applied in the analysis of triplet states [52,53] antibunching and rotational
correlation [15,32] as well as in kinetic studies of dye–nucleotide interactions
[54] and of conformational transitions in GFP [22,55,56].

A different way of following single-molecule events is the use of 2-D charge-
coupled devices (CCD). In this situation each pixel constitutes an individual
volume element and images of single molecules can be recorded as well as

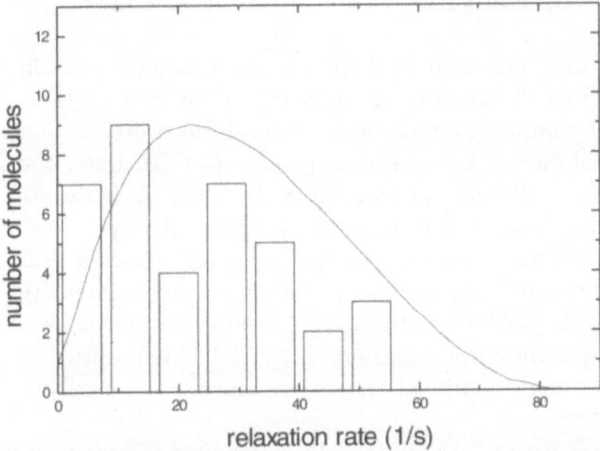

Fig. 10.5. Histogram of the relaxation rates of 37 single 217 bp DNA molecules as compared to the distribution of relaxation rates of the ensemble with $k = 20$ s^{-1} and $\beta = 0.44$ (from [48] and [11])

their fluctuations. Compared to APDs the time resolution of CCDs is rather limited by their readout time, and relatively slow processes can take seconds. A striking example is GFP, where single-molecule transitions were observed at seconds-long time scales [8], while conformational relaxations linked to a protonization processes could be recorded in the μs to ms time regime by FCS [22,51,55,56]. In summary, scanning analysis using APDs as introduced by [46] offers heretofore unsurpassed time resolution for the analysis of single molecules.

10.7 Kinetic Analysis, Death Numbers, and Survival Times

Important parameters for the analysis of single molecules include the number of photons which can be turned over until a molecule breaks down (death number) and the time interval during which this takes place (survival time). For TMR a detailed study at the level of single molecules was performed [47], and suggests the existence of a population with a turnover (death) number of 15,000 detected photons, as well as a second population with a higher value. Depending on the excitation intensity and the counting rate, the breakdown time (survival time) will assume different values which can be calculated from death numbers and excitation rates. Thus for recording fast processes, high excitation intensities are needed in order to obtain adequate S/N ratios, while for the analysis of slow processes low excitation intensities are required.

10.8   The Fluctuating Enzyme

The notion that enzymes may fluctuate in their activity was addressed in a seminal paper by Careri et al. as early as 1975 [5]. It alluded to thermodynamic fluctuations of complex systems such as catalytic proteins, and the upcoming possibilities of fluctuation spectroscopy such as FCS. Later the problem was taken up by Yeung [58] and by Dovichi et al. [6,34]. Their results indicated that enzyme molecules may vary in their catalytic activity.

An important step was taken when Lu and Xie, in single molecule experiments could show spontaneous fluctuations in the fluorescence intensity of flavin adenosin dinucleotide (FAD) which acts as a cosubstrate for cholesterol oxidase in the catalytic turnover of cholesterol [26,27,57]. Their data also indicate the involvement of conformational transitions during the catalytic cycle.

In a complementary study we analyzed the catalytic turnover of horseradish peroxidase (HRP) using a nonfluorescent Rhodamine molecule as substrate, which by the action of HRP is oxidized to a fluorescent product. Based on experience from the study of DNA fragments, biotinylated HRP was attached to a strepatavidin-coated glass surface at concentrations which assure the attachment as single molecules (Fig. 10.6). By the turnover of the substrate into fluorescent product, photons are generated at each single HRP molecule. Intensity fluctuations recorded show a large variation in the activity of product formation (Fig. 10.7). In comparison to the time limitation imposed by the survival time of TMR linked to the DNA fragments, no such limitations exist, since new fluorescent substrates are always generated, which leave their point of generation by dissociation from the enzyme.

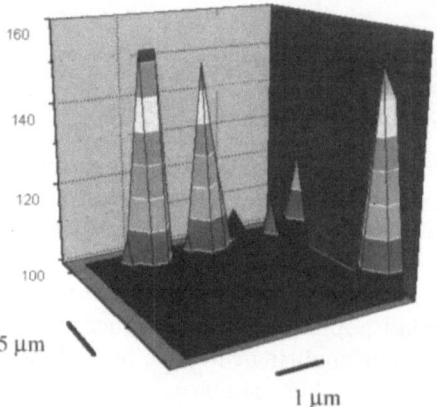

Fig. 10.6. Confocal image of single horseradish peroxidase molecules generating Rhodamin 6G as fluorescent product (from [14])

10.9   Multiple Conformational Transitions and Catalysis

Analysis of the intensity fluctuation traces show the diffusive part of the correlation function in the range of 100 μs and an extended (stretched) part from about 1 ms to the s range (Fig. 10.7).

Horseradish peroxidase interacts with the substrate in its oxidized form (5+ and 4+) and in either form can exchange an electron with the substrate, which is turned into a fluorecent product. The reduced enzyme form (3+) can be shifted into the oxidized form by hydrogen peroxide (H_2O_2) acting as

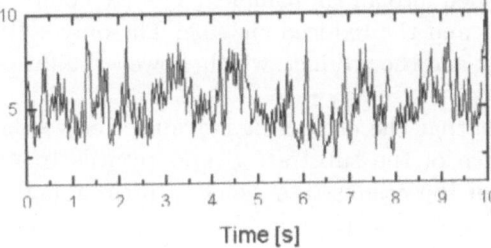

Time [s]

A

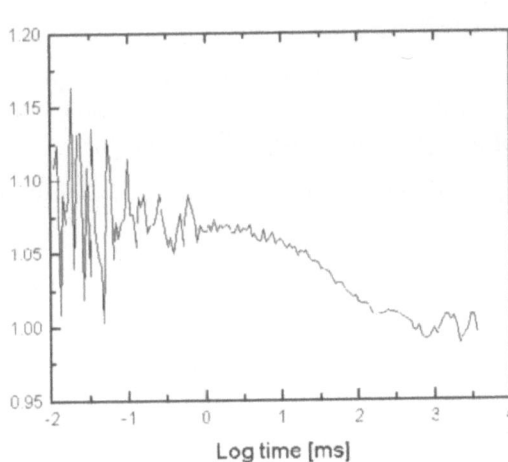

Log time [ms]

B

Fig. 10.7. a Fluctuation trace of single horseradish peroxidase molecules generating fluorescent Rhodamin 6G. **b** Autocorrelation function recorded simultaneously (from [14])

cosubstrate. A simple reaction scheme is as follows:

$$\text{S}\xrightarrow{\hspace{2cm}}\text{P} \qquad\qquad \text{S}\xrightarrow{\hspace{2cm}}\text{P}$$

$$\text{HRP(5+)} \xrightarrow{\hspace{2cm}} \text{HRP(4+)} \xrightarrow{\hspace{2cm}} \text{HRP(3+)} \qquad \text{(I)}$$

$$\xleftarrow{\hspace{2cm}} \quad \text{H}_2\text{O}_2 \quad \xleftarrow{\hspace{2cm}}$$

indicating the involvement of the heme iron in the oxidation–reduction process.

Using a simplistic model for the analysis of correlaton traces, we assume as minimal model the formation of a substrate enzyme complex (ES complex) which by catalysis will be transformed into an EP complex. The EP complex finally dissociates into the product and the reduced enzyme. The only state which is observed is the EP complex and the product, which, however, diffuses rapidly out of the VE.

The model starts from the fact that the enzyme E is transferred to the visible EP complex in the presence of the substrate S and returns to its starting position in the presence of the cosubstrate which is present under saturating conditions:

$$\begin{array}{ccc}
\text{ES} & \longleftrightarrow & \text{EP} \qquad\text{(II)}\\
\mid & & \mid \\
\text{S} + \text{E(5+)} & \longleftarrow & \text{E(3+)} + \text{P}
\end{array}$$

Calculating the correlation function $G(\tau)$ for the formation of the ES and the EP complex, which is the only observable form, gives rise to two exponential processes. We consider the case of a single immobilzed HRP molecule:

$$G(\tau) = 1 + \{[A\exp(-k_1\tau) + B\exp(-k_2\tau)] + 1\} \qquad (10.5)$$

with

$$A = K_1(1 + K_1 + K_1K_2)[Q_s - Q_{ES}]^2/$$
$$(K_1 + 1)(Q_S + Q_{ES}K_1 + Q_{EP}K_1K_2)^2$$

$$k_1 = k_{on}(E + S) + k_{off}$$

$$B = K_1K_2(Q_S + K_1Q_{ES} - (1 + K_1)Q_{EP})^2/$$
$$(K_1 + 1)(Q_S + Q_{ES}K_1 + Q_{EP}K_1K_2)^2$$

$$k_2 = k_{SP}k_{on}(E + S)/k_1 + k_{PS}$$

with K_1 and K_2 = equilibrium constants for first and second step

Q_S = quantum yield of S

Q_{ES} = quantum yield of ES

Q_{EP} = quantum yield of EP

For the realistic case that $Q_S = Q_{ES} = 0$, $A = 0$ and the first process is not visible. Only the formation of EP can be observed:

with $Q_{EP} \gg Q_S, Q_{ES}$, B reduces to

$B = (1 + K_1)/K_1/K_2$.

The amplitude of the visible step increases with increasing substrate concentration and reaches a maximum determined by $1/K_2$.

Equation (9) is valid for the case that the formation of the ES complex is fast as compared to the EP complex, i.e., $k_{on}, k_{off} \gg k_{SP}, k_{PS}$. It is also readily seen that Eq. (5) is a special case of Eq. (9) for the case that $K_2 = 0$.

Considering the whole catalytic cycle, we have reduced the reaction scheme (II) even further, indicating that the oxidized enzyme is transformed into the visible EP complex and vice versa [13]:

$$\begin{array}{ccc} & \xleftarrow{\hspace{1.5cm}k_f\hspace{1.5cm}} & \\ \mathrm{E} & \mathrm{EP} & \mathrm{(III)} \\ & \xrightarrow[\hspace{1.5cm}k_b\hspace{1.5cm}]{} & \end{array}$$

In this scheme k_f describes the formation of the EP complex, including all intermediate steps, while k_b describes the formation of the product, the essential catalytic step as well as the reoxidation of the enzyme. The correlation function $G(\tau)$ is

$$G(\tau) = 1 + \{1/K[\exp -(k_f + k_b)\tau] + 1\} , \tag{10.6}$$

with $K = k_f/k_b$ with $k_f = k_{EP}k_{on}(E + S)/k_1$.

If the formation of the EP complex can proceed via a manifold of pathways, k_f will have some distribution; then

$$G(\tau) = 1 + \{\Sigma p(k_f)\{1/K[\exp -(k_f + k_b)\tau] + 1\}dk_f$$

with $\Sigma p(k_f)dk_f = 1$ giving the normalization of the distribution function.

We assume two cases:

a) the catalytic step is much faster then the formation of the EP complex, $k_b \gg k_f$; then a single exponential correlation function will be observed.

$$G(\tau) = 1 + \{1/K[\exp(-k_b\tau)] + 1\}$$

b) the the EP formation step is much faster then the catalytic step i.e. $k_f \gg k_b$:

$$G(\tau) = 1 + \{a[\exp(-k_f\tau)^\beta] + 1\}$$

with $\exp -(k_f\tau)^\beta = \Sigma p(\lambda)\exp(-\lambda t)d\lambda$.

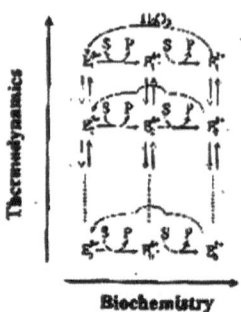

Fig. 10.8. Scheme of conformational transitions (vertical axis) and catalysis (horizontal axis) of horseradish peroxidase (from [13])

Thus for the case that $k_f \gg k_b$, k_f will be distributed and have a stretched appearance, but will assume a single exponential behavior which is domintaed by $\exp(-k_b\tau)$ as the catalytic rate exceeds the rate for the formation of the EP complex.

The normalized intensity autocorrelation function for a single immobilized HRP molecule can then be described by

$$G(\tau) = a \exp[-(k_f\tau)^\beta] + b \exp[-(k_b\tau)^\gamma] + c .$$

In this description the first term represents the formation of the EP complex, while the second terrm represents the catalytic step including generation of the product. For both steps stretch parameters have been introduced.

Stretched behaviour and conformational behavior

In the evaluation for high substrate concentrations (130 nM), k_f varied between 510–4100 s^{-1} with a β value between 0.66–0.16, while k_b was found to be between 5–11 s^{-1} with a γ value 1.2 to 0.8. From a comparison of the measured turnover rate in solution (50 s^{-1}) and the dependence of the substrate concentration, we assign the process characterized by a stretch parameter close to unity to the catalytic step of the reaction.

The formation of the enzyme product complex proceeds over a manifold of transitions as indicated by the low value of the stretch parameter. One interesting behavior is the observation that the stretch paramter of the EP formation decreases with increasing measurement time (Table I in [33]). This is best explained by the fact that given limited observation time the catalysis proceeds in a limited range of a landscape of activation barriers. This landscape is explored more extensively the longer the observation time is, leading to a broader distribution of the transition rates and a decrease of the stretch parameter.

The analysis of the enzyme catalysis at the single-molecule level has revealed properties which would not have been observed in an ensemble measurement. The results support very clearly Frauenfelder's model of conformational substates [19] as a basis for the functional behavior of proteins such as enzymes. The scheme in Fig. 10.8 is an example of the fact that HRP exists in a manifold of conformations of which only a few will lead to catalysis. The obvious behavior of times where the enzyme is active and other times where the enzyme is inactive as seen from the single-molecule traces is generated by the large variety of activation barriers which have to be crossed to lead to catalysis.

10.10   Higher-Order Correlations and Non-Markovian Behaviour

Enzymatic action may be linked to a complex network of conformational transitions that lead to properties which depend on the previous history, e.g.

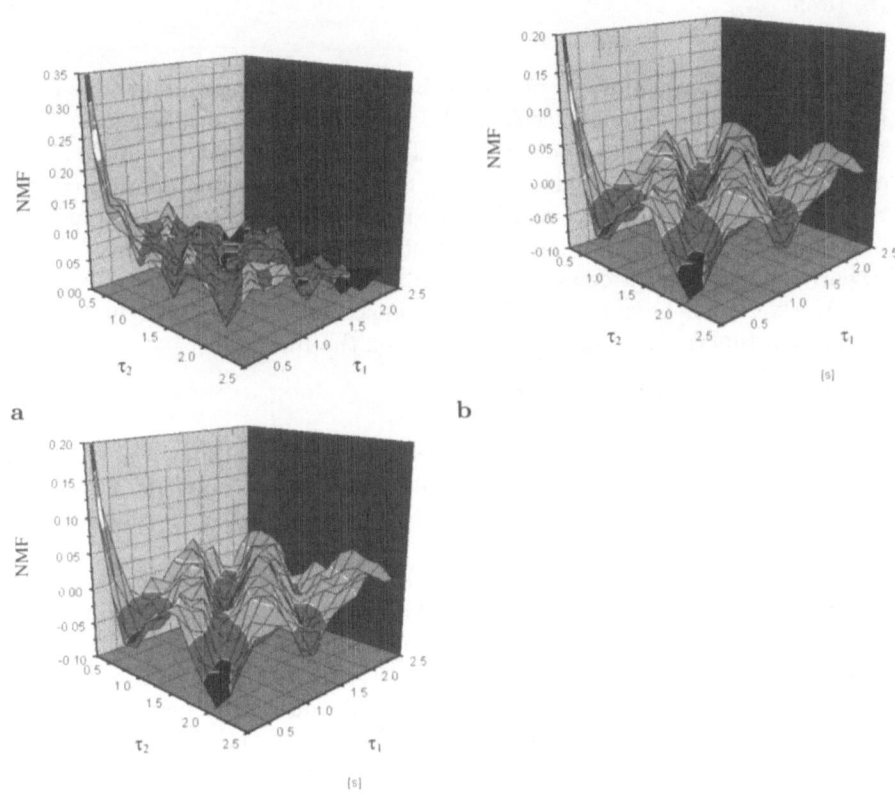

Fig. 10.9. Non-Markovian landscapes of 2 horseradish peroxidase molecules (**a** and **b**) generating fluorescent products (Rhodamin 6G). **c** Complete reaction system but without enzyme (from [13])

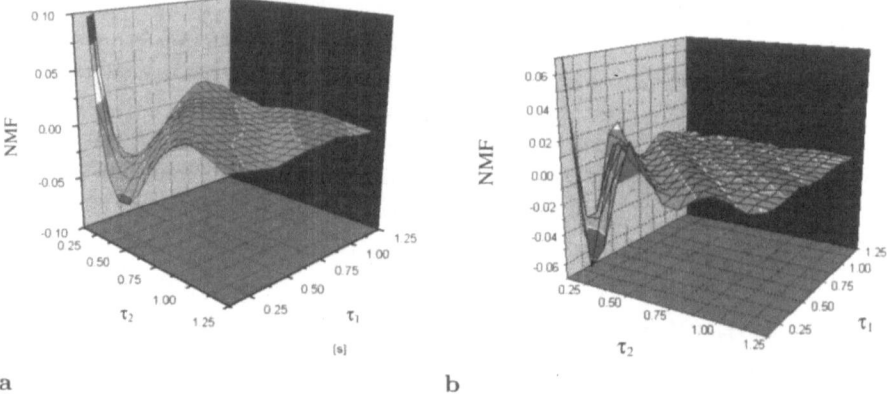

a   b

Fig. 10.10. Stochastic simulation of a closed reaction scheme of horseradish peroxidase generating the fluorescent product enzyme complex via several intermediates. Rates only in the forward direction are assigned with $k = 10\ s^{-1}$ (**a**) and $k = 22\ s^{-1}$ (**b**) (from [13])

hysteresis. From the analysis of the probability distribution of on time pairs, Lu and Xie [27] found evidence for nonexponential behavior during the catalysis of cholestrol by colcsterol oxidase. They also analyzed the applicability of classical kinetic models with coupling terms as well as the fluctuating barrier height model of Agmon and Hopfield [1,2] to their data set [42] in order to address the existence of memory effects in enzymatic pathways. We have recently suggested strategies which are able to discriminate Markovian from non-Markovian behavior [13]. They are based on the comparison between correlation functions of different order. Thus the function

$$\text{NMF} = G(\tau_1, \tau_2)/[G(\tau_2) - G(\tau_1)]\,, \tag{10.7}$$

will be zero for Markovian processes, for which only the immediate past is known. Analysis of the trajectory of product fluctuation from HRP (Fig. 10.9) gives a very reproducible picture indicating a non-Markovian relaxation on the time scale 20 ms to 2.5 s.

As can be shown from a detailed analysis of the spectrum of eigenvalues as well as from stochastic simulations, the NMF relaxation and the periodic oscillations (Fig. 10.10) are related to the catalytic cycle embedded in the network of state transitions. The periodic oscillations observed for the first time in a simple enzyme molecule (Figs. 10.9 and 10.10) originate from a stationary nonequilibrium state due to the catalytic cycle.

10.11   Conclusions

For understanding single-molecule behavior appropriate analysis of the molecular trajectories is required. Single-molecule detection and FCS analysis provide adequate tools for online analysis of the intensity fluctuations and their correlation functions. They provide the experimental complement to theoretical studies of single molecule processes [25,31,44,45]. Higher-order correlation functions provide information on non-Markovian processes. Stationary nonequilibrium systems such as substrate-driven enzymatic catalysis can be observed at the single-molecule level.

Acknowledgments. We acknowledge the continuous support of the Natural Science Research Council, the Technology Science Research Council, and the Knut and Alice Wallenberg Foundation.

References

1. N. Agmon and J. J. J. Hopfield, Chem. Phys. **78**, 6947 (1983)
2. N. J. Agmon, Phys. Chem. B 2000, **104**, 7830 (2000)
3. W. P. Ambrose, P. M. Goodwin, J. C. Martin, and R. A. Keller, Science **265**, 364 (1994)
4. A. M. Bereshkovskii, A. Szabo, and G. H. J. Weis, Chem. Phys. **110**, 9145 (1999)
5. G. Careri, P. Fasella, and E. Gratton, Crit. Rev. Biochem. **3**, 141 (1975)
6. D. B. Craig, E. A. Arriaga, C. Y. Wong, H. Lu, and N. J. J. Dovichi, J. Am. Chem. Soc. **118**, 5245 (1996)
7. J. Dapprich, Ü. Mets, W. Simm, M. Eigen, and R. Rigler, Exp. Tech. Phys. **46**, 259 (1995)
8. R. M. Dickson, A. B. Cubitt, R. Y. Tsien, and W. E. Moerner, Nature **355** (1997)
9. O. Edholm and C. Blomberg, Chem. Phys. **252**, 221–225 (2000)
10. L. Edman, Ü. Mets, and R. Rigler, Tech. Phys. **41**, 259 (1995)
11. L. Edman, Ü. Mets, and R. Rigler, Proc. Natl. Acad. Sci. USA **93**, 6710 (1996)
12. L. Edman, S. Wennmalm, F. Thamsen, and R. Rigler, Chem. Phys. Lett. **292**, 15 (1998)
13. L. Edman, Z. Földes-Papp, S. Wennmalm, and R. Rigler, Chem. Phys. **247**, 11 (1999)
14. L. Edman and R. Rigler, Proc. Natl. Acad. Sci. USA (2000)
15. M. Ehrenberg and R. Rigler, Chem. Phys. **4**, 390 (1974)
16. M. Eigen and L. DeMeyer, in: *Techniques in Org. Chemistry. Investigation of Rates Mechanism of Reactions*, S. L. Friess, E. S. Lewis, and A. Weissberger (eds.), vol. VIII, part II, pp. 895–1054 (Interscience Publishers, 1967)
17. M. Eigen and R. Rigler, Proc. Natl. Acad. Sci. USA **91**, 5740 (1994)
18. E. Elson and D. Magde, Biopolymers **13**, (1974)
19. H. Frauenfelder, S. G. Sligar, and P. G. Wolynes, Science **254**, 1598 (1991)
20. E. Geva and J. L. Skinner, Chem. Phys. Lett. **288**, 255 (1998)
21. A. Y. T. T. Ha, J. Liang, W. B. Caldwell, A. A. Deniz, S. Chemla, P. G. Schultz, S. and Weiss, Proc. Natl. Acad. Sci. USA **96**, 893 (1999)
22. U. Haupts, S. Maiti, P. Schwille, and W. W. Webb, Proc. Natl. Acad. Sci. 13573 (1998)

23. A. Ishijima, H. Kojima, T. Funatsu, Tokumanaga, and T. Yanagida, Cell **92**, 143 (1998)
24. Y. Jia, A. Sytnik, L. Li, S. Vladimirov, B. S. Cooperman, and R. M. Hochstrasser, Proc. Natl. Acad. Sci. USA **94**, 7932 (1997)
25. M. J. Karplus, Phys. Chem. B **194**, 11 (2000)
26. H. P. Lu and X. S. Xie, Nature **385**, (1997)
27. H. P. Lu, L. Xun, and X. S. Xie, Science **282**, 1877 (1998)
28. D. Magde, E. Elson, and W. W. Webb, Phys. Rev. Lett. **29**, (1972)
29. D. Magde, E. Elson, and W. W. Webb, Biopolymers **13**, 29 (1974)
30. U. Mets and R. Rigler, J. Fluorescence **4**, 259 (1994)
31. J. A. McCammon, B. R. Gelin, M. Karplus, and P. G. Wolynes, Nature **262**, 325
32. U. Mets, J. Widengren, and R. Rigler, Chem. Phys. **218**, 191 (1997)
33. W. E. Moerner and M. Orrit, Science, **283**, 1670 (1999)
34. R. Polakowski, D. B. Craig, N. Skelley, and N. J. J. Dovichi, Am. Chem. Soc. **112**, 4853 (2000)
35. R. Rigler, J. Biotechnology **41**, 1779 (1995)
36. R. Rigler and J. Widengren, J. Bio. Science, B. Klinge and Ch. Owman (eds.), p. 180 (Lund University Press, Lund, 1990)
37. R. Rigler and U. Mets, Soc. Photo-Opt. Instr. Eng. **1921**, 23 (1993)
38. R. Rigler, F. Claesens, and G. Lomakka, Springer Series in Chemical Physics **38**, D. H. Auston and K. B. Eisenthal (eds.), p. 472 (Springer-Verlag, 1984)
39. R. Rigler, J. Widengren, and U. Mets, in: *Fluorescence Spectroscopy*, O. S. Wolfbeis (ed.), p. 13 (Springer-Verlag, 1992)
40. R. Rigler, U. Mets, J. Widengren, and P. Kask, Eur. Biophys. J. **22**, 169 (1993)
41. E. B. Shera, N. K. Seitzinger, L. M. Davies, R. A. Keller, and S. A. Soper, Chem. Phys. Lett. **174**, 553 (1990)
42. G. K. Schenter, H. P. Lu, and X. S. Xie, J. Phys. Chem. A **103**, 10477 (1999)
43. J. K. Trautman, J. J. Macklin, L. E. Bruss, and E. Betzig, Nature **369**, 40–42 (1994)
44. J. Wang and P. G. Wolynes, Phys. Rev. Lett. **74**, 4317 (1995)
45. J. Wang and P. G. Wolynes, J. Chem. Phys. **110**, 4812 (1999)
46. S. Wennmalm, L. Edman, and R. Rigler, Proc. Natl. Acad. Sci. USA **94**, 10641 (1997)
47. S. Wennmalm and R. Rigler, J. Phys. Chem. B **103**, 2516 (1999)
48. S. Wennmalm, L. Edman, and R. Rigler, Chem. Phys. **247**, 61 (1999)
49. J. Widengren and R. Rigler, Bioimaging **4**, 149–157 (1996)
50. J. Widengren and R. Rigler, J. Fluorescence **7**, 211 (1996)
51. J. Widengren and R. Rigler, Cell. Mol. Bio. **44**, 57 (1998)
52. J. Widengren, R. Rigler, and U. Mets, J. Fluorescence **4**, 255 (1994)
53. J. Widengren, U. Mets, and R. Rigler, J. Phys. Chem. **3368**, (1995)
54. J. Widengren, J. Dapprich, and R. Rigler, Chem. Phys. **16**, 417 (1996)
55. J. Widengren, T. Terry, and R. Rigler, Chem. Phys. **249**, 259–271 (1999)
56. J. Widengren, U. Mets, and R. Rigler, Chem. Phys. **250**, 171 (1999)
57. X. S. Xie and J. K. Trautman, Ann. Rev. Phys. Chem. **59**, 441 (1998)
58. Q. F. Xue and E. S. Yeung, Nature **373**, 681 (1995)

11 The Characterization of a Transmembrane Receptor Protein by Fluorescence Correlation Spectroscopy

T. Wohland, K. Friedrich-Bénet, H. Pick, A. Preuss, R. Hovius, and H. Vogel

Fluorescence Correlation Spectroscopy (FCS) is a powerful technique for the *in vivo* and *in vitro* investigation of biomolecular interactions. It was first described almost three decades ago [8,13,14] and has found, especially in the last ten years, increased interest as a tool for screening of ligand–receptor interactions in fundamental research [18,12,24,31] and in drug development [19,20,1,30]. In this contribution, we use FCS for the characterization of a transmembrane receptor protein. Determined parameters include the molecular mass of the protein, the stoichiometry of binding, equilibrium constants of receptor–ligand interactions, and the lateral diffusion and organization of proteins on the surface of living cells.

The transmembrane protein characterized in this work is the 5-hydroxy-tryptamine receptor of type 3 ($5HT_{3A}$-R). This receptor belongs to the family of ligand-gated ion channels which include the muscle and neuronal nicotinic acetylcholine receptors, the γ-aminobutyric acid type A receptor, and the glycine receptor. The three-dimensional structure of the receptor is not known to atomic resolution, but electron microscopy has delivered low-resolution images of the protein. It has fivefold symmetry, and can be represented as a cylinder 11 nm in length and 8 nm in diameter [3]. The $5HT_{3A}$-R subunits can assemble into a homopentamer. A subunit has a molecular mass of 54 kDa. This structure implies for symmetry reasons that there are either one or five binding sites for ligands. Dose response curves for the activation of the receptor have given Hill coefficients from 1.8 to 2.8 [10], suggesting cooperative binding, but the evidence from such measurements was not conclusive.

The $5HT_{3A}$-R is of medical and pharmacological interest because several of its ligands show therapeutic effects. They have an influence on depression, anxiety, and migraine. Antagonists of the receptor prevent nausea and emesis in cancer patients undergoing chemotherapy or radiotherapy [10]. Therefore it is of great interest to find new methods, such as FCS, to characterize these important proteins both to better understand their function, and to screen compounds for potential new medicines.

11.1   Materials and Methods

11.1.1   Principle

In an FCS experiment, fluctuations of the fluorescence $\delta F(t)$ around the average fluorescence $\langle F(t) \rangle$ are measured, yielding information on molecular processes or molecular motions. The fluctuations of the fluorescence signal stem either from changes in the number of fluorescent particles, or from variations of the fluorescence quantum yield of the particles in the open probe volume, which is defined by the focal volume of a tightly focused laser beam. To analyze these fluctuations the autocorrelation function (ACF) of the fluorescence intensity fluctuations is calculated by

$$G(\tau) = \frac{\langle \delta F(t) \delta F(t+\tau) \rangle}{\langle F(t) \rangle^2}. \tag{11.1}$$

The angular brackets $\langle \rangle$ indicate a time average, $F(t)$ is the fluorescence signal as a function of time, and τ is the correlation time. Analytic solutions for this correlation function have been calculated and published elsewhere [8,27,22,29,17]. These solutions include systems with one or several components, and comprise possible triplets.

11.1.2   FCS Setup

The experimental setup schematically shown in Fig. 11.1 has been described in detail elsewhere [31].

The position of the laser focus inside the sample was controlled by an xy-translation microscope table (Scan IM 100x100/MCL-2, Maerzhaeuser, 35579 Wetzlar, Germany) and a piezoelectric microscope focusing drive for the z direction (PIFOC 721.10, Physik Instrumente, 76337 Waldbronn, Germany). All measurements were done at constant temperature using a thermostated perfusion chamber (Carl Zeiss AG, Oberkochen, Germany).

11.1.3   Receptor and Ligands

For measurements in solution, we used the purified murine $5HT_{3A}$-R solubilized in 0.4 mM of the detergent $C_{12}E_9$. All experiments were performed using a buffer containing 10 mM 4-(2-hydroxyethyl)piperazine-1-ethanesulfonic acid (HEPES, pH=7.4). The ligand GR-H (GR119566X obtained from the former Geneva Biomedical Research Institute: Glaxo Wellcome, Geneva, Switzerland) was labeled with one of the following fluorophores (Fig. 11.2): fluorescein, N-[7-nitrobenz-2-oxa-1,3-diazol-4-yl] (NBD), Rhodamine 6G, and the cyanine dye Cy5. The resulting compounds are referred to as GR-Flu, GR-NBD, GR-Rho, and GR-Cy5, respectively (for details on the labeling and purification procedures see [31]).

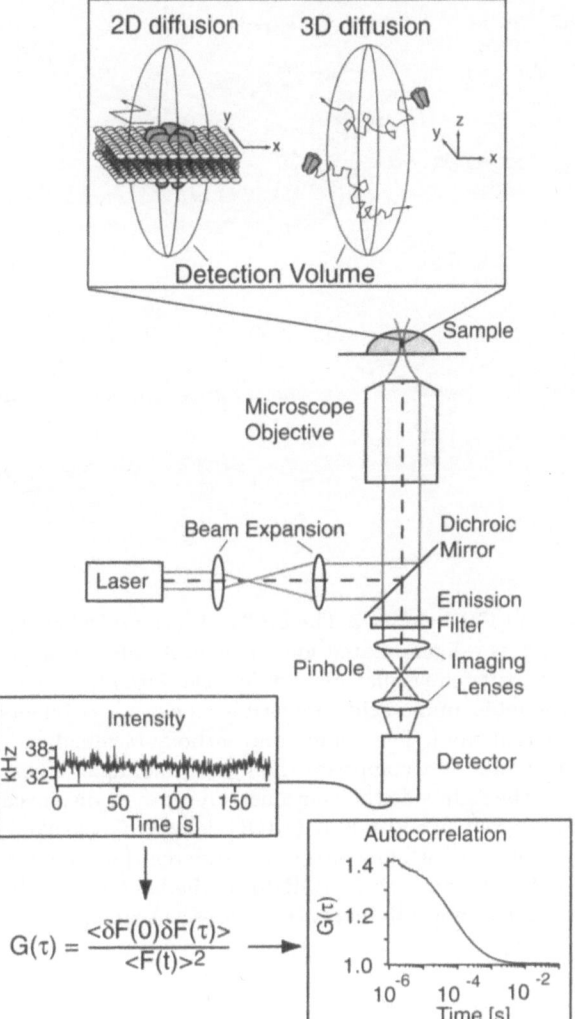

Fig. 11.1. A typical FCS experiment is shown. A laser is focused by a microscope objective and excites fluorophores in the sample. The emitted fluorescence is separated from the excitation light by a dichroic mirror and is spatially filtered by a pinhole. This is important to assure a small and defined detection volume. The light is then focused onto a detector which counts the incoming photons in fixed time intervals. This fluorescence signal is autocorrelated as indicated in the lower part of the figure. From the autocorrelation function, one can determine different parameters depending on the underlying physical process that causes fluctuations in the fluorescence intensity. By translational diffusion, fluorescent particles enter and leave the focal volume leading to fluctuations in the intensity signal. Thus lateral diffusion in membranes (2D diffusion) or in solution (3D diffusion) can be measured (top panel)

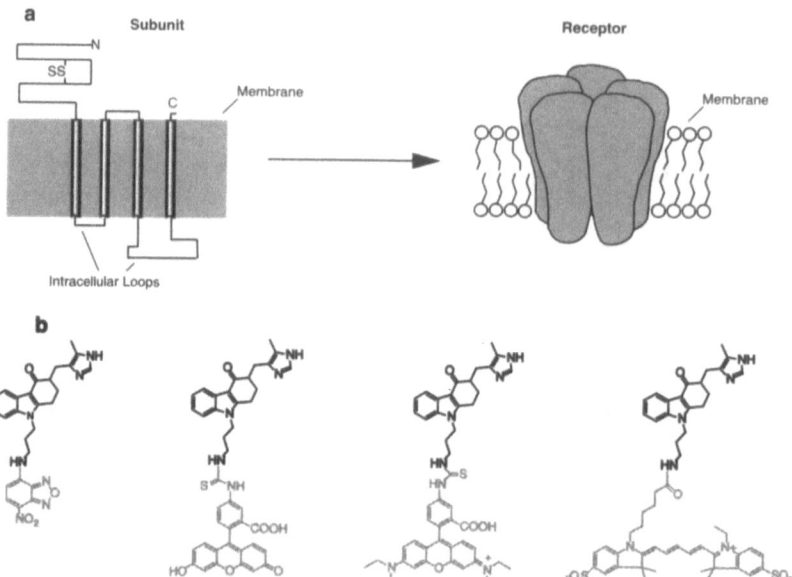

Fig. 11.2. 5HT$_{3A}$-R and fluorescent GR-ligands. **a** The 5HT$_{3A}$-R is a typical exam-ple of a neuroreceptor functioning as a ligand-gated ion channel. A subunit of the homopentameric receptor has a relative molecular mass of 54 kDa. From the distri-bution of hydrophilic and hydrophobic amino acids, four transmembrane segments are predicted within one subunit (left side). The amino- and carboxy-terminal ends are both extracellular. The C-terminal part comprises the ligand binding site near the indicated disulfide bridge. On the right side the pentameric receptor is displayed in a lipid bilayer. **b** The nonfluorescent ligand GR-H (=GR119566X, Glaxo Well-come, Geneva, Switzerland) was labeled with one of the four different fluorophores (from left to right): NBD (GR-NBD), fluorescein (GR-Flu), rhodamine B (GR-Rho), or Cy5 (GR-Cy5). The ligand is shown in black, the fluorescent labels in grey

The 5HT$_{3A}$-R was expressed either transiently in HEK293 cells or in a stable manner in a cell population created using the HEK293EBNA cells and the Epstein–Barr virus expression system (pCep4 plasmid, [2]) for measure-ments in biological cells.

11.1.4   Measurements in Solution

The setup was first calibrated using solutions of the fluorophores. Then, either 50 μl droplets of ligand solution (GR-Flu, GR-NBD, and GR-Rho) were deposited on a glass slide and receptor was titrated in small amounts to the droplet, or separate samples of particular ligand–receptor concentrations were prepared 1 h before the measurement, and then a 50 μl droplet of a particular solution was deposited on the glass slide (GR-Cy5). For every condition, 10 different measurements were performed and the results were averaged.

11.1.5   Measurements on Living Cells

The glass slides with the adherent cells were mounted on the microscope table. The cells were held at a temperature of 30–35°C if not stated otherwise. Depending on the measurement, the cells were either pre-incubated with a fluorescently labeled ligand for 15 minutes (GR-Cy5, 1.2 - 12 nM), or a membrane probe (Cy5-C18, R18: Cy5 or Rhodamine B linked to stearyl amine) was added directly to the solution. Because these membrane probes are hydrophobic, they integrate spontaneously into the cell membranes. The final probe concentration (between 1 and 10 nM) was optimal for FCS measurements avoiding too many fluorophores in the focal spot (numbers were usually held between 0.5 and 10 fluorescent particles per focal spot on the plasma membrane).

Keeping the laser blocked by a shutter to prevent photobleaching, cells which were not, or hardly, in contact with other cells were selected and placed at the center of the field of view, where the laser focus is located (the position of the laser focus can be visually determined by a highly concentrated solution of fluorophores). After defocusing, the shutter was opened, so that the actual laser focus was below the glass slide outside of the sample chamber. Raising the microscope objective scans the laser focus through the glass slide (both reflections from the first and second surface of the glass slide can usually be seen) and enters the sample chamber (Fig. 11.3). Following the second reflex from the glass slide a peak from the cell membrane close to the glass slide can be seen. Scanning further into the sample chamber a second peak can be observed. We attribute this second peak to the cell membrane facing the solution. All measurements were made on this membrane to avoid any influence of the glass slide on the diffusion of proteins in the membrane via nonspecific interactions.

11.2   Results and Discussion

11.2.1   Measurements in Solution

In FCS, several parameters have to be considered to fit the experimental ACFs correctly. First, the number of different components in solution must be addressed. Three components are present in the receptor/ligand solutions: the free ligand, the ligand–receptor complex, and the ligand partitioned into detergent micelles. In each case, the fluorophore is in a different environment and can therefore have different diffusion and fluorescence characteristics. Therefore, each component is defined in addition to its mole fraction by its diffusion coefficient and its fluorescence yield. The change of the fluorescence yield of the fluorophore in each different component has a decisive influence because the contribution of each component to the ACF depends quadratically on the fluorescence yield. The differences in the fluorescence yield can be measured with static fluorescence techniques ([18,31], and Table 11.1). When

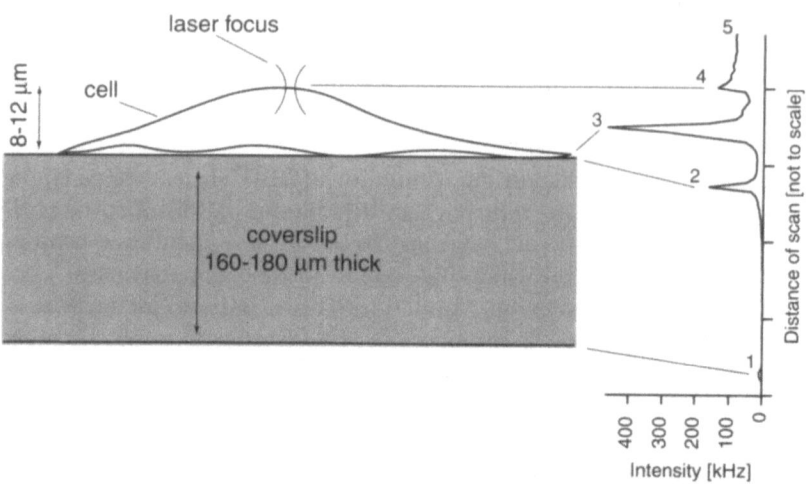

Fig. 11.3. Cells expressing the 5HT$_{3A}$-R are incubated with a fluorescent ligand. A scan of the laser focus through the glass slide and cell reveals different fluorescence intensity maxima. First the two reflections from the interface water/glass from the glass slide can be seen (1 and 2). Then the signal from the cell membrane close to the glass slide is detected (3). Inside the cell the fluorescence signal is low, then the cell membrane facing the solution comes into focus (4). Finally, the ligand in solution is detected (5). The distance between the lower and upper cell membrane was usually between 8 and 12 µm. The drawing is not to scale. While the difference between point 1 and point 2 is about 180 µm, the distance between points 2 and 3 is less than 1 µm

these factors are taken into account in the evaluation of the ACF, the equilibrium constants can be determined from the fractions of each component in the solution, and the molecular mass of the ligand–receptor complex can be estimated from their diffusion times.

The results of the binding measurements for the ligands are summarized in Table 11.2. The determination of equilibrium constants was not possible for GR-Rho for two reasons. Firstly, the ligand aggregated in the presence of the protein, leading to large peaks in the intensity trace and to very slow diffusion components in the ACF; secondly, the ligand adsorbed to the glass slide and was thus depleted from solution, making any determination of equilibrium constants unreliable. Measurement times were 30 s for GR-Flu, GR-Rho, and GR-Cy5, and 60 s for GR-NBD because of its low fluorescence yield.

Competition binding experiments between GR-Flu and granisetron (Fig. 11.5a) shows that granisetron has an IC_{50} of 10.9 ± 6.6 nM with a Hill coefficient of 0.61 ± 0.25, in agreement with published data [21].

To investigate the stoichiometry of ligand binding to the homopentameric 5HT$_{3A}$-R we monitored the number of fluorescent particles in solution as a function of 5HT$_{3A}$-R concentration. No change in the number of particles was seen for GR-Flu and GR-Cy5 upon addition of increasing amounts of recep-

Table 11.1. Experimental conditions and properties of the labeled ligands

Ligand	λ [nm][a]	P [mW][b]	Count rate[c] [kHz/particle]	τ_{exp} [µs][d]	ΔI [%][e]
GR-NBD	488	0.08	2.5	–	−50
GR-Flu	488	0.08	8	2–3	−70
GR-Rho	514	0.08	10	–	+25
GR-Cy5	633	7	120	1.2–2.5	–

[a] λ is the excitation wavelength used for the particular fluorophore.
[b] P is laser power used for excitation in mW. It was measured in front of the back aperture of the microscope objective and is therefore higher than in the focal plane due to incomplete transmission of the laser light through the objective.
[c] Fluorescence intensity measured as count rate per second and particle.
[d] τ_{exp} is the time constant of a fast molecular process that contributes to the ACF at short correlation times and shows an exponential correlation curve; it is a triplet in the case of GR-Flu, and a cis-trans transformation for GR-Cy5 [26].
[e] ΔI is the change of the fluorescence intensity in percent of the count rate when the ligand is bound to the receptor.

Table 11.2. Molecular masses and dissociation constants of the ligand–5HT$_{3A}$-R complex

probe	$K_d^{FCSa}(nM)$[a]	$K_d^{RBA}(nM)$[b]	$M_{C12E9}(MDa)$[c]	$M_{5HT3}(kDa)$[d]
R18	–	–	72 ± 15	–
GR-NBD	0.52 ± 0.77	0.5 ± 0.2	–	0.7 ± 0.7
GR-Flu	0.51 ± 0.24	0.32 ± 0.19	230 ± 75	0.7 ± 1.0
GR-Rho	–	0.8 ± 0.2	74 ± 18	0.7 ± 0.4
GR-Cy5	15.7 ± 8.0	18.0 ± 2.0	74 ± 33	0.5 ± 0.3

[a] K_d^{FCS} is the K_d determined by FCS.
[b] K_d^{RBA} is the K_d determined by radioligand binding assay.
[c] M_{C12E9} is the relative molecular mass of the $C_{12}E_9$ micelles determined by FCS.
[d] M_{5HT3} is the relative molecular mass of the 5HT$_{3A}$-R in micelles determined by FCS.

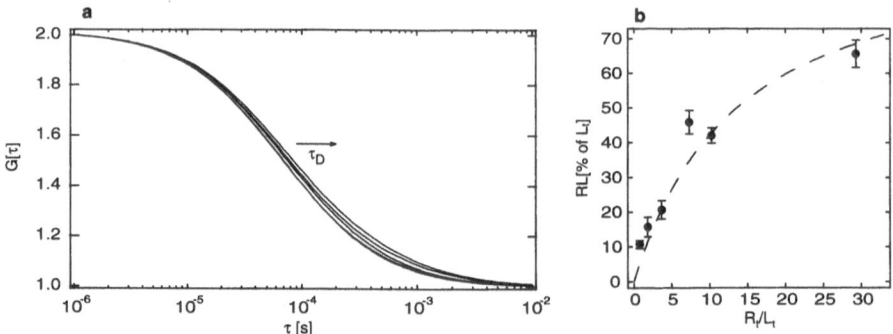

Fig. 11.4. a Three-component fits to the experimental ACFs (experimental data not shown) for GR-NBD (1.0-1.2 nM) with increasing concentrations of $5HT_{3A}$-R (0 nM receptor: *black* to 0.46 nM receptor: *light grey*). All fit functions were normalized for clarity. The average correlation time increased with the receptor concentration: 75.0 μs (0.0 nM, *black*) to 76.2 μs (0.05 nM), 84.9 μs (0.11 nM), 103.6 μs (0.25 nM), and 120.8 μs (0.46 nM, *light grey*), respectively. The inset shows an enlarged view of the ACFs, and how the correlation time changes with the increase of ligand–receptor binding. **b** Binding of the ligand GR-Cy5 to the $5HT_{3A}$-R. Solutions with a concentration of 1.2 nM of GR-Cy5 and various receptor concentrations between 0.9 and 35.2 nM were prepared and each measured separately. The ratio of the concentration of ligand–receptor complex RL to the total ligand concentration is calculated from the individual ACFs and plotted versus the ratio of total receptor and total ligand concentration (R_t/L_t). The experimental data were fitted to a simple binding equilibrium ($R + L \rightleftharpoons RL$) yielding a dissociation constant of $K_d = 15.7 \pm 8.0$ nM

tor (Fig. 11.5b). When the same experiment is performed with streptavidin and a fluorescein-labeled biotin, the concentration of fluorescent particles decreased by a factor of 4 due to the capacity of streptavidin to bind four biotin molecules. These data clearly show that only one fluorescent ligand binds per receptor homopentamer.

The measurements show that one has to choose fluorescent labels so that they are adapted to the particular experiment, and very likely several labels have to be used to yield a complete characterization of the protein. While rhodamine 6G and Cy5 both yield high count rates per residence time in the focus, have low non-diffusive contributions to the ACF, and are very photostable, they either induce receptor aggregation and make measurements unreliable or decrease the binding constant of the ligand, respectively. NBD yields very few photons per residence time, and thus necessitates longer measurement times. Furthermore, it was not possible to determine the stoichiometry reliably because of too low a signal-to-noise ratio. Fluorescein is in between the two extremes mentioned above. Its fluorescence yield is high enough to keep measurements short, but the error in the diffusion time is still very high (50%) and thus makes mass determinations unreliable.

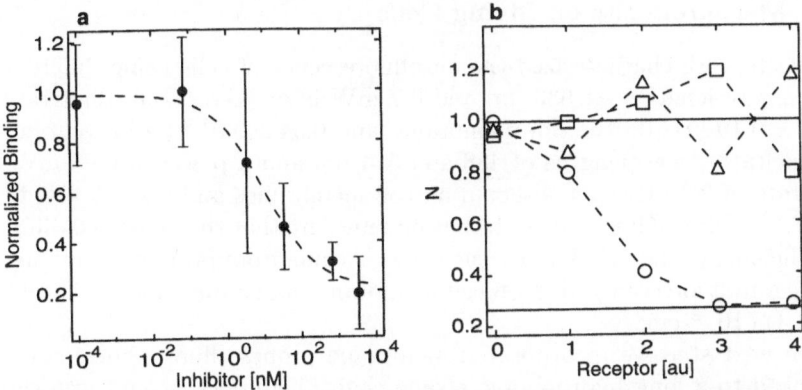

Fig. 11.5. a Competitive inhibition of granisetron for the binding of GR-Flu to the 5HT$_{3A}$-R. The concentration of granisetron was varied between 0.1 pM and 3.5 μM. The binding data was fitted to a binding isotherm [5], yielding an $IC_{50} = 10.9 \pm 6.6$ nM and a Hill coefficient of 0.61 ± 0.25 (*dashed line*). The 5HT$_{3A}$-R and GR-Flu concentrations were 8.8 and 1.75 nM, respectively. **b** Determination of the stoichiometry of ligand binding to the 5HT$_{3A}$-R. The concentration of fluorescent particles was measured for samples of GR-Flu or GR-Cy5 with increasing concentrations of 5HT$_{3A}$-R. For GR-Flu (*squares*), the concentration of the ligand varied from 1.9 to 1.5 nM, the receptor concentration increased from 0.0 to 0.85 nM, in turn the bound ligand increased from 0 to 40% of total ligand concentration. Relative errors are smaller than 15% for all measurements. In the case of GR-Cy5 (*triangles*), the ligand was held constant at 1.2 nM, and the receptor concentration varied from 0.0 to 35.2 nM, with a concomitant increase of ligand binding from 0.0 to 65.7% of total ligand concentration. Relative errors are smaller than 3% for all points. No significant decrease in the particle number of the fluorescent ligands was observed. In a control experiment (*circles*), biotin–fluorescein (3.6 nM) was titrated with increasing concentrations of streptavidin (0 to 1.8 nM), resulting approximately in a fourfold decrease of the number of fluorescent particles in solution at 0.9 nM streptavidin. The relative error was in the 20% range. For GR-Flu and biotin–fluorescein the number of particles is corrected for changing ligand concentrations during titration. The abscissa shows the number of titration steps, with increasing receptor concentration from left to right. The *solid lines* give the reference values for the stoichiometric ratio between receptor and ligand N=1 and N=0.25. All data points are an average of 10 experiments. All curves were normalized to their initial value. Error bars were omitted for clarity

The molecular mass determined for the receptor is too high because we use a very simple model assuming the receptor to be a sphere. As already mentioned in the introduction, the receptor more resembles a cylinder with a channel along its symmetry axis. Because such a structure diffuses slower than a sphere of the same mass [7] our estimate is an upper limit but yields the right order of magnitude.

11.2.2   Measurements on Living Cells

As a first step, we characterized the autofluorescence of cells using different excitation wavelengths. At 633 nm and 0.7 mW laser power, the count rate was 0.57 ± 0.19 kHz on the cell membranes and 0.34 ± 0.01 kHz in solution. At the excitation wavelengths of 488 and 514 nm and a power of 0.08 mW, a count rate of 2.6 kHz was observed on cell membranes and 0.2–0.5 kHz in solution. A stable cell line was used, thus making sure that the measured diffusion coefficients of the autofluorescence always come from protein-expressing cells. The autofluorescence seen on cell membranes had a diffusion coefficient of $(2.6\text{–}4.4) \cdot 10^{-8}$ cm^2/s.

In the next step we incorporated membrane probes (fluorophores covalently linked to a long hydrophobic alkane chain, here octadecane) into the cell membrane to measure the diffusion coefficient of lipids and lipid-sized molecules. We used two membrane probes, either R18 (octadecyl rhodamine B chloride, Molecular Probes, Eugene, OR, USA) for excitation at 514 nm, or Cy5-C18 for excitation at 633 nm. In the case of R18 we measured a diffusion coefficient $D = (2.6 \pm 2.1) \cdot 10^{-8}$ cm^2/s at 35°C. For Cy5-C18 measurements yielded a value of $D = (0.8 \pm 0.5) \cdot 10^{-8}$ cm^2/s at 35°C. To test whether small differences in the diffusion coefficient could be reliably measured, we cooled the cells down to 15°C. At this temperature it was noted earlier [6] that diffusion was slowed down by a factor of 2 in soybean lipids. Measurements with Cy5-C18 showed at 15°C a diffusion coefficient $D = (0.4 \pm 0.2) \cdot 10^{-8}$ cm^2/s.

A comparison of the diffusion coefficients of the autofluorescence and of the membrane probe shows that the autofluorescence stems either from small molecules that have integrated into the lipid membrane or from molecules that are membrane-associated. The diffusion constant of the autofluorescence is too fast to be related to transmembrane proteins which move much slower, as we will see below.

FCS measurements of GR-Cy5 bound in the presence of HEK293 cells resulted in ACFs best fit by 3 components (Table 11.3). Confocal images showed that the ligand bound specifically to the cell surface and could be inhibited with an excess of the competitor quipazine (data not shown). The fastest correlation time was of the order of one to a few hundred µs. The corresponding diffusion coefficient ($D_1 \sim 10^{-7} - 10^{-6}$ cm^2/s) corresponds to the ligand free in solution. It can be ruled out that this contribution to the ACF stems from intracellular compartments because as shown in Fig. 11.6, inside the cell no correlation could be observed at an excitation wavelength of 633 nm.

The second correlation time was in the range 1–10 ms ($D_2 \sim 10^{-9}$–10^{-8} cm^2/s). This is very close to the range of the membrane probes and is probably caused by nonspecific binding of some of the GR-Cy5 ligands to lipids or membrane-associated molecules. Because of its low intensity at an excitation wavelength of 633 nm, autofluorescence of the cells can be excluded as the source of this correlation time. For both components, D_1 and D_2, the

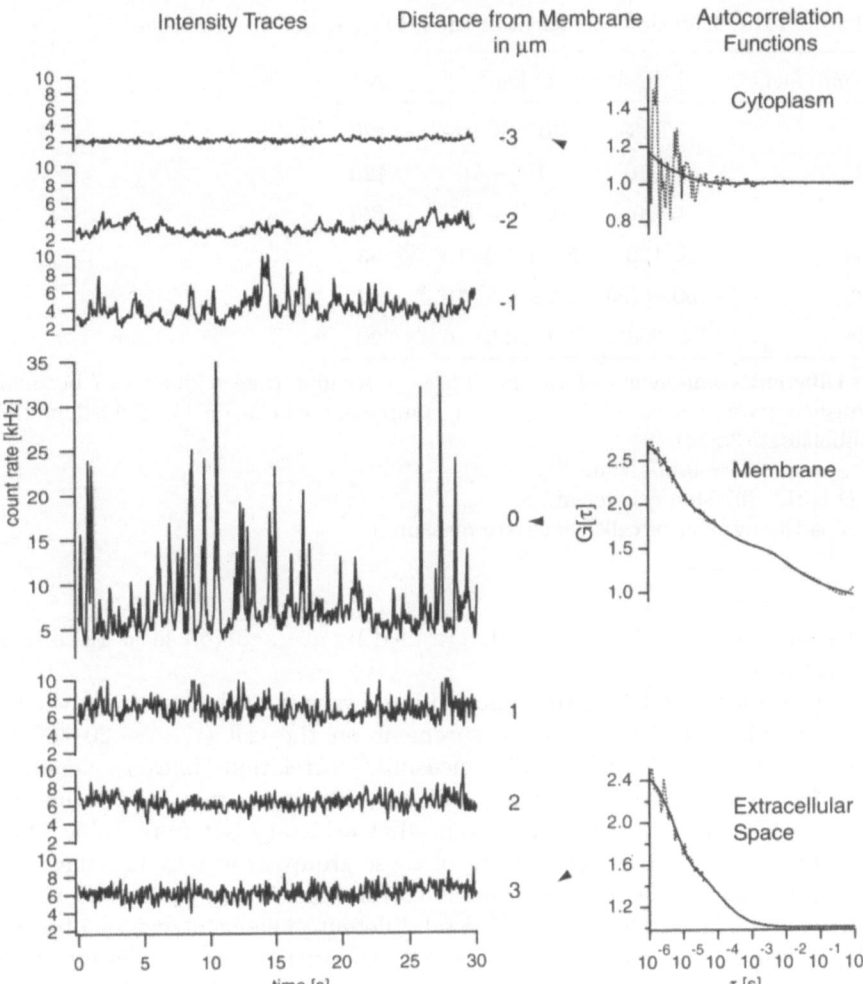

Fig. 11.6. The focus of a laser beam is scanned in micrometer steps across the plasma membrane of a HEK293 cell which expresses the 5HT$_{3A}$-R. The receptors were labeled by binding of GR-Cy5 (1.2 nM in solution). Starting with the focus inside the cell, the fluorescence is low (2 kHz) and no correlation can be seen (-3 µm). When the focus approaches the membrane the intensity rises and the fluctuations become larger (-2 to -1 µm). With the focus on the membrane (0 µm) photon bursts that originate from single receptor molecules or small receptor aggregates can be clearly detected. The correlation shows a fast part corresponding to the diffusion of the ligand in solution and a slower one corresponding to the membrane diffusion. When moving the focus further into the solution (1–3 µm), the diffusion of the free ligand can be seen and the autocorrelation function only contains the fast diffusing component. The excitation wavelength was 633 nm and the laser power 0.7 mW

Table 11.3. Diffusion coefficient of the $5HT_{3A}$-R in cell membranes

component[a]	τ_D^b [ms]	D [cm^2/s]^c	N^d
1	< 0.3	$10^{-7} - 10^{-6}$	120
2	$1 - 10$	$10^{-9} - 10^{-8}$	120
3	> 20	$10^{-10} - 10^{-9}$	120
3a	< 100	$5.5 \pm 1.0 \cdot 10^{-9}$	83
3b	$100 - 200$	$2.6 \pm 0.5 \cdot 10^{-9}$	17
3c	> 200	$0.9 \pm 0.5 \cdot 10^{-9}$	20

[a] Different components of the fit. Three particular time regimes can be distinguished (components 1-3). The third component can again be divided in three subclasses (3a-3c).
[b] τ_D is the correlation time.
[c] D is the diffusion coefficient.
[d] N is the number of cells that were measured.

relative error was 50% or less, determined by averaging at least 10 different measurements.

The third correlation time showed large variations depending on the particular cell and the site of measurement on the cell ($\tau_{D3} \sim$ 20–400 ms, $D_3 \sim 10^{-10}$–10^{-9} cm^2/s). The measured correlation times τ_{D3} could be divided in three groups: $\tau_{D3a} \leq 100$ ms, 100 ms $\leq \tau_{D3b} \leq 200$ ms, and $\tau_{D3c} \geq 200$ ms. This division is somewhat arbitrary but from Table 11.3 it can be seen that the relative error of these groups is usually less than 43%, while the averaging of all measured correlation times τ_{D3} lead to relative errors much larger than 100%. Several different phenomena can be observed, depending on the measurement. 1) For observations on a single membrane spot, τ_{D3} did not change from measurement to measurement, but could be attributed to one of the three regimes (τ_{D3a}, τ_{D3b}, or τ_{D3c}). 2) For measurements on an individual cell with the laser focused on different sites on the cell membrane, τ_{D3} varied strongly and could not be attributed to one single time regime. We attribute this third diffusion coefficient to receptor-bound GR-Cy5, because it is too slow to be explained by diffusion of the ligand in solution or diffusion of lipids. Analyzing different parts of the cellular membrane, τ_{D3} ranged over one order of magnitude, 0.6–6.9·10^{-9} cm^2/s. Nonetheless, the measurement errors on single spots are around 50% and can therefore not explain this large range of diffusion coefficients. In addition, measurements taken at different spots on one cell or taken randomly over several cells (one measurement per cell) can clearly be assigned to one of the different categories. While this classification does not mean that there are three different kinds of domains in the cell membranes, it clearly shows

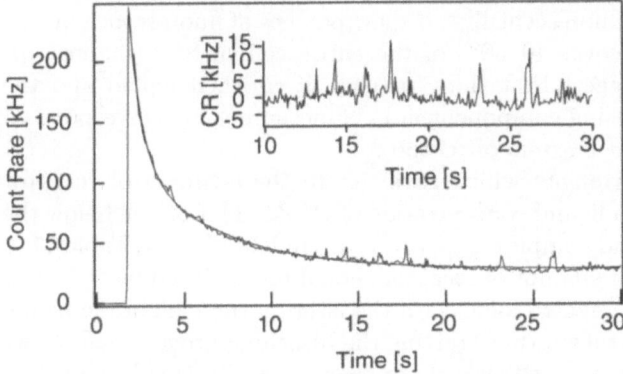

Fig. 11.7. Bleaching curve on the cell membrane. Strong photobleaching can be seen on some sites on the cell membrane. The decay can be fitted by a two exponentials. After subtracting the calculated double exponential curve from the experimental intensity trace, there remains a flat curve showing typical fluctuations as expected for diffusing receptors on a cell membrane (inset). The GR-Cy5 concentration was 12 nM and the laser power was set to 0.7 mW. The time constants of the fit were $t_1 = 0.5$ s and $t_2 = 3.4$ s, and the fraction of the fast component was 6 times higher than the slow one

that the diffusion of the 5HT$_{3A}$-R is not homogeneous over the whole cellular membrane.

Apart from these normal diffusion measurements in which the count rate did not change over the measurement time, there were other experiments which clearly showed photobleaching (Fig. 11.7). In these measurements a double-exponential decay of the count rate was observed which had time constants of 0.4 ± 0.5 s and 3.7 ± 1.7 s, the short-lived species being more abundant by a factor 6–10. Superimposed on this decay were distinct but correlated fluctuations of the fluorescence signal. The two time constants seen for the photobleaching might be due to the environment of the fluorescent ligand [25]. Different conformations of the receptor could change the accessibility of the fluorophore to the surrounding medium.

The maximum count rates registered on these sites of photobleaching were about 2700 kHz, much higher than the count rates obtained in experiments in which no photobleaching could be seen (5–90 kHz). Considering that GR-Cy5 molecules yield about 6–16 kHz (at 0.7 mW) in the membrane, a count rate of 2700 kHz corresponds to 169–450 bound GR-Cy5 molecules in the focal spot. This leads to a number of 500–1400 labeled receptors per µm^2.

In addition to the photobleaching, diffusion is still observed during these experiments, so we conclude that at least some of the receptors are immobile or diffuse so slowly that they are bleached. We attribute this immobile fraction to the clustering of the 5HT$_{3A}$-R, as supported by other observations. Confocal microscopy images of 5HT$_{3A}$-R expressing HEK293 cells incubated with the same fluorescent ligand show patches of fluorescence on the cell membranes which cover 40–50% of the entire cell surface (manuscript in preparation). We suggest that these patches of enriched ligand and the immobile or slowly diffusing component in FCS measurements correspond to each other and represent clusters of receptor.

The bleaching experiments which gave rise to the estimate of receptor density were done with a ligand concentration of 12 nM. This is well below the K_d of the receptor–ligand complexes, which is around 18 nM; see Table 11.2. Therefore, the estimated amount of receptor should be increased by at least a factor of 2. Moreover we had to focus with the laser on the membrane before measurements could be taken, thus starting the bleaching process well before the first intensity was measured. We think that especially this last process shifts the apparent receptor number to much lower values and our estimate is therefore a lower limit. Assuming that this process photobleaches at least half of the labels before measuring, the receptor density in the clusters can be higher than 2000–6000 receptors per μm^2.

11.3 Conclusions

From this report, the following general conclusions can be drawn for the characterization of membrane proteins by FCS:

FCS is suitable for studying ligand–receptor interactions in detergent solution and under physiological conditions, i.e. on surfaces of biological cells. The equilibrium constants of such binding reactions can be determined by FCS with accuracy comparable to that of other traditional methods, but with much higher efficiency: typical measurements are performed in seconds to minutes using few microliters of sample volume. In addition, FCS delivers important information on molecular interactions which can at best only be obtained with great difficulty, e.g. the stoichiometry of ligand–receptor complexes or the lateral organization of proteins on cellular membranes.

In the case of the homopentameric $5HT_{3A}$-R it was particularly interesting to obtain by a model-free approach a 1:1 stoichiometry for the complex between an antagonist and the receptor. This implies that either the antagonist binds to the symmetry axis of the ion channel, or if it binds to one subunit, the protein undergoes an allosteric conformational change that strongly affects the affinity of the other binding sites.

Receptor clustering on cell surfaces was determined *in vivo*, yielding lower limits of the receptor density in these patches in the range of 2000–6000 receptors per μm^2. These values are close to the experimentally determined receptor densities of other neuroreceptors, such as the nicotinic acetylcholine receptor in neuromuscular junctions, with about 9000 receptors per μm^2 [15].

Using novel technologies for the specific *in vivo* incorporation of fluorescent labels into key molecules [23,9,4], FCS makes it possible to investigate particular molecular interactions under physiological conditions like those in living cells.

Acknowledgments. We thank Horst Blasey (Serono Pharamaceutical Research Institute, Geneva, Switzerland) for delivering the stably expressing cell population and Zeptosens AG (Witterswill, Switzerland) for the gift of the Cy5-C18. The financial support by the Swiss National Science Foundation and the EPFL is gratefully acknowledged.

References

1. M. Auer, K. J. Moore, F. J. Meyer-Almes, R. Guenther, A. J. Pope, and K. A. Stoeckli, "Fluorescence correlation spectroscopy: Lead discovery by miniaturized HTS," Drug Discovery Today **3**, 457–465 (1998)
2. H. D. Blasey, R. Hovius, H. Vogel, and A. R. Bernard, "Transient-expression technologies, their application and scale-up: 5HT$_3$ serotonin receptor case study," Biochem. Soc. Trans. **27**, 956–960 (1999)
3. F. G. Boess, R. Beroukhim, and I. L. Martin, "Ultrastructure of the 5-hydroxytryptamine 3 receptor," J. Neurochem. **64**, 1401–1405 (1995)
4. A. Cha, G. E. Snyder, P. R. Selvin, U. Meseth, and F. Bezanilla, "Atomic scale movement of the voltage-sensing region in a potassium channel measured via spectroscopy," Nature **402**, 809–813 (1999)
5. Y. C. Cheng and W. H. Prusoff, "Relationship between the inhibition constant (K_1) and the concentration of inhibitor which causes 50 per cent inhibition (I_{50}) of an enzymatic reaction," Biochem. Pharmacol. **22**, 3099–3108 (1973)
6. R. M. Clegg and W.L.C. Vaz, in: *Progress in Protein-Lipid Interactions*, vol. 1, A. Watts and J. H. M. DePont (eds.) (Elsevier Science, Amsterdam, 1985)
7. H. G. Elias, *Macromolecules*, 2nd ed. (Plenum Press, New York, 1984)
8. E. L. Elson and D. Madge, "Fluorescence correlation spectroscopy. i. Conceptual basis and theory," Biopolym. **13**, 1–27 (1974)
9. B. A. Griffin, S. R. Adams, and R.Y. Tsien, "Specific covalent labeling of recombinant protein molecules inside live cells," Science **281**, 269–272 (1998)
10. M. Hamon (ed.), *Central and Peripheral 5-HT3 Receptors*, Neuroscience Perspectives (Academic Press Limited, London, 1992)
11. P. Kask, R. Gunther, and P. Axhausen, "Statistical accuracy in fluorescence fluctuation experiments," Eur. Biophys. J. **25**, 163–169 (1997)
12. J. Klingler and T. Friedrich, "Site-specific interaction of thrombin and inhibitors observed by fluorescence correlation spectroscopy," Biophys. J. **73**, 2195–2200 (1997)
13. D. Madge, E. L. Elson, and W. W. Webb, "Fluorescence correlation spectroscopy. ii. Experimental realization," Biopolym. **13**, 1–27 (1974)
14. D. Madge, W. W. Webb, and E. L. Elson, "Fluorescence correlation spectroscopy. iii. Uniform translation and laminar flow," Biopolym. **17**, 361–376 (1978)
15. J. A. Mathews-Bellingers and M. M. Salpeter. "Fine structural distribution of acetylcholine receptors at developing mouse neuromuscular junctions," J. Neurosci. **3**, 644–657 (1983)

16. R. M. McKernan, N. P. Gillard, K. Quirk, C. O. Kneen, G. I. Stevenson, C. J. Swain, and C. I. Ragan, "Purification of the 5-hydroxytryptamine 5HT$_3$ receptor from NCB20 cells," J. Biol. Chem. **265**, 13572–13577 (1990)

17. U. Meseth, T. Wohland, R. Rigler, and H. Vogel, "Resolution of fluorescence correlation measurements," Biophys. J. **76**, 1619–1631 (1999)

18. B. Rauer, E. Neumann, J. Widengren, and R. Rigler, "Fluorescence correlation spectrometry of the interaction kinetics of tetramethylrhodamin α-bungarotoxin with *Torpedo californica* acetylcholine receptor," Biophys. Chem. **58**, 3–12 (1996)

19. M. V. Rogers, "Light on high-throughput screening: fluorescence-based assay technologies," Drug Discovery Today **2**, 156–160 (1997)

20. S. Sterrer and K. Henco, "Minireview: Fluorescence correlation spectroscopy (FCS) – A highly sensitive method to analyze drug/target interactions," J. Recept. Signal Transduct. Res. **17**, 511–520

21. A.-P. Tairi, R. Hovius, H. Pick, H. Blasey, A. Bernard, A. Surprenant, K. Lundström, and H. Vogel, "Ligand binding to the serotonin 5HT$_3$ receptor studied with a novel fluorescent ligand," Biochemistry **37**, 15850–15864 (1998)

22. N. L. Thompson, in: *Topics in Fluorescence Spectroscopy, Volume 1: Techniques*, J. R. Lakowicz (ed.) (Plenum Press, New York, 1991)

23. G. Turcatti, K. Nemeth, M. D. Edgerton, U. Meseth, F. Talabot, M. Peitsch, J. Knowles, H. Vogel, and A. Chollet, "Probing the structure and function of the tachykinin Neurokinin-2 receptor through biosynthetic incorporation of fluorescent amino acids at specific sites," J. Biol. Chem. **271**, 19991–19998 (1996)

24. E. Van Craenenbroeck and Y. Engelborghs, "Quantitative characterization of the binding of fluorescently labeled colchicine to tubulin in vitro using fluorescence correlation spectroscopy," *Biochem.* **38**, 5082–5088 (1999)

25. S. Wennmalm and R. Rigler, "On death numbers and survival times of single dye molecules," J. Phys. Chem. B **103**, 2516–2519 (1999)

26. J. Widengren, *Fluorescence Correlation Spectroscopy, Photophysical Aspects and Applications*, Ph. D. thesis, Karolinska Institut, Department of Medical Biophysics, Stockholm, Sweden, 1996

27. J. Widengren, Ü. Mets, and R. Rigler, "Fluorescence correlation spectroscopy of triplet states in solution: A theoretical and experimental study," J. Phys. Chem. **99**, 13368–13379 (1995)

28. J. Widengren, R. Rigler, and Ü. Mets, "Triplet-state monitoring by fluorescence correlation spectroscopy," J. Fluorescence **4**, 255–258 (1994)

29. J. Widengren and R. Rigler, "Review - Fluorescence correlation spectroscopy as a tool to investigate chemical reactions in solutions and on cell surfaces," Cell. Mol. Biol. **44**, 857–879 (1998)

30. T. Winkler, U. Kettling, A. Koltermann, and M. Eigen, "Confocal fluorescence coincidence analysis – An approach to ultra high-throughput screening," Proc. Natl. Acad. Sci. USA **96**, 1375–1378 (1999)

31. T. Wohland, K. Friedrich, R. Hovius, and H. Vogel, "Study of ligand receptor interactions by fluorescence correlation spectroscopy with different fluorophores: Evidence that the homopentameric 5-hydroxytryptamine type 3As receptor binds only one ligand," Biochem. **38**, 8671–8681 (1999)

12 Applications of Dual-Color Confocal Fluorescence Spectroscopy in Biotechnology

A. Koltermann, U. Kettling, J. Stephan, M. Rarbach, T. Winkler, and M. Eigen

Classical applications of fluorescence spectroscopy detect emission which is collected from comparatively large ensembles of fluorescent particles, i.e. the signal is averaged over space and time. In contrast to this, confocal fluorescence methods restrict the probe volume to a tiny spot of less than one femtoliter, which is the size of a typical bacterial cell. The high spatial resolution can be accomplished by epi-illumination of a microscope objective with appropriate laser beams and confocal imaging of the collected emission photons onto the detector. A sufficiently high temporal resolution (down to the range of nanoseconds) is enabled through the use of avalanche photo diodes as sensitive single-photon detectors in combination with suitable data processing. Employing fluorophore concentrations of one nanomole per liter and below, on average less than one emitter resides in the femtoliter focal volume, and the temporal behavior of single molecules can be followed. The technique of dye-tagging molecules of interest then allows one to investigate biomolecular processes at the single molecular level.

In the early 1970s, fluorescence correlation spectroscopy (FCS) was introduced [4,5,8,13]. This method mainly focuses on the analysis of statistical fluctuations of fluorescent particles and determines concentrations and diffusion characteristics. During the last decade, its combination with the aforementioned confocal techniques and the synthesis of a great variety of dyes as efficient fluorescent probes resulted in a widely-used analytical tool which permits the observation of the dynamics of single biomolecules in real time [7,14]. Nowadays, FCS is applied by several laboratories and companies all over the world in basic research as well as for industrial applications like drug screening [1,7,18].

Recently, single-color FCS was extended to a dual-color cross-correlation scheme which shows several advantageous characteristics [7,17]. Dual-color FCS allows the tracing of two spectrally distinguishable fluorophores at the same time, and the cross-correlation analysis of the two signals gives access to number and time constants of correlated fluorescence fluctuations in the different emission ranges. The cross-correlation function of signals from freely diffusing fluorophores in an ideally calibrated system reads [17]:

$$G(\tau) = \frac{\langle C_{12}\rangle\, Diff(\tau)}{V_{eff}\,(\langle C_1\rangle + \langle C_{12}\rangle)\,(\langle C_2\rangle + \langle C_{12}\rangle)} \tag{12.1}$$

C_1, C_2, and C_{12} denote the concentrations of fluorophore 1, fluorophore 2, and double fluorescent particles (1 and 2), respectively. Here,

$$Diff(\tau) \equiv (1 + \tau/\tau_d)^{-1}\,\left(1 + r_0^2\tau/z_0^2\tau_d\right)^{-1/2} \tag{12.2}$$

describes the temporal decay due to molecular diffusion of the double fluorescent species with a diffusion time τ_d. Structural parameters r_0 and z_0 are determined by the experimental geometry, and V_{eff} is the effective measurement volume. Compared to single-color autocorrelation measurements, signal specificity and accuracy is strongly improved in this new method, i.e. even at large excess of background fluorescence in each channel, dual-color FCS makes it possible to specifically detect double fluorescent molecules. In this regard, a pioneering biochemical application will be summarized in the following section, a real-time kinetic study of the cleavage of a double-stranded DNA substrate by the endonuclease EcoRI [9].

The typical assay format investigated by the dual-color cross-correlation method is based on the distinction between double-labeled and single-labeled molecules. The special features of the technique enable the experimenter to examine every reaction that can be attributed to either formation or destruction of a linkage between two fluorophore-tagged (molecular) fragments. Consequently, its universality suggests adapting it to screening applications in biochemistry and cellular biology. This was implemented for the first time in Rapid Assay Processing by Integration of Dual-color Fluorescence Cross-correlation Spectroscopy (RAPID FCS) [11]. While conventional autocorrelation methods identify molecules by their diffusion properties, requiring a considerable amount of analysis time, RAPID FCS simply *counts* double-labeled molecules. Data collection times for precise determination of endonucleolytic activity lie in the range of one second and below, which corresponds to a screening throughput up to 10^5 samples per day. This will be briefly described in a later section.

Besides cross-correlation FCS, alternative algorithms for data processing of multicolor fluorescence signals emanating from single molecules are currently examined. Special attention is given to the extraction of accurate signals at shortest analysis times possible, thereby enhancing the throughput rate. Confocal Fluorescence Coincidence Analysis (CFCA) uses a coincidence algorithm, which is explained schematically in Fig. 12.1 [19]. This approach was combined with technical improvements, like modifications concerning the laser source and a controlled external enhancement of concentration fluctuation. As a result, sampling times in the range of 100 ms were achieved, which would allow sampling of approximately 10^6 variants per day. This work will be summarized in a later section.

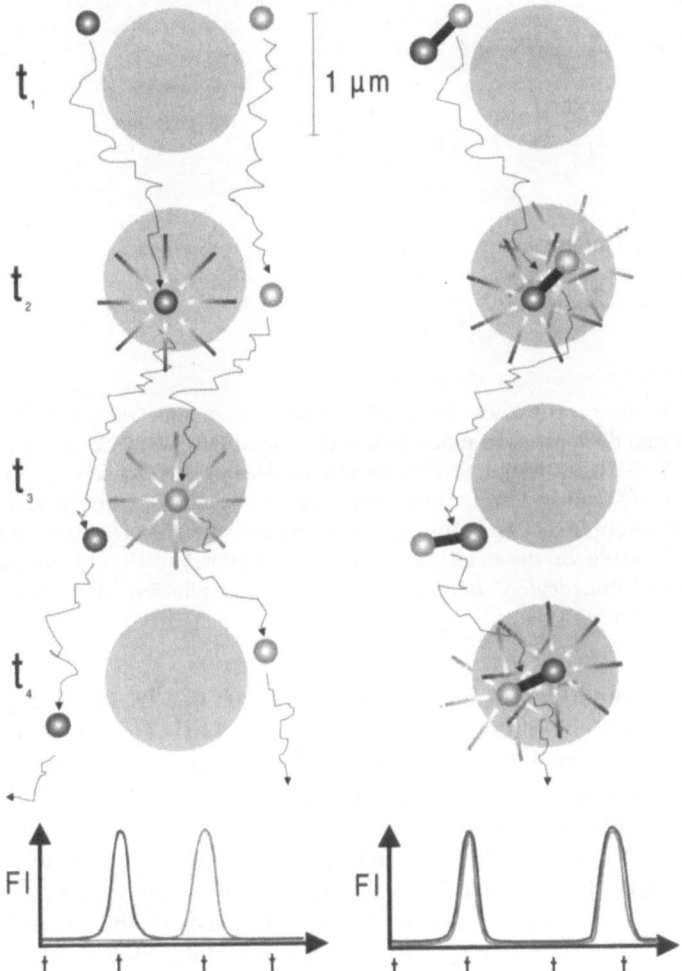

Fig. 12.1. The principle of CFCA analysis. *Left side:* Single-labeled molecules (light gray and dark gray) diffuse through the illuminated volume element. The occurrence of a green dye molecule is independent of the occurrence of a red one, and vice versa. Thus the photon events in one time trace are statistically distributed and completely independent from the events in the other time trace ($K = 1$). *Right side:* Both of the labeling dyes are connected to the same molecule, accordingly they jointly diffuse through the laser-illuminated area. In doing so, two kinds of spectrally different photons are emitted and detected in both time traces at the same time. The quota of these coincident events compared to the statistical value of independent signals of both colors increase the coincidence value ($K > 1$)

I. Covalent bond

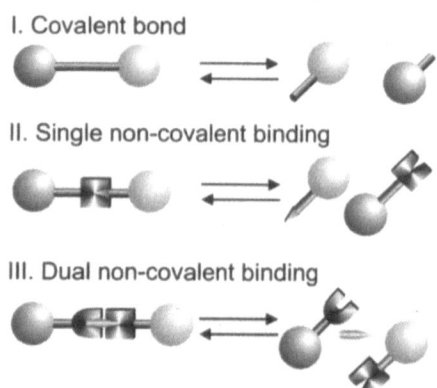

II. Single non-covalent binding

III. Dual non-covalent binding

Fig. 12.2. Different assay principles. Principles of employing spectrally separated fluorophores (light and dark gray) to monitor reactions by dual-color FCS or CFCA. *I. Covalent bond:* Substrate design for proteases, nucleases, esterases, and other cleavage or ligation reactions. The bond of the assay molecule can be broken or built between the fluorophores, depending on the enzyme employed. *II. Single non-covalent binding:* Binding or dissociation of receptor–agonist, antibody–antigen, and others. *III. Dual noncovalent binding:* Binding or dissociation of antibody-antigen–antibody and others

Figure 12.2 presents different assay principles which can be investigated by confocal dual-color techniques like dual-color FCS, RAPID FCS, or CFCA. Depending on the requirement of the application, the respective assay can be used for kinetic studies and/or (ultra-)high-throughput screening (HTS). In this article, the application of the methods used will be described in more detail for the example of endonucleolytic activity of the specific endonuclease EcoRI, one of a large variety of different enzymatic cleavage and ligation reactions which we have carried out thus far. On the basis of the presented results, the great potential of confocal dual-color methodologies for selection by screening strategies in evolutionary optimization processes is discussed.

12.1   Kinetic Studies of Enzymatic Activity in Real Time by Dual-Color FCS

Kinetic studies on enzymes are very significant for the understanding of biological interactions at the molecular level. In combination with new approaches in genetic engineering and structure determination, in recent years major efforts have been made to develop more sensitive and precise techniques for characterizing the kinetics of enzymatic reactions. With the help of these techniques, it was possible to develop efficient assays for analyzing catalytic parameters such as turnover rates, substrate specificity, and regio- and stereospecificity; they constitute a major part of the biochemical and pharmaceutical research. The dual-color cross-correlation method circumvents the

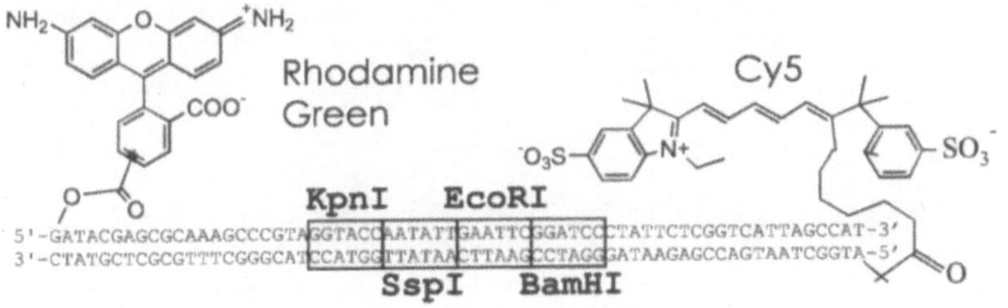

Fig. 12.3. The double-stranded DNA substrate. Sequence of the double-stranded DNA substrate with fluorophore labels Cy-5 and Rhodamine Green at its 5′-ends. The recognition sequences for different specific endonucleases are indicated in rectangles

necessity of evaluating the diffusion characteristics of the product fractions; therefore, simple mathematical evaluation is adequate [15,17]. The instrumental setup makes it possible to sensitively detect molecules which bear two spectrally separated fluorescent labels. The experiment is designed in such a way that the reaction step to be investigated converts double-labeled educts into single labeled products or vice versa. In the past few years, different assays have been designed such as hybridization kinetics [17] or prion aggregation [2]. Monitoring of enzymatic reactions by dual-color FCS was first applied by Kettling et al. [9]. In this work, the cleavage reaction of a double-stranded DNA molecule catalyzed by the restriction endonuclease EcoRI for analyzing enzyme kinetics was investigated. The DNA molecule was labeled with a red and a green dye at opposite ends (Fig. 12.3), and the catalyzed reaction was monitored online using a dual-color fluorescence cross-correlation spectrometer prototype, which was described in detail in earlier work [15,17].

Cleavage of the double-stranded DNA substrate by EcoRI breaks the chemical linkage between the two different fluorophores, resulting in loss of the cross-correlation signal. The time course of this enzymatic reaction can be monitored in solution precisely and homogeneously by dual-color FCS (Fig. 12.4).

The fluorescence signals can be measured continuously, and the cross-correlation analysis is usually carried out at a rate of one per minute. During the cleavage reaction, the fraction of the cross-correlation curves corresponding to the concentration of the double-labeled DNA substrate decreased successively; the second important parameter – the average diffusion time of cross-correlating entities – remained constant, indicating a highly specific detection of the double-labeled DNA substrate by this method. FCS measurements usually require pico- to nanomolar concentrations of fluorescent molecules for optimal correlation analysis. Here, a kinetic analysis with broader ranges of substrate concentrations was achieved by adding unlabeled substrate to the reaction sample. Cleavage kinetics using different ratios of labeled to unlabeled substrate ranging from 0.05 to 0.5 were obtained by plot-

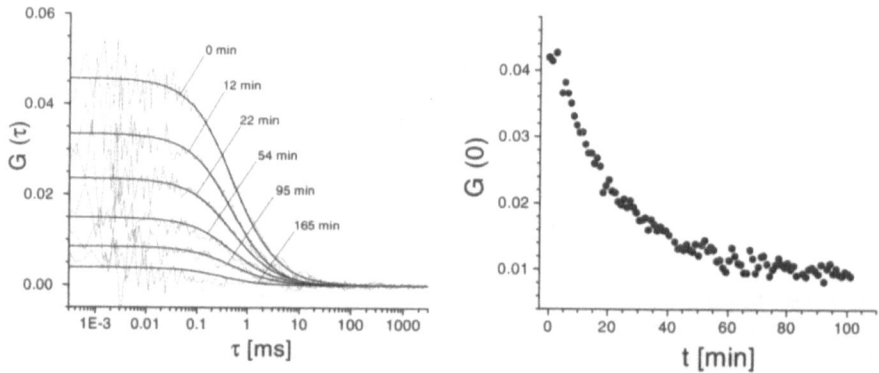

Fig. 12.4. Cross-correlation curves at different times during the endonucleolytic EcoRI cleavage reaction. 10 nM labeled DNA, 80 nM unlabeled DNA, and 1.6 nM EcoRI were incubated in the reaction buffer at 27°C. *Left side:* The dotted lines represent the original data, the fitted curves are given by solid lines. During the reaction, the cross-correlation amplitude G(0), which is a measure of the reaction progress, gradually decreases. *Right side:* Plot of the cross-correlation amplitude at $\tau = 0$ (G(0)) over the macroscopic time of the enzymatic cleavage reaction

ting the evaluated substrate concentrations versus time (results not shown here). These measurements revealed identical kinetics, providing evidence that the fluorophores attached to the ends of the DNA molecule did not interfere with the enzyme's catalytic action; therefore, the labeled substrate served as a one-to-one indicator.

The ability of cross-correlation analysis to sensitively quantify enzyme activity is shown in Fig. 12.5. With an initial substrate concentration of 0.8 nM (below the K_M), a linear relation between initial reaction rates and enzyme concentrations was obtained over a concentration range of more than one order of magnitude (Fig. 12.5, inset). Enzyme activity was detected down to 1.6 pM. The extremely low enzyme activities detected by dual-color FCS demonstrate the enormous sensitivity of this technique.

The class II restriction endonuclease EcoRI catalyzes the cleavage of the phosphodiester bond between the guanosine and the adenosine monomer of the palindromic recognition sequence GAATTC in each of the two strands. Dual-color FCS detects cleavage reactions by observing the separation of two different fluorescent labels, resulting from scissions in both strands of a single molecule; consequently, the kinetics that is observed represents an overall reaction rate. Cleavage reactions were monitored over a wide range of substrate concentrations (from 1 to 130 nM) by adding a defined amount of unlabeled to a constant amount of labeled substrate, the latter serving as an indicator. From a linear regression of initial slopes at different substrate concentrations it was confirmed that EcoRI catalysis obeys the Michaelis–Menten equation (Fig. 12.6). From an Eadie–Hofstee plot (Fig. 12.6, inset)

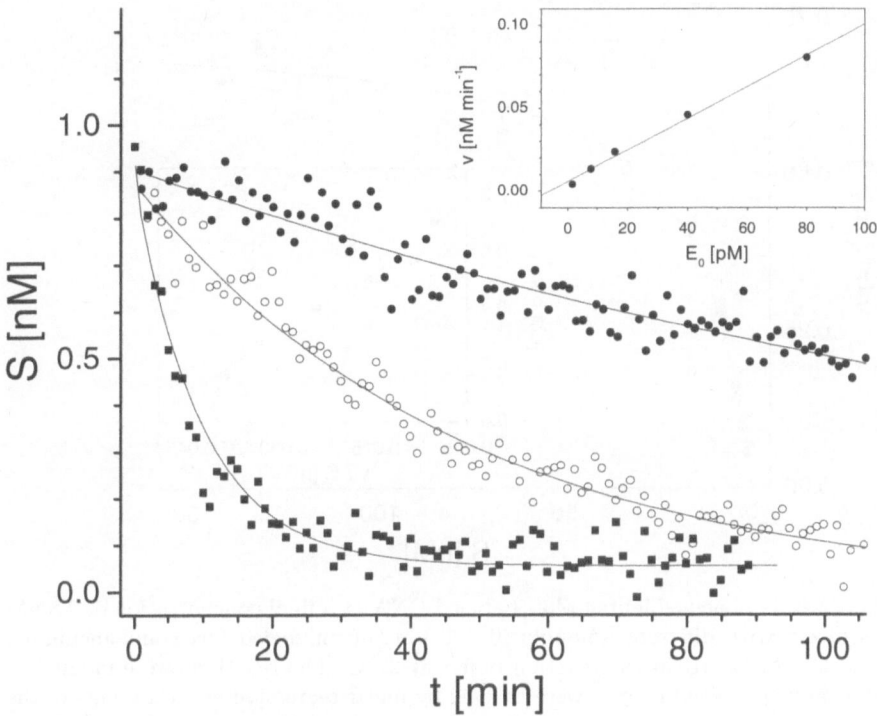

Fig. 12.5. Reaction rates at different enzyme concentrations. Endonucleolytic reactions were carried out in the reaction buffer at 27°C with 0.8 nM double-labeled DNA substrate and different enzyme concentrations ranging from 1.6 to 80 pM. Time courses are shown for $E_0 = 1.6$ pM (solid circles), 8 pM (open circles), and 80 pM (solid squares). *Inset:* Initial rates v were calculated from linear regression of the data points of the first 5–50 minutes; the plots of these initial rates versus enzyme concentrations E_0 indicate a clearly linear relationship between v and E_0

a K_M of (14 ± 1) nM and a k_{cat} of (4.6 ± 0.2) min^{-1} were derived, and the suitability of dual-color FCS for measuring kinetics in the nanomolar range was confirmed.

12.2   Dual-Color Confocal Fluorescence in High-Throughput Screening

Dual-color cross-correlation spectroscopy has several outstanding characteristics, like signal specificity and precision. These attributes, combined with its general applicability and compatibility with biological environments, make it a well suited tool for high-throughput screening in the biological disciplines. Termed Rapid Assay Processing by Integration of Dual-color Fluorescence Cross-correlation Spectroscopy (RAPID FCS), short analysis time of samples

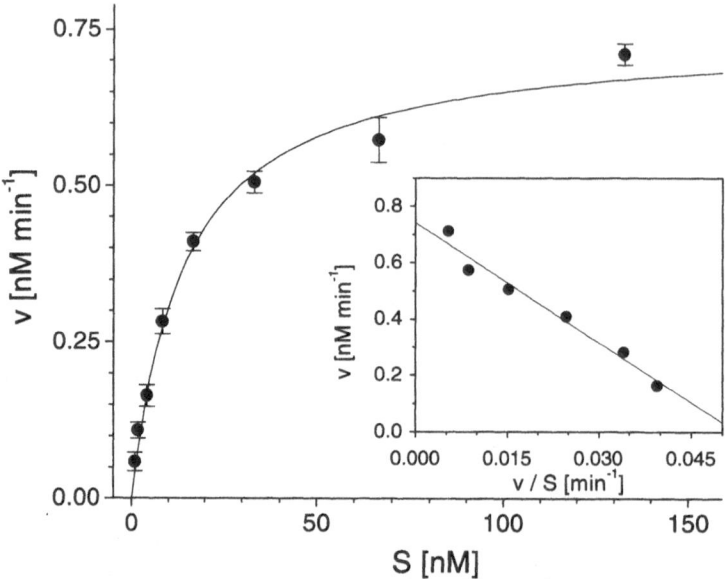

Fig. 12.6. Michaelis–Menten plot. Labeled DNA at a final concentration of 0.8 nM was mixed with different amounts (0–130 nM) of unlabeled DNA and incubated with 160 pM EcoRI in the reaction buffer at 27°C. The reactions were monitored online and the initial rates v were derived by linear regression of data points of the first 5–20 minutes. *Inset:* Calculations from an Eadie–Hofstee plot lead to a K_M value of (14 ± 1) nM and v_{max} of (0.74 ± 0.03) nM min^{-1}

was demonstrated with the endonucleolytic cleavage assay described in the preceding section [11]. A simulated HTS with homogeneous assays for restriction enzymes (EcoRI, BamHI, and SspI) which cleave the double-stranded DNA substrate (Fig. 12.3), as well as a restriction enzyme without a recognition site in the substrate (HindIII), was carried out with sample volumes of a few microliters. Several experimental parameters, such as substrate concentration or instrument parameters of the confocal set-up, were optimized in order to achieve shortest sample analysis times, and endpoint determinations for enzymatic activity were screened with different analysis times. The lower limit of analysis times was between 1 and 2 seconds, depending on the chosen tolerance for false positive or false negative precise yes-or-no decisions. This corresponds to a screening throughput rate of 10^4 to 10^5 samples per day. Due to the listed features, RAPID FCS is a suitable tool for screening applications with moderate throughput rates. Furthermore, Koltermann et al. [11] indicate that a combination of dual-color FCS with nanotechnology exhibits great potential for progressive selection strategies in evolutionary biotechnology, where rare and specific binding or catalytic properties must be screened from large numbers.

In contrast to conventional confocal correlation methods which evaluate diffusion parameters, dual-color FCS exploits the amplitude $G(0)$ of the cross-correlation curve (see (12.1)); i.e. RAPID FCS "counts" double-labeled molecules. In order to extract this number in a more efficient fashion, which leads to shorter sample analysis times, several alternative data processing algorithms were investigated in our group. Winkler et al. [19] presented the Confocal Fluorescence Coincidence Analysis (CFCA). This method utilizes a normalized dot product algorithm which is applied to the simultaneously recorded time traces of the two detector channels:

$$K(n) = \frac{\sum_m N_1(m)N_2(m)}{\sum_m N_1(m) \sum_m N_2(m)} * n \,. \tag{12.3}$$

$K(n)$ represents the coincidence value as a measure of the frequency of coincidence events in the two detection channels. $N_1(m)$ and $N_2(m)$ are the number of counts in the different emission ranges in time channel m, and n is the total number of time channels in the trace. The analysis times are determined by the time-channel width and the number of channels n. This data evaluation procedure was implemented online, and the readout parameter was a single number per measurement without any further fitting routine (as required in FCS diffusion analysis). The data recording technique was combined with improvements in the experimental setup: a high-speed motion of the sample volume relative to the focal volume was applied, and a single laser in multiline mode emitted both excitation wavelengths simultaneously. The CFCA setup is shown in Fig. 12.7.

Winkler et al. [19g] investigated and optimized essential experimental parameters of the CFCA method, e.g. time-channel width and frequency of oscillation. The performance of the method was demonstrated by means of the above described homogeneous assay for restriction endonuclease *EcoRI*. The minimum readout times achieved for precise yes-or-no decisions with regard to enzymatic cleavage activity amounted to 100 ms or less. Compared to RAPID FCS, a tenfold increase in sample analysis speed was obtained. Figure 12.8 shows the improvements using the CFCA setup and the applied relative motion between sample and focal volume.

These achievements break the ground for throughput rates as high as 10^6 samples per day using small amounts of sample substance, and therefore constitute a solid base for screening applications in drug discovery and evolutionary biotechnology.

12.3   Dual-Color FCS in Evolutionary Biotechnology

Designing biomolecules by evolutionary approaches opens up a new area in biotechnology. The idea to transfer nature's principles of Darwinian evolution, i.e. variation and selection, to the molecular level and to apply it to the directed evolution of molecules is the underlying concept of *evolutionary*

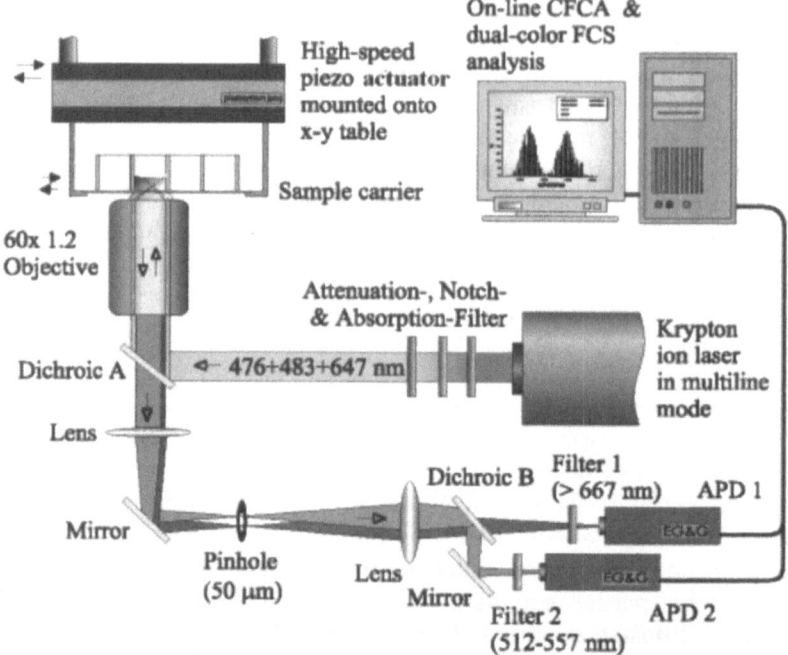

Fig. 12.7. The Confocal Fluorescence Coincidence Analysis (CFCA) setup. Fluorescence excitation was achieved by epi-illumination of a water-immersion objective with the output of a multiline laser. The sample holder was connected to a high-speed, two-dimension piezo actuator which in turn was mounted onto a high-precision x-y mechanical positioning table. Fluorescence photons were separated by dichroic beamsplitter B and, after filtering in the red and the green channel, were imaged onto two avalanche photodiodes (APD). Digital pulses from the APD were recorded by an on-board processor PC card which was programmed to perform online data processing

biotechnology [6]. Important prerequisites are a comprehensive understanding of the mode of molecular evolution as well as the ability to apply its principles to experimental systems in order to create and optimize molecular functions with scientific or economic value. In recent years, evolutionary approaches turned out to be the most successful techniques for molecular design if one deals with biopolymers of more than a few monomers [10,12]. Selection of variants with properties more similar to the intended phenotype than the average is one of the most crucial steps. The most common strategy for selection is to couple the desired molecular feature directly to the amplification rate of a replicator unit (*in vitro*) or the growth rate of a host system (*in vivo*). Many successful examples are reported in detail, which have shown the power of this nature-like strategy, where the desired molecular feature influences the amplification rate. Unfortunately, this strategy is strongly de-

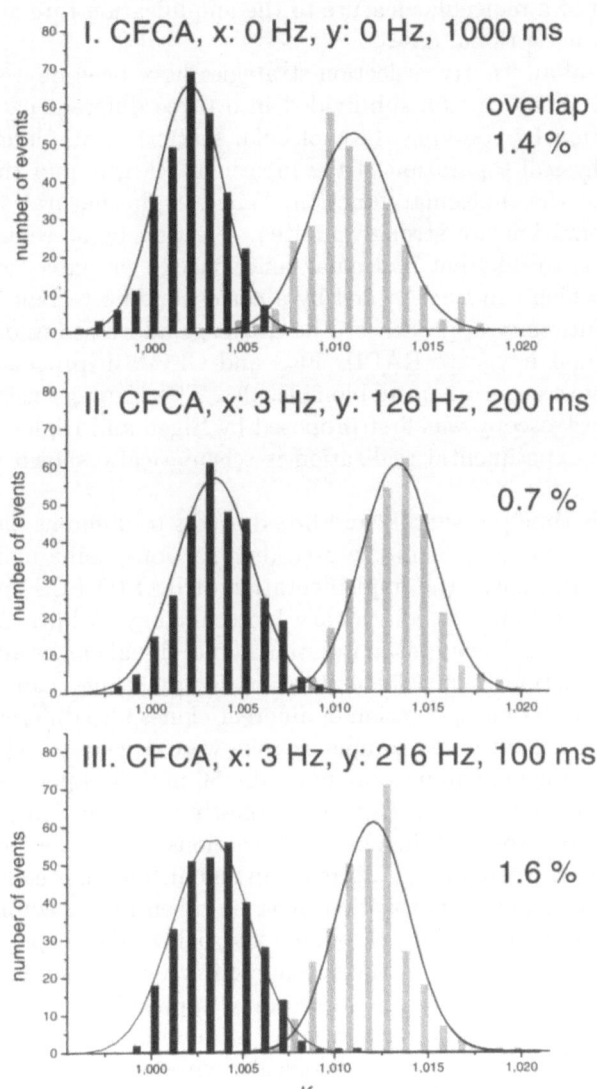

Fig. 12.8. Histogram plots of CFCA. Histogram plots of 2 × 300 coincidence measurements each at sample oscillation frequencies of 0 Hz (y) and 0 Hz (x) (upper plot), 126 Hz (y) and 3 Hz (x) (middle plot), and 216 Hz (y) and 3 Hz (x) (lower plot). Samples containing pure substrate (10-nM double-labeled dsDNA, black bars) and cleaved products (endonuclease-treated 10-nM double-labeled dsDNA, gray bars), respectively. The distributions were fitted with Gaussian functions and the respective overlap in percent between corresponding fit functions was calculated

pendent on the coupling of a molecular feature to the amplification rate and thereby limited to a few exceptional cases.

In recent years, several alternative selection strategies have been devised and technically realized. They can be subdivided into three different categories: (i) natural selection by coupling the molecular feature to amplification, (ii) selection by physical separation of the molecular feature, and (iii) selection by screening of the molecular function. Whereas the natural selection and physical separation are strongly limited, selection by screening offers the broadest access to different molecular functions. In principle, every molecular function which can be screened by a corresponding technique can be subjected to evolutionary approaches. The challenge addressed in our laboratory is to set up and integrate RAPID FCS and CFCA in processes for selection by screening to generate novel biomolecules. This strategy using confocal fluorescence spectroscopy was first proposed by Eigen and Rigler in the early 1990s [7]. The experimental realization is schematically shown in Fig. 12.9.

Besides short analysis time per sample and broad access to different biochemical reactions, which is discussed in the preceding sections, some more major issues must be addressed to the implementation of RAPID FCS and CFCA for selection by screening in evolutionary biotechnology. Additional steps, as pointed out in Fig. 12.9, are spatial isolation of individual clones and expression of molecular functions. Spatial isolation of individual clones can be carried out by simple dilution of a population of different clones into different compartments. Therefore, miniaturized sample carriers were developed. Different designs were realized using volumes ranging from 50 nl to 5 µl per well as shown in Fig. 12.10 (upper pictures). Because of the tiny volume element of confocal fluorescence spectroscopy, biochemical reactions can be carried out in these sample carriers without any decrease in signal, but provisions against evaporation and nonspecific adsorption must be taken into account. The standard samples volume is one µl per sample and below. For dispensing bacterial cultures or assay buffers in the sub-microliter range, standard piezo driven pumps or magnetic valves are successfully applied (Fig. 12.10, lower plot). The expression of molecular functions is monitored by dual-color confocal fluorescence spectroscopy. In biological environments, even in crude samples where, besides the double-labeled substrate, bacteria, culture media, and metabolic products are present, dual-color confocal fluorescence spectroscopy shows great sensitivity in monitoring biochemical reactions, as shown in Fig. 12.11. The only alteration compared to the experiments described above was the application of two different clones. One clone carries a plasmid which encodes the subtilisin E gene for secretion whereas the other clone carries a reference plasmid lacking the subtilisin E gene. The monitoring of the precise degradation of low amounts of double-labeled peptide substrate after several hours shows the enormous sensitivity of this method in biological environments.

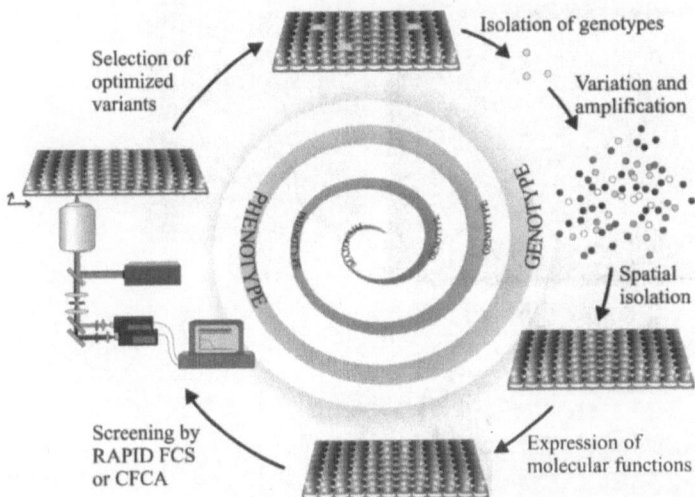

Fig. 12.9. Evolutionary biotechnology is the cyclic interaction of variation, amplification and selection in a defined technical environment. All steps are separated by technical means. Expression of function is only required when carriers of phenotype and genotype are separate molecules. A library of genotypes is dispensed into compartments of a sample carrier. After expressing the molecular phenotype, individual variants are screened by RAPID FCS or CFCA. The integration of RAPID FCS or CFCA is needed for the screening of individual molecular variants. Corresponding genotypes are isolated and subjected to the next round of optimization in order to increase the level of intended functions

12.4   Outlook

The present developments and innovations in the field of dual-color confocal fluorescence spectroscopy show the great potential of this single-molecule based technology. The broad application to a variety of different biochemical reactions as well as the short analysis times per sample are first steps towards the development of a full-grown detection technology. It is assumed that analysis times per sample can be reduced by at least one order of magnitude. The intrinsic features of miniaturized dimensions and extreme sensitivity may lead away from the statistical analysis of an ensemble of molecules towards the analysis of a single molecule. Integration of pulsed laser systems for two-photon excitation [16] and time-resolved analysis [3] will be the next milestones on the way to improving dual-color based confocal techniques.

On the other hand, confocal fluorometric techniques have already been successfully applied in life science industries as high-throughput screening tools for drug discovery and screening [1,7,18]. Nowadays, dual-color confocal techniques are implemented, or will be implemented for high-throughput screening in the near future.

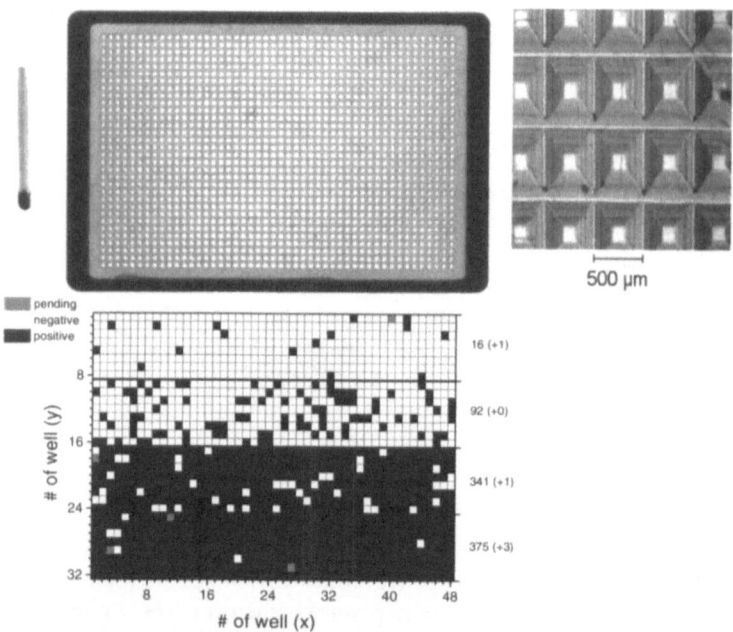

Fig. 12.10. Sample carrier and dispensing. In our sample carriers designed for confocal fluorescence spectroscopy, reactions can be monitored through a coverglass which is fixed to the substrate. Therefore no loss in signal intensity can be observed, which leads to high quality of the measurement. *Upper left side:* Designed sample carrier in a 1536er standard microtiterplate format. Reaction volume can be choosen from 1 to 4 µl. *Upper right side:* Designed nanotiterplate with volumes of 50 nl per well. *Lower side:* Spatial isolation of individual clones by dispensing small amounts of a bacterial culture into compartments of a sample carrier. A decreasing decadic dilution series at each 384-well-quarter of a 1536er microtiterplate shows the ability of single clone separation. Culture conditions are described in Fig. 12.11

In evolutionary biotechnology, the integration of dual-color confocal techniques such as CFCA for selection by screening of biomolecular functions was recently established with throughput rates of more than 100,000 samples in one day. They even have the potential of more than one million samples per day. The application relevance of these techniques can be outlined by the flexibility for precise monitoring of different biochemical reactions in short analysis times. Further improvements in different fields such as miniaturization, detection efficiency, parallelization, improved optics, evaluation of collected data and assay developments will push this limit up to several millions of samples per day in the near future. The combination with nanotechnology will further reduce sample volumes, and applications to flows in micro- and nanostructures will lead to ultrasensitive analysis tools. Nowadays, the problem is no longer single-molecule detection, which can "easily" be achieved

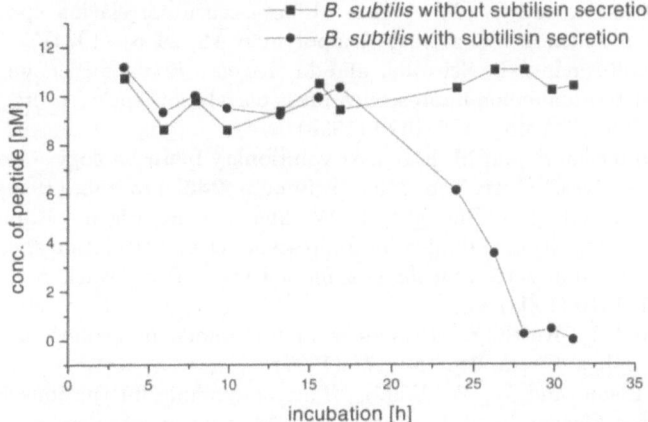

Fig. 12.11. Monitoring the degradation of 10 nM double-labeled peptide in a growing *B. subtilis* culture. Optimized synthetic culture media was innoculated with approx. 100 cells of *B. subtilis* and incubated at 37°C. Two genetic constructs of *B. subtilis* (with and without secretion of subtilisin, an unspecific alkaline protease) were cultured and analyzed by dual-color FCS as described above. The change in concentration of 10 nM double-labeled peptide in both cultures was monitored for a period of more than 30 h

below nanomolar concentrations. The challenge is to search single molecules at dramatically low concentrations of femto- to attomolar (10^{-15} to 10^{-18} M) and to screen for rare events among large numbers of samples.

References

1. M. Auer, K. J. Moore, F. J. Meyer-Almes, R. Günther, A. J. Pope, and K. A. Stoeckli, "Fluorescence correlation spectroscopy – lead discovery by miniaturized HTS," Drug Discovery Today **3**(10), 457–465 (1998)
2. J. Bieschke and P. Schwille, "Aggregation of prion protein investigated by dual-color fluorescence cross-correlation spectroscopy," Fluorescence Microscopy and Fluorescent Probes **2**, 81–86 (1998)
3. C. Eggeling, J. R. Fries, L. Brand, R. Günther, and C. A. M. Seidel, "Monitoring conformational dynamics of a single molecule by selective fluorescence spectroscopy," Proc. Natl. Acad. Sci. USA **95**, 1556–1561 (1998)
4. M. Ehrenberg and R. Rigler, "Rotational brownian motion and fluorescence intensity fluctuations," Chem. Phys. **4**, 390–401 (1974)
5. M. Ehrenberg and R. Rigler, "Fluorescence correlation spectroscopy applied to rotational diffusion of macromolecules," Quart. Rev. Biophys. **9**, 69–81 (1976)
6. M. Eigen and W. Gardiner, "Evolutionary molecular engineering based on RNA replication," Pure. Appl. Chem. **56**, 967–978 (1984)
7. M. Eigen and R. Rigler, "Sorting single molecules: Application to diagnostics and evolutionary biotechnology," Proc. Natl. Acad. Sci. USA **91**, 5740–5747 (1994)

8. E. L. Elson, D. Magde, and W. W. Webb, "Fluorescence correlation spectroscopy. II. An experimental realization," Biopolymers **13**, 29–61 (1974)

9. U. Kettling, A. Koltermann, P. Schwille, and M. Eigen, "Real-time enzyme kinetics monitored by dual-color fluorescence cross-correlation spectroscopy," Proc. Natl. Acad. Sci. USA **95**, 1416–1420 (1998)

10. U. Kettling, A. Koltermann, and M. Eigen, "Evolutionary biotechnology – Reflections and perspectives," Curr. Top. Microb. Immun. **243**, 173–185. (1999)

11. A. Koltermann, U. Kettling, J. Bieschke, T. Winkler, and M. Eigen, "Rapid assay processing by integration of dual-color fluorescence cross-correlation spectroscopy: High throughput screening for enzyme activity," Proc. Natl. Acad. Sci. USA **95**, 1421–1426 (1998)

12. A. Koltermann and U. Kettling, "Principles and methods of evolutionary biotechnology," Biophys. Chem. **66**, 159–177 (1997)

13. D. Magde, E. L. Elson, and W. W. Webb, "Thermodynamic fluctuations in a reacting system – Measurement by fluorescence correlation spectroscopy," Phys. Rev. Lett. **29**, 705–708 (1972)

14. R. Rigler, U. Mets, J. Widengren, and P. Kask, "Fluorescence correlation spectroscopy with high count rate and low background analysis of translational diffusion," Eur. Biophys. J. **22**, 169–175 (1993)

15. P. Schwille, *Fluoreszenz-Korrelations-Spektroskopie: Analyse biochemischer Systeme auf Einzelmolekülebene,* Thesis, Technische Universität Braunschweig (1996)

16. P. Schwille, U. Haupts, S. Maiti, and W. W. Webb, "Molecular dynamics in living cells observed by fluorescence correlation spectroscopy with one- and two-photon excitation," Biophys. J. **77**, 2251–2265 (1999)

17. P. Schwille, F. J. Meyer-Almes, and R. Rigler, "Dual-color fluorescence cross-correlation spectroscopy for multicomponent diffusional analysis in solution," Biophys. J. **72**, 1878–1886 (1997)

18. S. Sterrer and K. Henco, Fluorescence correlation spectroscopy (FCS) – A highly sensitive method to analyze drug/target interactions'," J. Recept. Signal. Transduct. Res. **17**(1–3), 511–520 (1997)

19. T. Winkler, U. Kettling, A. Koltermann, and M. Eigen, "Confocal fluorescence coincidence analysis: An approach to ultra high-throughput screening," Proc. Natl. Acad. Sci. USA **96**, 1375–1378 (1999)

13   Single-Molecule Enzymology*

X. S. Xie** and H. P. Lu

Viewing a movie of an enzyme molecule made by molecular dynamics simulation, we see incredible details of molecular motions, be they changes of the conformation or actions during a chemical reaction. Molecular dynamics simulations have advanced our understanding of the dynamics of macromolecules in ways that would not be deducible from the static crystal structures [1,2]. Unfortunately, these "virtual movies" do not run long enough, compared to the time scale of milliseconds to seconds in which most enzymatic reactions take place. In recent years, rapid advances in the patch clamp technique [3], atomic force microscopy [4,5], optical tweezers [6,7], and fluorescence microscopy [8-11] have permitted making single-molecule "movies" *in situ* at the millisecond to second time scale. These techniques do not have time resolutions as high as that of molecular dynamics simulations, but their single-molecule sensitivities allow probing of conformational motions, which are otherwise masked in ensemble-averaged experiments. Moreover, chemical reactions can now be observed on a single-molecule basis. For example, enzymatic turnovers of a few motor proteins [12–17] and a few enzymes [18–20] have been monitored in real time.

Our knowledge of enzyme kinetics has come primarily from experiments conducted on large ensembles of enzyme molecules, in which concentration changes over time are measured. In a single-molecule experiment, the concentration of the molecule being studied becomes meaningless in discussing chemical kinetics. However, this does not negate the fundamental principles of chemical kinetics. Chemical kinetics can be cast in terms of single-molecule probabilities. Thinking of chemical kinetics in terms of single molecules is not only pertinent to the ever increasing single-molecule studies, but is also insightful and very often more informative.

Such "single-molecule" thinking is also useful in understanding chemistry in living cells. In a living cell, the number of enzyme molecules in a cellular

* Part of the text appeared in a minireview in Journal of Biological Chemistry **274**, 15967–15970 (1999). Courtesy of the copyright owner, the American Society for Biochemistry and Molecular Biology.

** To whom correspondence should be addressed.

component is not large. In this situation, the concentration in a small probe volume is no longer a constant but a fluctuating quantity, as molecules react or diffuse in and out of the probe volume. In fact, the reaction rate (and diffusion rate) can be extracted from analyses of concentration fluctuations [21,22]. This approach is referred to as Fluctuation Correlation Spectroscopy (FCS) [23], and has recently been conducted with single-molecule sensitivity [24]. A typical FCS trace, however, is averaged over a large number of molecules diffusing one or a few at a time in and out of a fixed probe volume. In studying biochemical reactions in living cells, there are situations in which we need to focus on the behavior of a single molecule. For example, DNA exists as a "single molecule" inside a bacteria cell. The behaviors of a DNA–enzyme complex can be tracked. In another example, a single receptor protein at a particular spot in a membrane can be interrogated by optical or scanning probe microscopy. Studies in a similar line have been extensively carried out on ion channel proteins with the patch clamp technique [3].

In this article, we use our recent work on a flavoenzyme [18] to discuss the underlying principles of single-molecule kinetics and the information obtainable from single-molecule studies.

13.1   Why Single-Molecule Real-Time Studies?

What does one gain by doing single-molecule enzymatic studies? The stochastic events of individual molecules are not observable in conventional measurements, and the steady state concentrations of transient intermediates are usually too low to detect. The single-molecule experiments allow direct observations of individual steps or intermediates of biochemical reactions. The trajectories of motor proteins serve as good examples of the visualization of individual steps [12–17].

Perhaps less obviously, single-molecule experiments allow determination of static and dynamic disorder. Seemingly identical copies of biomolecules often have broad distributions of molecular properties because of static heterogeneity and fluctuations. In studying rate processes, static disorder is the static heterogeneity of reaction rates within a large ensemble of molecules. Dynamic disorder [25] is the time-dependent fluctuation of the reaction rate of an individual molecule. Both static and dynamic disorder result in distributions of rates and multiexponential kinetics in conventional experiments. Ensemble-averaged measurements can hardly determine distributions of rates, nor can they distinguish between static and dynamic disorder.

A recent single-molecule enzymatic assay by Xue and Yeung has revealed static disorder in enzymatic turnover rates of genetically identical and electrophoretically pure enzyme molecules [26]. In a capillary tube containing a solution of highly diluted enzyme molecules (lactate dehydrogenase, LDH-1) and concentrated substrate molecules (lactate and nicotinamide adenine dinucleotide, NAD^+), each enzyme molecule produced a discrete zone of thousands of NADH molecules after 1 hour of incubation. The zones were then eluted by capillary electrophoresis and monitored by natural fluorescence of NADH. The enzyme molecules had a broad and asymmetrical distribution of activity, which was otherwise masked by ensemble-averaged measurements. The heterogeneity was found to be static at the hour time-scale because the same enzyme molecule produces the same zone intensity after another incubation period. Using a similar approach, Craig et al. studied single alkaline phosphatase molecules and found a broad and multipeak distribution of activities [27]. The microscopic origin of the static disorder observed is an interesting subject that deserves future research.

Yeung's experiment is capable of determining static disorder but not dynamic disorder. Dynamic disorder in enzymatic turnover rates has been observed by real-time single-molecule experiments [18], as discussed below. We note that a homogenous and ergodic system may exhibit "static disorder" because the time scale of a fluctuation is longer than the time scale of a measurement. In reality, however, static disorder also arises from the intrinsic heterogeneity of biological systems, and often creates a major complication in conventional experiments. With the separation of static and dynamic disorder in single-molecule experiments, relations between the dynamics and functions of enzymes can be better interrogated.

13.2 Viewing Single-Molecule Enzymatic Reactions by Fluorescence

Consider the example of cholesterol oxidase, a 53-kDa flavoprotein that catalyzes the oxidation of cholesterol by oxygen (Fig. 13.1). The active site of the enzyme (E) involves a flavin adenine dinucleotide (FAD), which is naturally fluorescent in its oxidized form but not in its reduced form. The FAD is reduced by a cholesterol molecule to FADH2 and is then oxidized by molecular oxygen. As shown in Fig. 13.1, fluorescence turns on and off as the redox state of the FAD toggles between the oxidized and reduced states, respectively. Each on–off cycle corresponds to an enzymatic turnover.

The single-molecule fluorescence measurements are carried out with an inverted fluorescence microscope, as described elsewhere [18,28,29]. It is desirable to study immobilized molecules to avoid the complications of the diffusion process. Our samples are thin films of agarose gel of 99% water. The single enzyme molecules are confined in the gel with no noticeable translational diffusion. In contrast, small substrate molecules still diffuse freely. Though confined in the polymer matrix, the enzyme molecules freely rotate within the gel, which was evidenced by a polarization modulation experiment, as previously described [30,31]. This means that the enzyme molecules do not bind to the polymer matrix. To prepare the sample, 10^{-9} M cholesterol oxidase was mixed with 0.2–6 mM steroid substrate in buffer solution (pH 7.4). Because of the low solubility of cholesterol, detergent was needed

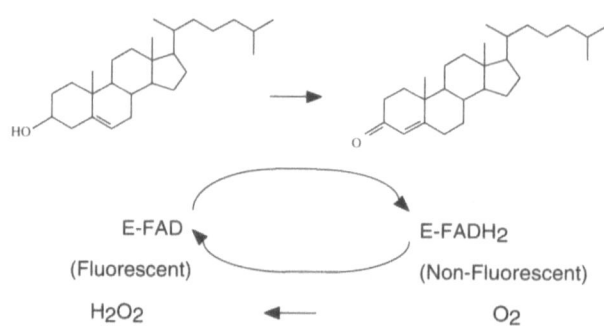

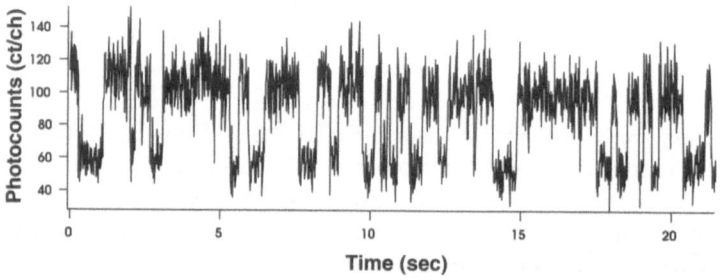

Fig. 13.1. Enzymatic cycle of cholesterol oxidase and real-time observation of enzymatic turnovers of a single cholesterol oxidase molecule. Each on–off cycle in the emission intensity trajectory corresponds to an enzymatic turnover

for this preparation. A 10^{-2} M steroid substrate was first dissolved into 1.7 M Triton X-100 (Sigma, St. Louis, MO), and was further diluted to 0.2–6 mM steroid concentration. 1% agarose in buffer solution was kept at a temperature slightly higher than the gelling temperature ($>30°$C). A 1:1 mixture of the agarose solution and the enzyme/steroid solution lowered the gel temperature and caused gelling. Two glass cover slips previously coated with agarose were used to sandwich the transparent sample gel of 10- to 20-μm thickness. Control experiments were done to ensure that the conventional enzymatic assays gave the same results in gel and in solution.

The turnover trajectories contain detailed dynamic information, which is extractable from statistical analyses. Good statistical analyses require long trajectories. The lengths of the trajectories are limited by photo-bleaching through photochemistry on the excited state [8]. We observed better photostability for the FAD chromophore in protein than for dye molecules, most likely because of the protection by the protein. Trajectories with more than 500 turnovers and 2×10^6 detected photons (detection efficiency 10%) have been recorded. Similar photostability has been seen for other natural fluorophores, such as those in GFP [32]. In the case of GFP, emission of a single molecule blinks off due to photoinduced chemical reactions [32]. We did not observe such photoinduced blinking of cholesterol oxidase. We have also done control experiments to make sure that repetitive excitation/de-excitation does not perturb the enzymatic reactions.

13.3   Chemical Kinetics in Terms of a Single Molecule

Many two-substrate enzymes, such as cholesterol oxidase, follow the ping-pong mechanism for the two-substrate binding processes, obeying the Michaelis–Menten mechanism [33]:

$$\text{E-FAD} + \text{S} \;\underset{k_{-1}}{\overset{k_1}{\rightleftarrows}}\; \text{E-FAD*S} \;\overset{k_2}{\longrightarrow}\; \text{E-FADH}_2 + \text{P} \tag{13.1}$$

$$\quad\;\text{on} \qquad\qquad\quad \text{on} \qquad\qquad \text{off}$$

$$\text{E-FADH}_2 + \text{O}_2 \;\underset{k_{-1}}{\overset{k_1}{\rightleftarrows}}\; \text{E-FADH}_2^*\text{O}_2 \;\overset{k_2}{\longrightarrow}\; \text{E-FAD} + \text{H}_2\text{O}_2 \tag{13.2}$$

$$\quad\;\text{off} \qquad\qquad\qquad \text{off} \qquad\qquad \text{on}$$

The emission on-times and off-times correspond to the "waiting time" for the FAD reduction and oxidation half reactions, respectively. The simplest analysis of the trajectory is the distribution of the on- or off-times. We limit our discussion below to the on-time distributions, although similar analyses can be done for off-times as well.

First, take a simple case in which k_2 is rate-limiting. This situation can be created with a slowly reacting substrate (derivative of cholesterol) and a high concentration of the substrate. The FAD reduction reaction follows a simple kinetic scheme:

$$\text{E-FAD*S} \;\overset{k_2}{\longrightarrow}\; \text{E-FADH}_2 + \text{P} \tag{13.3}$$

$$\quad\;\text{on} \qquad\qquad \text{off}$$

For this scheme, the probability density of the on-time, τ, for a single-molecule turnover trajectory is an exponential function, $p_{on}(\tau) = k_2 \exp(-k_2\tau)$, with the average of the on-times being $1/k_2$, the time constant of the exponential. The exponential function follows from the fact that Eq. (3) is a Poisson process. One caveat is that this does not mean that zero on-time has the highest probability: $p_{on}(\tau)$ is the probability density, whereas the probability for the on-time to be between τ and $\Delta\tau$ is given by $p_{on}(\tau)\Delta\tau$, with the integral of $p_{on}(\tau)$ from zero to infinity being one.

A simple example of a Poisson process is the case of a telephone being turned on and off. The Poisson process corresponds to a situation in which all phone calls are independent. This results in an exponential probability density distribution of the lengths of the calls (on-times). In such an analogy, our real-time experiment determines the on-time distribution of a single phone, Yeung's experiment determines the average on-time of a single phone, and an ensemble-averaged measurement determines the average on-time of all the phones in the world.

Figure 13.2a shows the histogram of on-times derived from a trajectory with a cholesterol derivative of 2-mM concentration, which was fitted with a single exponential.

The inset of Fig. 13.2a shows the distribution of k_2 among 33 molecules examined, reflecting the large static disorder for the activation step. However, we did not observe static disorder for k_1, k_1', and k_2'.

The on-time histogram (or probability density) takes a more complex form when k_2 is not rate-limiting. Assuming that $k_{-1} = 0$ in Eq. (1), the probability density of the on-times is expected to be the convolution of the two exponential functions (time constants of $1/k_1$ and $1/k_2$) with an exponential rise (the faster of k_1 and k_2) and an exponential decay (the slower of k_1 and k_2) [18]. Figure 13.2b shows an on-time histogram for an enzyme molecule with 2-mM cholesterol, with the solid line being the convolution of two exponentials. The time lag in the histogram arises from the fact that there is an intermediate, E-FAD*S. There is no E-FADH$_2$ generated until the intermediate emerges. In the telephone analogy, this corresponds to the caller having to wait for an operator to connect. If there is more than one fluorescent intermediate, the on-times are expected to have a narrower distribution. This is discussed in detail in [34].

13.4 What is the New Information?

So far, the analyses of on-time distributions illustrate at the single-molecule level the validity of chemical kinetics, in particular the Michaelis–Menten mechanism (Eq. (13.1)). What new information can we obtain from the trajectory analyses? Chemical kinetics holds for Markov processes, implying that an enzyme molecule undergoing a turnover exhibits no memory of its preceding turnovers. Dynamic disorder is beyond the scope of chemical kinetics. Figure 13.2 does not have a good enough signal-to-noise ratio to reveal multiexponential decay because of dynamic disorder. Furthermore, being a scrambled histogram, Fig. 13.2 is not sensitive to memory effects. What we need is a way to examine how a particular turnover is affected by its predecessors.

We evaluated the conditional probability $p(X, Y)$ for a pair of on-times (X and Y) separated by a certain number of turnovers. Figures 13.3a and b are the two-dimensional histograms of a pair of on-times adjacent to each other and those separated by 10 turnovers, respectively. In the absence of dynamic disorder, $p(X, Y)$ should be independent of the separation of turnovers. However, Figs. 13.3a and b are clearly different. For the separation of 10

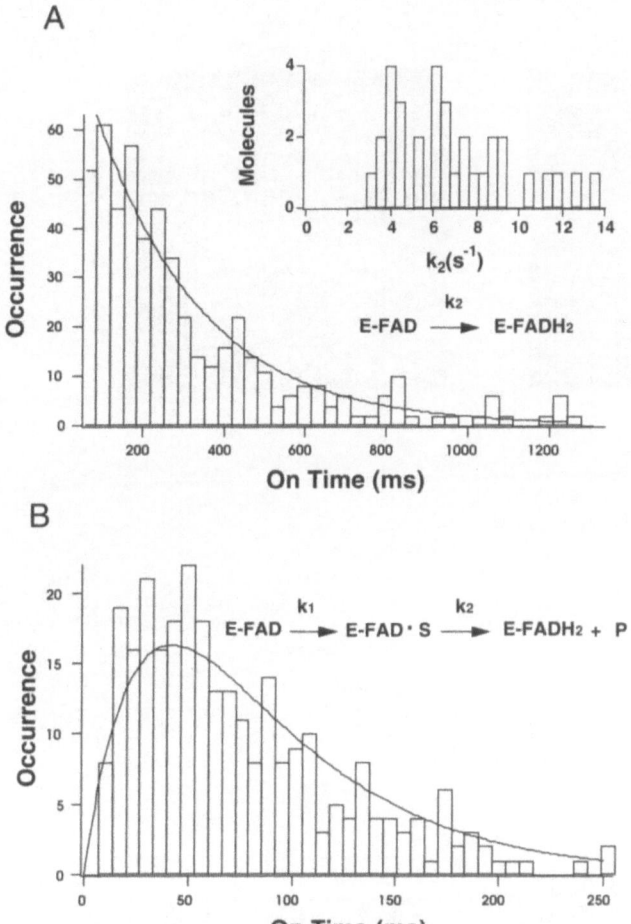

Fig. 13.2. a Histogram of the on-times in a single-molecule turnover trajectory taken with a derivative of cholesterol (inset) at 2-mM concentration (k_2 being rate-limiting). The dotted line is a single exponential fit with $k_2 = 3.9 \pm 0.5$ s^{-1}. The inset shows the distribution of k_2 derived from 33 molecules in the same sample. **b** Histogram of the on-times in a single-molecule turnover trajectory taken with cholesterol at 2-mM concentration. The solid line is the convolution of two exponentials with time constants $k_1 = 33 \pm 6$ s^{-1} and $k_2 = 17 \pm 2$ s^{-1}

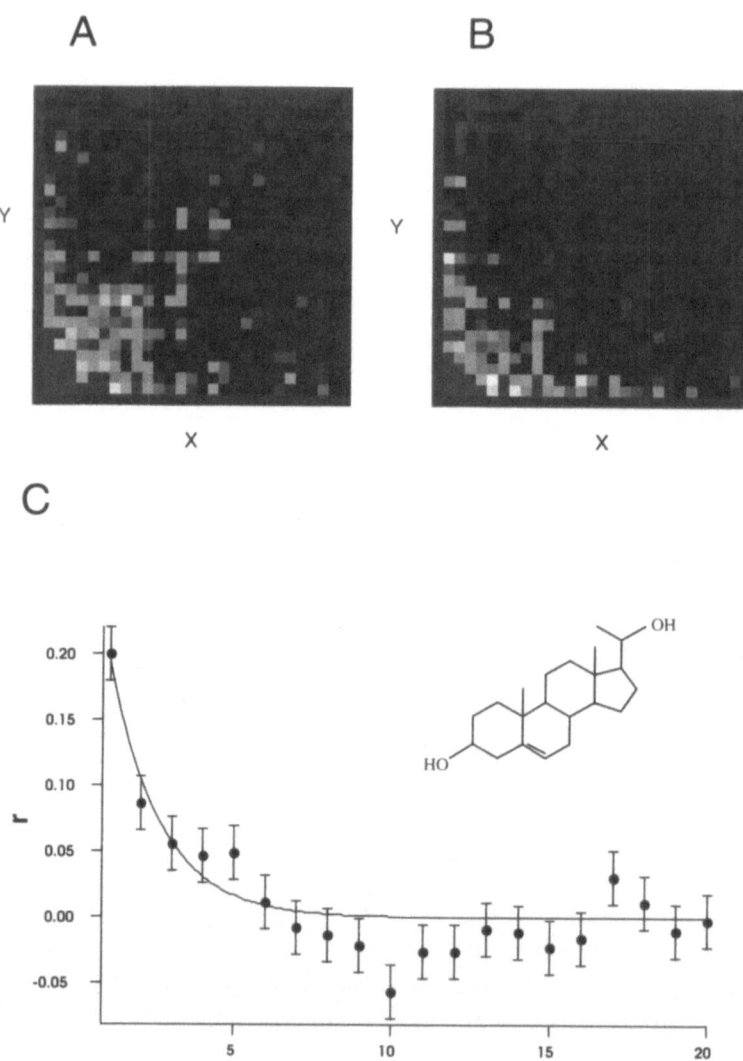

Fig. 13.3. a Two-dimensional histogram of pairs of on-times adjacent to each other. A diagonal feature indicates a memory effect. **b** Two-dimensional histogram of pairs of on-times separated by 10 turnovers. The on-times that are 10 turnovers apart become independent as the memory is lost. **c** Autocorrelation function of the on-times for a turnover trajectory, $r(m) = \langle \Delta\tau(0)\Delta\tau(m)\rangle/\langle\Delta\tau^2\rangle$, m being the index number of the turnovers and $\Delta\tau(m) = \tau(m) - \langle\tau\rangle$. The fact that r(m) is not a spike at $m = 0$ indicates dynamic disorder. The time constant of the decay gives the time scale of the k_2 fluctuation. The inset shows the structure of the substrate used, which is a derivative of cholesterol

turnovers (Fig.13.3b), the loss of memory leads to $p(X, Y) = p(X)p(Y)$, where $p(X)$ and $p(Y)$ are the same as in Fig. 13.3a. For pairs of adjacent on-times (Fig. 13.3a), there is a diagonal feature, indicating that a short on-time tends to be followed by another short on-time, and a long on-time tends to be followed by another long on-time. This means that an enzymatic turnover is not independent of its previous turnovers. The memory effect arises from a slowly varying rate (k_2). Coming back to the telephone analogy, this corresponds to the average lengths of phone calls varying over the course of a day.

Although the two-dimensional conditional probability plot provides a clear visual illustration, it needs to be constructed from a large number of turnovers of many molecules. A more practical and quantitative way to evaluate the dynamic disorder is the autocorrelation function of the on-times, $r(m) = \langle \Delta\tau(0)\Delta\tau(m)\rangle/\langle\Delta\tau^2\rangle$, where m is an index number for the turnovers in a trajectory and $\Delta\tau(m) = \tau(m) - \langle\tau\rangle$, and where the bracket denotes the average along the trajectory. The physical meaning of $r(m)$ is as follows: In the absence of dynamic disorder, $r(0) = 1$ and $r(m) = 0$ ($m > 0$). In the presence of dynamic disorder, $r(m)$ decays, with the initial ($m = 1$) amplitude reflecting the variance of k_2 and the decay time yielding the time scale of the k_2 fluctuation. Figure 13.3c shows the $r(m)$ derived from a single-molecule trajectory, with the decay constant being 1.6 ± 0.5 turnover (1.0 ± 0.3 s).

We attribute the dynamic disorder behavior to a slow fluctuation of protein conformation, which was independently observed by spectral fluctuation of FAD [18]. Figure 13.4a shows a trajectory of the spectral mean of the emission spectra of a single enzyme molecule in the absence of cholesterol molecules. The slow fluctuation of the emission spectrum (Fig. 13.4b) reflects conformational changes around the FAD, a phenomenon that is otherwise hidden in ensemble-averaged measurements. Similar room-temperature spectral fluctuation of single dye molecules in polymer has been studied in detail, providing information regarding the energy landscape [29]. Conformational fluctuations on a similar slow time scale have been observed on other biomolecules with single-molecule experiments [4,35–38]. The autocorrelation function of the spectral mean trajectory is shown in Fig. 13.4c. The decay curve was found to be independent of excitation rate, indicating a spontaneous (rather than photo-induced) conformational fluctuation. Interestingly, the decay constant is 1.3 ± 0.3 sec, which is in the same time scale as the correlation time of k_2. This provides strong evidence that a fluctuation in conformation results in variation of the enzymatic rate k_2.

The simplest model we proposed for the dynamic disorder involved two slowly converting conformational states (E and E'):

$$E \underset{k_{E'}}{\overset{k_E}{\rightleftarrows}} E'$$

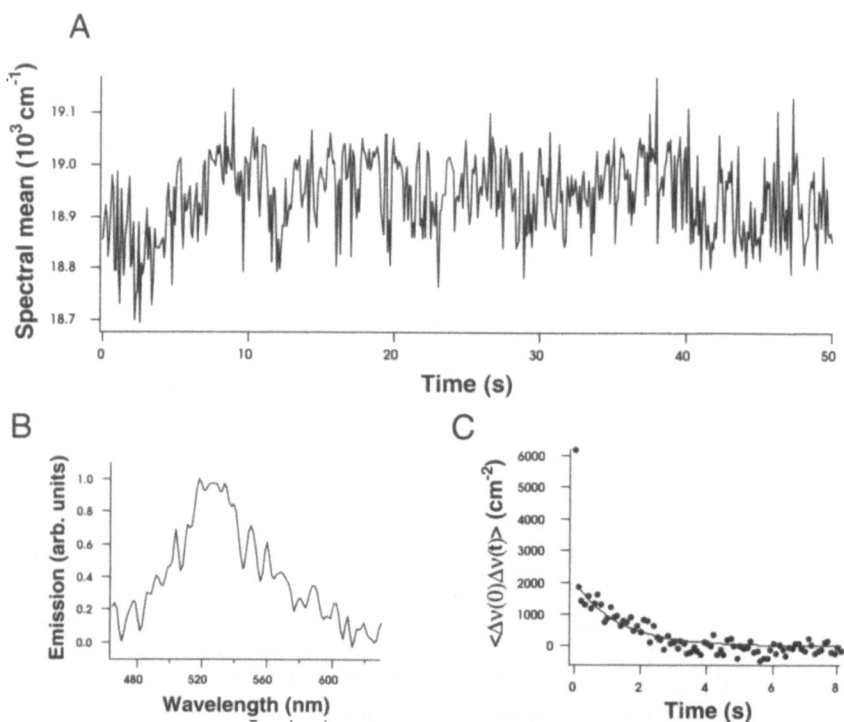

Fig. 13.4. a Trajectory of the spectral mean of a cholesterol oxidase molecule in the absence of steroid substrate. The spectral fluctuation is attributed to a spontaneous conformational fluctuation. **b** Emission spectrum of the FAD in the enzyme molecule taken in 100 ms. The spectral mean of the spectrum is the first data point in **a**. **c** The correlation function (dots) of spectral mean for the trajectory in **a**. The first data point, a spike at zero time, is due to uncorrelated measurement noises and spectral fluctuations faster than the 100-ms time resolution. The solid line is a fit by a single exponential with a time constant of 1.3 ± 0.3 sec, which coincides with the correlation time for k_2 fluctuation

$$
\begin{array}{ccccc}
\text{E-FADH} + \text{S} & \xrightarrow{k_1} & \text{E-FAD*S} & \xrightarrow{k_{21}} & \text{E-FADH}_2 + \text{P} \\
\Big\updownarrow & & \Big\updownarrow & & \Big\updownarrow \\
\text{E-FADH} + \text{S} & \xrightarrow{k_1} & \text{E-FAD*S} & \xrightarrow{k_{22}} & \text{E-FADH}_2 + \text{P}
\end{array}
\tag{13.4}
$$

The dynamic disorder of k_2 for the Michaelis–Menten mechanism (Eq. 13.2) can be accounted for by the more complicated kinetics scheme (Eq. 13.4) with time-independent rates k_{21} and k_{22}. A data fit based on this kinetic scheme indicates $k_{21}/k_{22} \sim 7$ – a significant variation of the activation rate [39].

Of course, the activation rate can take more than two discrete values. Another limiting case is that there is a continuous distribution of k_2. In this situation, the time dependent $k_2(t)$ can be described by a diffusive process using the Langevin equation [39,40,41]. Such a slow fluctuation in reactivity has been termed "intermittency" [42]. Our data fit based on the diffusive model indicates that the standard deviation of k_2 is 150–200% of k_2 [39].

In summary, although the Michaelis–Menten mechanism provides a good description for the averaged behaviors of many molecules and for the averaged behaviors of many turnovers of a single molecule, it does not provide an accurate picture of the real-time behavior of a single molecule. On a single molecule basis, the rate for the activation step is fluctuating. We conclude that there are at least two slowly interconverting conformers with significantly different activation rates. This is a not a small effect!

13.5 Open Questions

The influence of conformational dynamics on protein functions has been a subject of extensive studies [42–50]. It has been well established that proteins have many conformational substates [49,50]. Our observations have revealed that the metastable conformations have significantly different rates of chemical reactions. The question is, what are the functionally important conformational states and the distribution of their rates? We have presented two models for two limiting cases, with either two conformations or a continuous distribution of conformations. We are not yet in a position to determine experimentally which one of the two models or an intermediate between the two models is the underlying mechanism. The answer to this question requires longer trajectories and/or more sophisticated analyses of trajectories. The single-molecule approach will provide detailed dynamical information about energy landscapes and their influence on protein functions.

We do not yet understand microscopic details of the conformational states, except that the barrier heights between the states are relatively high (~15 kcal/mole assuming an Arrhenius process), comparable to breaking and reforming a few hydrogen bonds. The conformational fluctuations influencing the reaction rate can be either global protein conformational changes or local conformational changes at the active site. Similar slow conformational changes have been inferred from other monomeric enzyme systems, which were postulated to be associated with physiological enzymatic regulation [51–53]. The physiological relevance of the dynamic disorder phenomenon [54], if any, will be best probed by single-molecule experiments in cellular environments.

Not assumed in Eq. (4) is the possibility of conformational changes induced by substrate binding and/or the redox reactions, which can either lead to new conformational states or shift the equilibrium between the existing conformational states [51–53]. Conformational changes associated with the "induced fit" idea have been illustrated [55]. A study of cross-correlations between simultaneous spectral and turnover trajectories is underway to investigate this possibility.

13.6   In the Future

We have shown that statistical analyses of turnover trajectories of single enzyme molecules can unequivocally reveal mechanistic information hidden in ensemble-averaged measurements. The analyses described here are generally useful as more and more single-molecule *in vitro* enzymatic assays are being developed. Single-molecule real-time enzymology will allow investigation of molecular interactions and reaction mechanisms at a level of great detail.

For living cells where biochemical reactions take place in small volumes, often on a single-molecule basis, thinking of chemical kinetics in terms of individual molecules is becoming a necessity as new tools of microscopy become available. In order to image single protein and DNA molecules in living cells, we have developed two new nonlinear optical imaging methodologies based on high-resolution near-field microscopy [56] and coherent anti-Stokes Raman scattering microscopy [57]. Our knowledge of biological processes, such as cell signaling and gene expression, will be greatly enhanced through real-time single-molecule experiments carried out under physiological conditions.

Acknowledgments. We thank Luying Xun and Greg Schenter for fruitful collaborations. This work was conducted at the Environmental Molecular Sciences Laboratory at Pacific Northwest National Laboratory and was supported by the Chemical Sciences Division of the Office of Basic Energy Sciences and the Office of Biological and Environmental Research within the Office of Science of the U. S. Department of Energy.

References

1. For a review, see C. L. Brooks III, M. Karplus, and B. M. Pettitt, *Proteins: A Theoretical Perspective of Dynamics, Structure, and Thermodynamics*, Advances in Chemical Physics, vol. LXXI (John Wiley & Sons Inc., New York, 1998)
2. For a review, see J. A. McCammon and S. C. Harvey, *Dynamics of Proteins and Nucleic Acids* (Cambridge Univ. Press, Cambridge, 1987)
3. For a review, see *Single-Channel Recording*, 2nd ed., B. Sakmann and E. Neher (eds.) (Plenum Press, New York, 1995)
4. M. Radmacher, M. Fritz, H. G. Hansma, and P. K. Hansma, Science **265**, 1577–1579 (1994)
5. W. A. Rees, R. W. Keller, J. P. Vesenka, and C. Bustamante, Science **260**, 1646–1649 (1993)
6. A. Ashkin and J. M. Dziedzic, Science **235**, 1517–1520 (1987)
7. T. T. Perkins, S. R. Quake, D. E. Smith, and S. Chu, Science **264**, 822–826 (1994)

8. For a recent review, see X. S. Xie and J. K. Trautman, Ann. Rev. Phys. Chem. **59**, 441–480 (1998)
9. For a recent review, see S. Nie and R. N. Zare, Ann. Rev. Biophys. Biomol. Struct. **26**, 567–596 (1997)
10. For a recent review, see W. E. Moerner and M. Orrit, Science, **283**, 1670–1676 (1999)
11. For a recent review, see S. Weiss, Science, **283**, 1676–1683, (1999)
12. T. Funatsu, Y. Harada, M. Tokunaga, K. Saito, T. Yanagida, Nature **374**, 555–559 (1995)
13. R. D. Vale, T. Funatsu, D. W. Pierce, L. Romberg, Y. Harada, and T. Yanagida, Nature **380**, 451–453 (1996)
14. A. Ishijima, H. Kojima, T. Funatsu, M. Tokunaga, H. Higuchi, H. Tanaka, and T. Yanagida, Cell **92**, 161–171 (1998)
15. H. Noji, R. Yasuda, M. Yoshida, and K. Kinosita, Nature **386**, 299–302 (1997)
16. R. Yasuda, H. Moji, K. Kinoshita, and M. Yoshida, Cell **93**, 1117–1124 (1997)
17. M. J. Schnitzer, S. M. Block, Nature **388**, 386–390 (1997)
18. H. P. Lu, L. Xun, and X.S. Xie, Science, **282**, 1877–1882 (1998)
19. T. Ha, A. Y. Ting, J. Liang, W. B. Caldwell, A. A. Deniz, D. S. Chemla, P. G. Schultz, and S. Weiss, Proc. Natl. Acad. Sci. USA **96**, 893–898 (1999)
20. L. Edman, Z. Foldes-Papp, S. Wennmalm, and R. Rigler, Chem. Phys. **247**, 11–22 (1999)
21. D. Magde, E. Elson, and W. W. Webb, Phys. Rev. Lett. **29**, 705–708 (1972)
22. G. Feher and M. Weissman, Proc. Nat. Acad. Sci. USA **70**, 870–875 (1973)
23. For a recent review, see S. Maiti, U. Haupts, and W. W. Webb, Proc. Natl. Acad. Sci. USA **94**, 11753–11757 (1997)
24. M. Eigen and R. Rigler, Proc. Natl. Acad. Sci. USA **91**, 5740–5747 (1994)
25. R. Zwanzig, Acc. Chem. Res. **23**, 148–152 (1990)
26. Q. F. Xue and E. S. Yeung, Nature **373**, 681–683 (1995)
27. D. B. Craig, E. A. Arriaga, J. C. Y. Wong, H. Lu, and N. J. Dovichi, J. Am. Chem. Soc. **118**, 5245–5253 (1996)
28. J. J. Macklin, J. K. Trautman, T. D. Harris, and L. E. Brus, Science **272**, 255–258 (1996)
29. H. P. Lu and X. S. Xie, Nature **385**, 143–146 (1997)
30. X. S. Xie and R. C. Dunn, Science **265**, 361–364 (1994)
31. M. T. Ha, T. Enderle, D. S. Chemla, P. R. Selvin, and S. Weiss, Phys. Rev. Lett. **77**, 3979–3982 (1996)
32. E. J. G. Peterman, S. Brasselet, and W. E. Moerner, J. Phys. Chem. A **103**(49), 10553–10560 (1999)
33. T. Palmer, *Understanding Enzymes*, 4th ed. (Prentice Hall, New York, 1991) Chap. 9
34. M. J. Schnitzer, S. M. Block, "Protein kinesis: The dynamics of protein trafficking and stability," Cold Spring Harbor Symp. Quant. Biol. **60**, 793–802 (1995)
35. J. Dapprich, U. Mets, W. Simm, M. Eigen, and R. Rigler, Exp. Tech. Phys. **41**, 259–264, (1995)
36. C. Eggeling, J. R. Fries, L. Brand, R. Gunther, and C. A. M. Seidel, Proc. Natl. Acad. USA **95**, 1556–1561 (1998)
37. Y. Jia, A. Sytnik, L. Li, S. Vladimarov, B. S. Coopeman, and R. M. Hochstrasser, Proc. Natl. Acad. Sci. USA **94**, 7932–7936 (1997)

38. E. Geva and J. L. Skinner, Chem. Phys. Lett. **288**, 225–229 (1998)
39. G. Schenter, H. P. Lu, and X. S. Xie, J. Phys. Chem. A **103**, 10477–10488 (1999)
40. N. Agmon and J. J. Hopfield, J. Chem. Phys. **78**, 6947–6959 (1983)
41. J. Wang and P. Wolynes, J. Chem. Phys. **110**, 4812 (1998)
42. J. Wang and P. Wolynes, Phys. Rev. Lett. **74**, 4317–4320 (1995)
43. J. R. Knowles, Nature **350**, 121–124 (1991)
44. W. R. Cannon, S. F. Singleton, and S. J. Benkovic, Nat. Struct. Biol. **3**, 821–833 (1996)
45. M. Karplus, J. Phys. Chem. B **104**, 3721–3743 (2000)
46. J. A. McCammon, *Simplicity and Complexity in Proteins and Nucleic Acids*, H. Frauenfelder, J. Deisenhofer, and P. Wolynes (eds) (Dahlem Univ. Press, Berlin, 1999)
47. W. N. Lipscomb, Acc. Chem. Res. **15**, 232 (1982)
48. S. J. Hagen and W. A. Eaton, J. Chem. Phys. **104**, 3395–3398 (1996)
49. H. Frauenfelder, F. Parak, and R. D. Young, Ann. Rev. Biophys. Chem. **17**, 451–479 (1988)
50. H. Frauenfelder, S. G. Sligar, and P. G. Wolynes, *Science* **254**, 1598–1603 (1991)
51. K. E. Neet and G.R. Anislie, Method Enzym. **64**, 192–226 (1980)
52. C. Frieden, Ann. Rev. Biochem. **48**, 471–489 (1979)
53. J. Ricard, J. Meunier, and J. Buc, Eur. J. Biochem. **49**, 195–208 (1974)
54. P. Stange, A. S. Milhailov, and B. Hess, J. Phys. Chem. **102**, 6273–6289 (1998)
55. D. E. Koshland, Nat. Med. **4**, xii–xiv (1998)
56. E. J. Sanchez, L. Novotny, and X. S. Xie, Phys. Rev. Lett. **82**, 4014 (1999)
57. A. Zumbusch, G. R. Holtom, and X. S. Xie, Phys. Rev. Lett. **82**, 4142 (1999)

14 Single-Molecule Enzymology

N. J. Dovichi, R. Polakowski, A. Skelley, D. B. Craig, and J. Wong

14.1 Introduction

The concept of the atom, the fundamental particle of chemistry, was understood by some of the 4th century BC Greeks, who philosophized that all matter is reducible to indivisible particles. Experiments done 200 years ago convinced the chemists of that era that these atoms combine to form molecules, and that all molecules of a substance are identical. The development of powerful spectroscopic tools over the past three decades allows us to characterize individual atoms and molecules, and to test fundamental assumptions of chemistry, physics, and biology.

14.1.1 Single-Atom Detection

The detection of individual atoms was reported in 1977 by Greenlees and Kaufman at the University of Minnesota [1] and in 1980 by Fairbank and She at Colorado State University [2]. In both cases, an atom beam intersected a CW laser beam. A fluorescent burst was generated as an atom passed through the laser beam. This burst was detected above the background signal associated with scattered laser light; successful detection of single atoms required that the background signal be reduced to the lowest possible levels.

14.1.2 Single-Molecule Detection

In 1976, Hirschfeld, working at Block Engineering, detected single polyethylene molecules that had been labeled with ~1,000 fluorescein dye molecules [3]. These molecules were spread as a thin film on a microscope slide and were detected with total internal reflection laser-induced fluorescence. In his experiment, a photon burst arose when a fluorescently labeled polyethylene molecule was scanned through the focused laser beam, which was observed against a background signal. The awkward sample preparation and limited sample throughput made Hirschfeld's work a historical curiosity; his approach has seldom been used again for single-molecule detection.

In 1983, Dovichi and Keller at Los Alamos reported the use of a sheath-flow cuvette for ultrasensitive laser-induced fluorescence detection [4,5]. In the initial work, a single molecule of rhodamine 6G, on average, was present in the probe volume. However, the signal from several thousand molecules was averaged during the integration time of the experiment. In 1987, Keller reported the first detection of single molecules in a flowing stream [6]. This experiment was analogous to the photon-burst technology used in the single-atom experiments. A stream of highly dilute B-phycoerythrin solution was passed through a focused laser beam, and single molecules were detected based on their fluorescence burst signature.

Detection of single molecules relies on minimizing the probe volume [7]. When no dye molecule is present in the probe volume, the analyte concentration is, by definition, zero. When a single molecule is present, the dye concentration can be large. For example, a single dye molecule present in a 1-micrometer diameter droplet results in a nanomolar dye concentration. The signal from this solution can be distinguished from the background signal in the absence of an analyte molecule.

14.1.3 Single-Molecule Chemical Analysis

We used the sheath-flow cuvette as a detector in capillary electrophoresis analysis of fluorescein thiocarbamyl amino acids [8]. The combination of ultrasensitive detection with the powerful separation technique resulted in a very useful tool for the analysis of many different types of biological samples, including DNA, proteins, and carbohydrates [9–11]. A capillary-array version of this instrument has been commercialized by Applied Biosystems and is widely used in large-scale DNA sequencing efforts [12,13].

We used our capillary electrophoresis/laser-induced fluorescence instrument for single-molecule analysis of B-phycoerythrin, based on excitation with a 2 mW helium–neon laser [14]. B-phycoerythrin is a protein that exists in several isoforms, and analysis of 30,000 analyte molecules generated a reproducible peak consisting of five components. However, injection of 3,000 or fewer molecules led to a noisy and irreproducible peak. This irreproducibility resulted from molecular shot noise due to inevitable fluctuations in the number of injected molecules. We demonstrated that the relative standard deviation of peak area, peak center, and peak width were inversely proportional to the square root of the number of injected molecules. At least 10,000 analyte molecules were required to define peak area and width with 1% relative precision. Fluctuation in the number of molecules taken for chemical analysis is a fundamental source of uncertainty.

14.1.4 Single-Molecule Physics

Moerner at IBM reported the spectra of single molecules in solids (15). Blinking, frequency shifts, and trapping of molecules in dark states have been observed for a number of systems. In each case, the phenomena arose from interaction of the molecule with its environment. In those single-molecule physics experiments, differences in the behavior of chemically identical molecules were associated with differences in their environment.

14.2   Single-Molecule Enzymology

Enzymes are proteins that catalyze a reaction. Like all catalysts, they speed the rate of a reaction but do not shift the equilibrium between reactant and product. By monitoring the products generated by a single enzyme molecule, we study the details of this catalysis, and we probe several fundamental assumptions of chemistry.

In 1961, Rotman at the Syntex Institute measured the activity of single β-galactosidase molecules [16]. A very dilute solution of enzyme was mixed with a high concentration of fluorogenic substrate. The enzyme converted the weakly fluorescent substrate to a highly fluorescent product. Rotman dispersed this mixture into a fine aerosol, coated the droplets with silicon oil to prevent evaporation, and determined the fluorescence from individual microdroplets after a 10–15 hour incubation period.

This aerosol-based method was cumbersome. Instead, Yeung's group at Iowa State University and our group at the University of Alberta developed a capillary-based method for monitoring the activity of single enzyme molecules [17–20]. Yeung studied lactate dehydrogenase from rabbit muscle while we studied alkaline phosphatase isolated from calf intestine. Both enzymes are commonly used reporter molecules for enzyme-linked immunoassays, and the enzymes and fluorogenic substrates are easily to obtain. In both cases, a dilute solution of enzyme and concentrated solution of fluorogenic substrate was used to partially fill a fused silica capillary. In our case, the capillary was 10 μm in inner diameter and 50 cm in length. Each molecule of enzyme catalyzed the conversion of the weakly fluorescent substrate into highly fluorescent product. After incubation, the fluorescent product was eluted from the capillary by use of an electric field and passed through a high sensitivity laser-induced fluorescence detector.

The use of the capillary for incubation and an electric field for elution of fluorescent product solved one important problem in single-molecule enzymology. Although fluorogenic substrates were relatively weakly fluorescent, they were present at millimolar concentration. The fluorescence from the high concentration substrate swamped the signal from the picomolar-concentration fluorescent product. By eluting the product with an electric field after incubation, differences in electrophoretic mobility could be used to separate the substrate from the product, Fig. 14.1. The pools of fluorescent product were detected in the absence of the signal from the fluorogenic substrate. Each pool generated a peak as the product was swept through the fluorescence detector.

Figure 14.2 presents the fluorescence signal generated by a typical experiment with calf-intestinal alkaline phosphatase. It took about nine minutes for

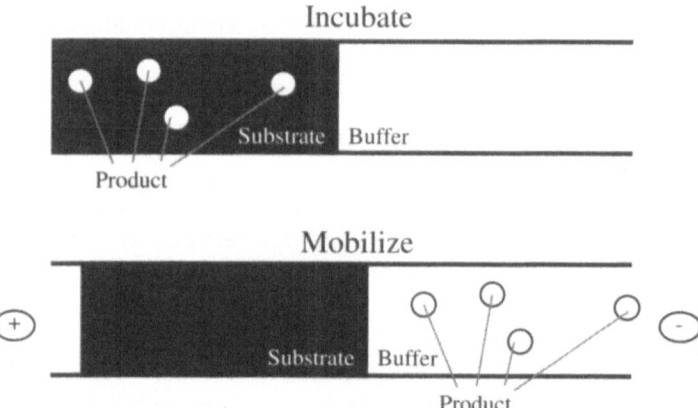

Fig. 14.1. Schematic of single molecule enzymology with on-column incubation. A highly dilute enzyme solution is mixed with substrate. A plug of this mixture is injected into a 15-μm diameter fused silica capillary. After incubation, electric field·is used to drive the fluorescent product to a high sensitivity detector. The net electrophoretic mobility of the product is higher than that of the substrate, and the fluorescent product generated by each enzyme molecule is detected as a peak in the absence of the background signal from the high concentration substrate

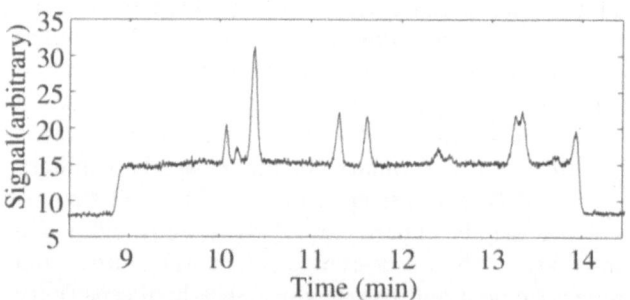

Fig. 14.2. Single-molecule enzymology of calf-intestinal alkaline phosphatase. The sample was incubated for 20 minutes at room temperature. The fluorescent substrate began to migrate from the capillary to the detector nine minutes after application of a 400 V/cm electric field. Autohydrolysis of the substrate resulted in a plug of fluorescent product within the capillary, which generated a five-minute plateau in the fluorescence signal. A set of ~12 peaks is observed on this plateau. Each peak was produced by the fluorescent product generated by a single enzyme molecule. The wide variation in peak height is characteristic of the molecule to molecule heterogeneity in activity for the mammalian enzyme

the fluorescent product to reach the detector. The substrate underwent autohydrolysis to produce the product, which formed a 5-minute wide plateau, corresponding to the volume of the capillary that was filled with substrate. A set of peaks was observed on this plateau; each peak was generated by the pool of fluorescent product generated by a single enzyme molecule.

Several tests were performed to confirm that these peaks were due to the pool of fluorescent product generated by single molecules. Most importantly, the number of peaks was Poisson distributed, with a mean value predicted by the nominal enzyme concentration and the sample volume. While the number of peaks was proportional to enzyme concentration, the area of each peak, which was proportional to the activity of each molecule, was independent of enzyme concentration.

14.2.1   Molecular Heterogeneity

There was large molecule to molecule variation in enzyme activity. This variation was surprising, and we invested significant effort to identify its source. We considered five possible sources. First, enzyme aggregates would cause heterogeneous activity, where each peak corresponds to several molecules of enzyme. However, zone electrophoresis showed that the enzyme migrates as several closely spaced bands with no evidence for aggregates. Second, the precision of the experiment may be poor, generating the large range of molecule to molecule activity. Third, enzymes could stick to the capillary walls; if the active site of a molecule is partially hidden, then its activity will be less than the activity of the free protein. Fourth, enzymes could denature during the assay; those molecules that denature early in the incubation period will have low apparent activity. Last, the enzyme may undergo posttranslational modifications that result in gross molecule to molecule structural differences, which generate the observed distribution in activity.

14.2.2   Repeated Incubations for a Single-Molecule

To investigate several of these experimental artifacts, we performed repeated incubations on a single enzyme molecule [19,20]. A sufficiently dilute solution of enzyme was used so that only one or two enzyme molecules were captured within the capillary. The enzyme was incubated for eight minutes, then a pulse of electric current was applied to drive the enzyme molecule into fresh substrate. This procedure was repeated for four-, two-, and one-minute incubations. After the last incubation, the electric field was used to drive the four pools of fluorescent product to the detector, which created a set of four peaks (Fig. 14.3).

Nonlinear regression analysis was used to fit the data with a set of four Gaussian peaks, which are shown as the dashed curve in the figure. The area of these peaks increased linearly with incubation time, $r = 0.998$, with zero

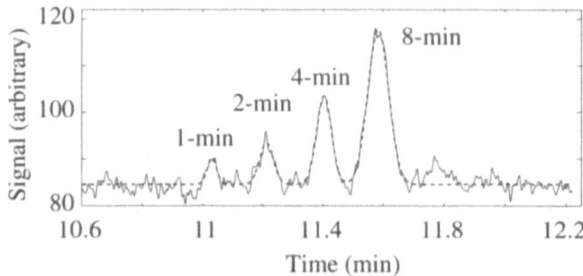

Fig. 14.3. Replicate incubations on the same enzyme molecule. A single enzyme molecule is captured within the capillary. After an 8-minute incubation, a brief electrophoretic pulse is applied to the capillary; this pulse separates the enzyme from the surrounding pool of product. The process is repeated for 4-, 2-, and 1-minute incubations. After the last incubation, the contents of the capillary are swept through the fluorescence detector by a 400 V/cm electric field. The dashed curve is the least-squares fit of a set of four Gaussian functions to the data

intercept. The peak width increased with incubation time due to diffusion of the fluorescent product during subsequent incubations.

Note that the last incubation, for one minute, corresponded to the first peak to elute from the capillary and that the first incubation, for eight minutes, corresponds to the last peak to elute. This reversal of peak elution order was only possible if the electrophoretic mobility of the enzyme was greater than the mobility of the fluorescent product.

This experiment eliminated most experimental artifacts as causes for the molecule to molecule variation in activity. The activity determination had better than 1% relative precision. The much larger molecule to molecule variation in activity observed in Fig. 14.2 was not due to poor experimental precision.

The data also eliminated the possibility that molecules stuck to the capillary walls. The inversion in migration order for the peaks in Fig. 14.3 was only possible if the mobility of the enzyme was higher than the mobility of the fluorescent product.

The data also demonstrated that the molecule did not denature during the period of the experiment. The peak corresponding to the last incubation shows no sign of the enzyme having denatured.

14.2.3   Activation Energy for the Reaction Catalyzed by One Molecule

To determine whether other properties of this enzyme also demonstrated molecule to molecule heterogeneity, we determined the activation energy for the reaction catalyzed by a single enzyme molecule. The instrument was modified so that the temperature of the capillary could be adjusted. Again, a dilute

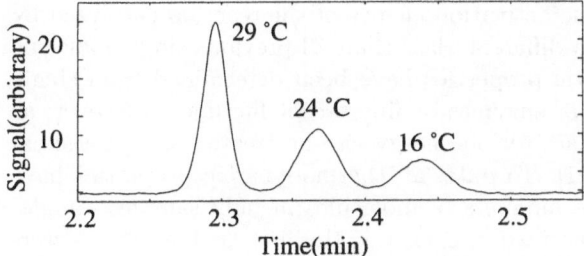

Fig. 14.4. Determination of the activation energy of the reaction catalyzed by a single enzyme molecule. A single enzyme molecule was captured within the capillary. The molecule was incubated for 15-minutes at 16°C. Brief electrophoretic pulses were applied to separate the enzyme molecule from the substrate, and incubations were repeated at 24 and 29°C. Following the third incubation, the sample was flushed past the detector by a 400 V/cm electric field

solution of calf-intestinal alkaline phosphatase was introduced into the capillary so that only one or two molecules were captured. In the data of Fig. 14.4, an incubation was performed at 16°C for 20 minutes. An electric field was applied to drive the molecule into fresh substrate, and a second incubation was performed at 24°C for 20 min. Last, the procedure was repeated to generate an incubation at 29°C for 20 min. After the last incubation, the contents of the capillary were flushed through the detector, and the fluorescence signal was recorded.

The peak area increased with reaction temperature, corresponding to a larger fraction of substrate molecules having sufficient energy to overcome the activation barrier and convert to product. The peak width increased for the low temperature data due to diffusion of those product molecules during the subsequent incubations.

A plot of ln(peak area) vs. $1/T$ yields a slope of $-E_a/R$, where E_a is the activation energy. A linear least-squares fit to the data was used to estimate the activation energy. The data were linear, with an average correlation coefficient of 0.994. The relative precision of the slope, estimated by the least-squares routine, ranged from 3 to 21%, with an average relative precision of 9%. Activation energies varied from 39 to 91 kJ mol^{-1}. The mean activation energy was 53 kJ mol^{-1} and the standard deviation of the distribution was 16 kJ mol^{-1} ($N = 8$). There was no correlation between activation energy and activity, $r = -0.26$. Activation energy and the preexponetial term of the Arrhenius equation were highly correlated, $r = 0.998$, with near-zero intercept.

A solution of 4.6×10^{-12} M calf-intestinal alkaline phosphatase was incubated at 23 and 32°C; fluorescent product was analyzed by capillary electrophoresis. The activation energy for this ensemble was 50 kJ/mol and was identical to the single-molecule result.

This determination of the activation energy of the reaction catalyzed by a single molecule falls into a different class than all previous single-molecule experiments. Classical kinetic properties have been determined from single molecules, such as activity of enzymes or fluorescent lifetimes. However, as Henry Eyring pointed out, activation energy can be treated as a statistical thermodynamic property [21]. To date, all thermodynamic properties have been determined from large numbers of molecules in bulk samples. Single-molecule determination of activation energy is the first time-average determination of a thermodynamic property, and is possible because we monitor the products generated from many reactions catalyzed by one molecule.

Although the average single-molecule determination of activation energy is identical to the ensemble determination, the molecule to molecule activation energy varies by nearly a factor of three. This molecule to molecule difference in activation energy is much larger than the experimental uncertainty, and it reflects the heterogeneous nature of this enzyme.

14.2.4   Causes of Heterogeneity

The central paradigm of chemistry states that chemical function is determined by chemical structure. These molecule to molecule differences in activity and activation energy must be due to molecule to molecule differences in structure.

If every molecule in a compound is identical, then these structural differences must be subtle indeed. Yeung has argued that these molecule to molecule differences are an extreme example of the energy landscape model of protein structure [17]. In this model, a protein explores a hierarchy of conformations. In Yeung's interpretation, the molecule to molecule differences in activity arise because the molecules are captured in different, long-lived conformations.

In contrast, we have argued that the molecule to molecule differences in activity and activation energy instead are the result of impure samples [19]. These impurities arise from differences in post-translational modification of the protein. Most eukaryotic enzymes are highly modified. Calf-intestinal alkaline phosphatase, like most eukaryote enzymes, is cleaved from an apoprotein [22]. Many enzymes are extensively glycosylated, have N-terminal acetylation, and other post-translational modifications. These modifications cause variations in both Km and Vmax in enzymes [23].

The post-translational modifications generated a range of products, which were revealed as a complex pattern during isoelectric focusing electrophoresis. Figure 14.5 presents the isoelectric focusing band generated by 1-microgram of calf-intestinal alkaline phosphatase that was visualized with a fluorogenic phosphatase reagent. The broad smear is associated with a complex mixture of enzyme isoforms.

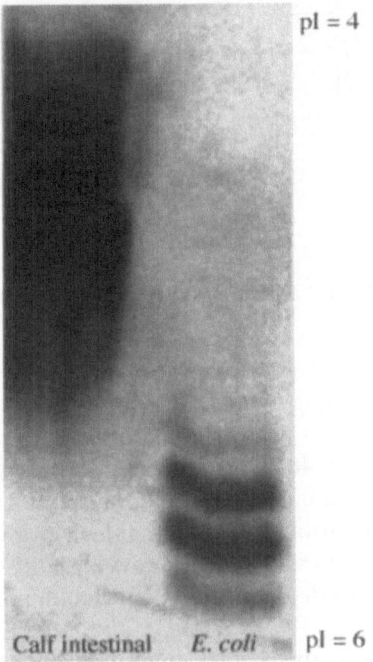

pl = 4

Calf intestinal   *E. coli*   pl = 6

Fig. 14.5. Isoelectric focusing gel of alkaline phosphatase. The left lane was generated by analysis of calf intestinal alkaline phosphatase while the right lane was generated by *E. coli* alkaline phosphatase. The gel was visualized by treatment with a fluorogenic reagent

14.3   Identical Molecules Behave Identically

While many eukaryotic enzymes undergo significant post-translational modification, prokaryotic enzymes tend to have fewer modifications and tend to generate much simpler isoelectric focusing patterns. *Escherichia coli* alkaline phosphatase is a non-glycosylated protein that is coded by a single PhoA gene. The enzyme also requires proteolysis of an apo-protein. The active enzyme consists of two 449 amino acid monomers. The enzyme has been found to exist in three forms, designated isozyme 1, 2, and 3. The isozymes differ by the presence of an N-terminal arginine, which is present in isozyme 1 but not isozyme 3. Isozyme 2 is a heterodimer of isozyme 1 and 3 monomers [24]. No other modifications are known of the native protein.

We analyzed this enzyme by isoelectric focusing, Fig. 14.5. The enzyme generated a set of three main bands with a fourth faint, lower pI band barely visible in the electropherogram. We purified each of the three main bands from the isoelectric focusing gel. Care was taken to preserve the purity of this enzyme. All procedures were performed at 4°C, and a protease inhibitor cocktail was added to the isolated enzyme to minimize its proteolytic degradation.

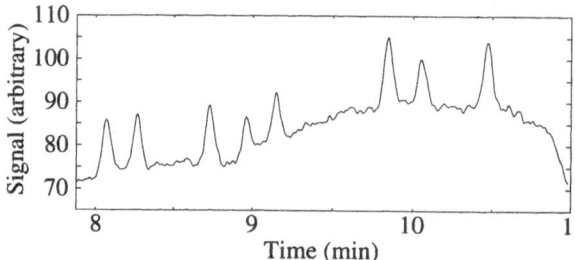

Fig. 14.6. Single-molecule enzymology of *E. coli* alkaline phosphatase. The mixture was incubated for 30 min at 40°C

We performed single-molecule enzymology with *E. coli* alkaline phosphatase [25]. This enzyme is much less active than the mammalian enzyme. We modified the instrument to operate at 40°C to increase the activity of the enzyme and we used relatively long incubation times in our experiments. Unfortunately, the high-temperature operation of our instrument introduced drift in the background signal. When combined with the low activity of the enzyme, the signal-to-noise ratio was degraded compared to the mammalian enzyme.

Figure 14.6 presents a single-molecule analysis of the highly purified *E. coli* enzyme. The molecule to molecule activity of the enzyme was much more homogeneous than observed for the calf-intestinal enzyme. The relative standard deviation in molecule to molecule activity was ~30% and was dominated by the precision of the measurement for this low activity enzyme.

There was no significant difference in the activity of molecules isolated from each isoelectric focusing band. This similarity was not surprising; the N-terminal arginine is far removed from the active site of alkaline phosphatase, and it should not modulate the activity of the enzyme.

14.3.1 Degraded *E. coli* Alkaline Phosphatase

There is a potential pitfall in single-molecule enzyme measurements. Samples of *E. coli* alkaline phosphatase were stored at 4°C for several days in the absence of protease inhibitor. During storage, the enzyme was degraded by endogenous proteases. An isoelectric focusing analysis of these samples revealed up to 12 fractions of enzyme that retained their phosphatase activity. The degraded sample was used for single-molecule enzymology. The molecule to molecule activity in this sample varied by a factor of five, which reflects the heterogeneous population produced during degradation. The data of Fig. 14.6 were generated with samples treated with a protease inhibitor, which helped preserve the enzyme's purity.

14.4   Time-Resolved Single-Molecule Enzymology

Xie at Pacific Northwest [26] and Rigler at the Karolinska Institute [27] measured time-resolved fluorescence from individual enzyme molecules undergoing reaction. In Xie's case, single molecules of cholesterol oxidase were trapped in a gel matrix. Fluorescence from a flavin prosthetic group was monitored. Reduction and oxidation of the group were used as a surrogate measure of the activity of the enzyme. Careful analysis of the kinetics of the reaction revealed that the enzyme did not function at a uniform rate; instead, the molecule fluctuated between active and inactive states. Similar results were obtained by Rigler for horseradish peroxidase that was tethered through a biotin/streptavidin linkage to a glass substrate. In addition to fluctuations in the time-resolved activity of a single molecule, both groups observed large molecule to molecule variation in activity.

14.5   Isoelectric Focusing Analysis of Other Enzymes

To see whether the molecular heterogeneity observed in earlier single-molecule experiments could be resolved by isoelectric focusing, we obtained samples from the manufacturer of the enzymes used for several of those studies. As shown in Fig. 14.5, calf-intestinal alkaline phosphatase generated a complex isoelectric focusing pattern and highly heterogeneous molecule to molecule activity.

Yeung reported large molecule to molecule variation in the activity of lactate dehydrogenase [17,18]. There is some confusion as to the identity of the enzyme actually used in the experiment; the enzyme was not purified before use. We analyzed Type I lactate dehydrogenase from rabbit muscle by isoelectric focusing and we observed four bands. These enzymes are known to be highly glycosylated [28]. The heterogeneity in single-molecule enzymatic activity may be due to post-translational modification of the enzyme, forming a series of isoenzymes with different activities.

We reported a wide range of activity for β-galactosidase [29]. Isoelectric focusing of this enzyme generated a smear from pH 4.6 to pH 5. The molecule to molecule variation in enzymatic activity likely reflects differences in structure due to proteolysis of this enzyme.

Xie reported the activity of single cholesterol oxidase molecules [26]. The observed rate constant was measured for 32 different molecules and varied by a factor of five. This enzyme was purified by gel filtration chromatography and generated a single band with SDS-PAGE. We obtained a sample of this enzyme from the manufacturer. Unfortunately, our isoelectric focusing ampholytes did a poor job of resolving components of this high-pI enzyme; the protein sample generated a set of two bands from pH 8.2 to pH 9.6, with most of the protein dispersed as a broad smear in the high pI region of the gel.

We have not obtained a sample of the horseradish peroxidase used by Rigler for his single-molecule studies. However, this enzyme is extremely complex. It is coded by at least two different genes and undergoes extensive post-transcriptional modification [30]. The main isozyme has eight different glycosylation sites and has a pyrrolidonecarboxyl amino terminus and heterogeneity at the carboxyl terminus [31]. In addition, the enzyme is further modified by the manufacturer by the chemical addition of a biotin group to lysine residues on the protein. This labeling chemistry is known to generate a complex mixture of products [32], which is undoubtedly responsible for the observed molecule to molecule variation in activity.

14.5.1   Molecule to Molecule Heterogeneity and the Energy Landscape Model

Gross structural differences are associated with molecule to molecule heterogeneity in enzyme activity. Conversely, highly purified enzyme generates homogeneous activity. There is no need to invoke the energy landscape model to explain molecule to molecule variations in enzyme activity. Instead, these molecule to molecule variations are associated with gross differences in the primary structure of the molecule. Post-translational modification of enzymes is particularly common in eukaryotes, and proteolytic digestion of all enzymes is a potential source of molecule to molecule structural differences. Great care is required to obtain pure enzyme and to preserve its purity during experiments. Isoelectric focusing is a useful tool to probe enzyme purity, and protease inhibitors should be used prophylactically to prevent degradation of the purified enzyme.

14.6   Thermal Denaturation of *E. coli* Alkaline Phosphatase

Enzymes, like all proteins, undergo irreversible denaturation when heated. The loss of activity is monitored from an ensemble of molecules. Aliquots are periodically withdrawn from a heated solution and quenched by dilution in cool buffer. The residual activity decreases with increasing heating time. Heating disrupts the hydrogen bonds that determine the secondary structure of the molecule. Prolonged heating disrupts more bonds, eventually causing the complete loss of activity. Single-molecule enzymology provides clues to the mechanism of thermal denaturation.

14.6.1   Cheshire Cat Model of Thermal Denaturation

Heating of the enzyme could drive the molecule through a series of active conformations, each with gradually lower activity. Disruption of bonds far from the active site would have a minor effect on the geometry of that site, leading to minor changes in activity. Disruption of bonds closer to the active site would lead to a more dramatic decrease in activity. In this Cheshire Cat model of thermal denaturation, heating causes the protein to form structures progressively further removed from the optimal conformation, with a concomitant reduction in activity.

The Cheshire Cat model of thermal denaturation can be used to predict the effects of heating on single-molecule analysis of enzyme activity. The loss of activity observed in bulk samples would be caused by a shift in the distribution of enzyme activity to lower values. However, the number of active enzyme molecules would not change dramatically with heating, at least until the bulk activity had decreased significantly.

14.6.2   Catastrophic Model of Thermal Denaturation

In contrast, if enzymes are relatively robust, the breakage of hydrogen bonds far removed from the active site should not significantly influence the molecule's activity. Only the disruption of bonds near the active site would be crucial, and the disruption of these bonds would likely cause an irreversible loss of activity.

If a sample is subjected to conditions that cause the loss of half the bulk activity, then this catastrophic denaturation model predicts that half the molecules will be irreversibly denatured while half will retain their activity.

14.6.3   Thermal Denaturation of *E. coli* Alkaline Phosphatase

A sample of *E. coli* alkaline phosphatase was heated to 96°C. Aliquots were periodically withdrawn from the sample and quenched by a two order of magnitude dilution in a borate buffer. These aliquots were then analyzed by single-molecule enzymology. There was no systematic change in the activity of the surviving, active molecules. Figure 14.7 presents the fraction of surviving molecules with heating time. The fraction of surviving molecules decreased exponentially ($X^2 = 6.6$, $D.F. = 15$, $P = 0.95$) with heating time, with a characteristic time constant of 360 s, but the distribution in activity of the surviving molecules was unchanged from the initial sample.

Single-molecule enzymology demonstrates that thermal denaturation is a catastrophic phenomenon. There is no evidence for a gradual loss of activity during thermal denaturation. Instead, the loss of activity due to denaturation results from the loss of active molecules, and the activity of the surviving molecules is independent of denaturation.

14.7   Conclusions

Single-molecule enzymology provides the study of the long-term behavior of individual molecules. This technology also allows us to test fundamental assumptions of chemistry.

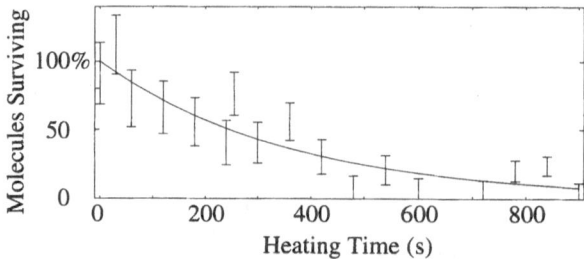

Fig. 14.7. Kinetics of thermal denaturation of *E. coli* alkaline phosphatase. *E. coli* was heated to 96°C. Aliquots were withdrawn and quenched to room temperature. The number of surviving molecules was determined by single-molecule enzymology. The fraction of surviving molecules decreased exponentially with heating time (*smooth curve*). Data are plotted ± one standard deviation, which was estimated based on Poisson statistics

Several conclusions may be drawn from these experiments.

- Chemically identical molecules behave identically.
- Chemically heterogeneous molecules generate a large molecule to molecule variation in activity and activation energy. The large molecule to molecule variation in enzyme activity reported by others is almost certainly an artifact due to the use of impure enzymes.
- The activation energy of the reaction catalyzed by a single molecule can be measured.
- Differences in glycosylation result in a nearly threefold difference in activation energy for calf-intestinal alkaline phosphatase.
- Thermal denaturation is a catastrophic phenomenon. The loss of activity due to thermal denaturation is due to the loss of active enzymes; there is no systematic change in the activity of the surviving molecules.

In addition to these molecule to molecule variations in activity, both Xie and Rigler have described a short-term fluctuation in the activity of an individual enzyme molecule [26,27]. Those fluctuations arise from the molecule switching between two states with a rate of ~ 1 s^{-1}. This dynamic variation in activity cannot be explained by post-translational modifications and instead must be due to short-term fluctuations in the molecule's structure.

Acknowledgment. This work was funded by a Research Grant from the Natural Sciences and Engineering Research Council.

References

1. G. W. Greenlees, D. L. Clark, S. L. Kaufman, D. A. Lewis, J. F. Tonn, and J. H. Broadhurst, "High resolution laser spectroscopy with minute samples," Opt. Commun. **23**, 236–239 (1977)

2. C. L. Pan, J. V. Prodan, W. M. Fairbank Jr., and C. Y. She, "Detection of individual atoms in helium buffer gas and observation of their real-time motion," Opt. Lett. **5**, 459–461 (1980)
3. T. Hirschfeld, "Optical microscopic observation of single small molecules," Appl. Opt. **15**, 2965–2966 (1976)
4. N. J. Dovichi, J. C. Martin, J. H. Jett, and R. A. Keller, "Attogram detection limit for aqueous dye samples by laser-induced fluorescence," Science **219**, 845–847 (1983)
5. N. J. Dovichi, J. C. Martin, J. H. Jett, M. Trkula, and R. A. Keller, "Laser-induced fluorescence of flowing samples as an approach to single-molecule detection in liquids," Anal. Chem. **56**, 348–354 (1984)
6. D. C. Nguyen, R. A. Keller, J. H. Jett, and J. C. Martin, "Detection of single molecules of phycoerythrin in hydrodynamically focused flows by laser-induced fluorescence," Anal. Chem. **59**, 2158–2161 (1987)
7. N. J. Dovichi, "Attogram detection limits using laser-induced fluorescence," Trends Anal. Chem. **3**, 55–57 (1984)
8. Y. F. Cheng and N. J. Dovichi, "Subattomole amino acid analysis by capillary zone electrophoresis and laser-induced fluorescence," Science **242**, 562–564 (1988)
9. H. Swerdlow, J. Z. Zhang, D. Y. Chen, H. R. Harke, R. Grey, S. Wu, C. Fuller, and N. J. Dovichi, "Three DNA sequencing methods using capillary gel electrophoresis and laser-induced fluorescence," Anal. Chem. **63**, 2835–2841 (1991)
10. Z. Zhang, S. Krylov, E. A. Arriaga, R. Polakowski, and N. J. Dovichi, "One-dimensional protein analysis of an HT29 human colon adenocarcinoma cell," Anal. Chem. **72**, 318–322 (2000)
11. J. Y. Zhao, N. J. Dovichi, O. Hindsgaul, S. Gosselin, and M. M. Palcic, "Detection of 100 molecules of product formed in a fucosyltransferase reaction," Glycobiology **4**, 239–242 (1994)
12. N. J. Dovichi, Development of DNA sequencer, Science **285**, 1016 (1999)
13. J. Z. Zhang, K. O. Voss, D. F. Shaw, K. P. Roos, D. F. Lewis, J. Yan, R. Jiang, H. Ren, J. Y. Hou, Y. Fang, X. Puyang, H. Ahmadzadeh, and N. J. Dovichi, "A multiple-capillary electrophoresis system for small-scale DNA sequencing and analysis," Nucleic Acids Res. **27**, E36 (1999)
14. D. Y. Chen and N. J. Dovichi, "Single-molecule detection in capillary electrophoresis: Molecular shot noise as a fundamental limit to chemical analysis," Anal. Chem. **68**, 690–696 (1996)
15. W. E. Moerner and L. Kador, "Optical detection and spectroscopy of single molecules in a solid," Phys. Rev. Lett. **62**, 2535–2538 (1989)
16. B. Rotman, "Measurement of activities of single molecules of β-galactosidase," Proc. Natl. Acad. Sci. **47**, 1981–1986 (1961)
17. Q. Xue and E. S. Yeung, "Differences in the chemical reactivity of individual molecules of an enzyme," Nature **373**, 681–683 (1995)
18. W. Tan and E. S. Yeung, "Monitoring the reactions of single enzyme molecules and single metal ions," Anal. Chem. **69**, 4242–4248 (1997)
19. D. B. Craig, E. Arriaga, J. C. Y. Wong, H. Lu, and N. J. Dovichi, "Studies on single alkaline phosphatase molecules: Reaction rate and activation energy of a reaction catalyzed by a single molecule and the effect of thermal denaturation – the death of an enzyme," J. Am. Chem. Soc. **118**, 5245–5253 (1996)
20. D. B. Craig, E. Arriaga, J. C. Y. Wong, H. Lu, and N. J. Dovichi, "The life and death of a single enzyme molecule," Anal. Chem. **70**, 39A–43A (1998)

256   N. J. Dovichi et al.

21. S. Gladstone, K. J. Laidler, and H. Eyring, *The Theory of Rate Processes*, p. 108 (McGraw-Hill, New York, 1941)
22. L. Engström, "Studies on calf-intestinal alkaline phosphatase. 1. Chromatographic purification, microheterogeneity and some other properties of the purified enzyme," Biochim. et Biophys. Acta **52**, 36–48 (1961)
23. A. Varki, "Biological roles of oligosaccharides: All of the theories are correct," Glycobiology **2**, 97–130 (1993)
24. R. A. Bradshaw, F. Cancedda, L. H. Ericsson, P. A. Neumann, S. P. Piccoli, M. J. Schlesinger, K. Shriefer, and K. A. Walsh, "Amino acid sequence of Escherichia coli alkaline phosphatase," Proc. Natl. Acad. Sci. USA **78**, 3473–3477 (1981)
25. R. Polakowski, D. Craig, A. Skelley, and N. J. Dovichi, "Single molecules of highly purified bacterial alkaline phosphatase," J. Am. Chem. Soc. **122**, 4853–4855 (2000)
26. H. P. Lu, L. Sun, and X. S. Xie, "Single-molecule enzymatic dynamics," Science 282, 1877–1882 (1998)
27. Edman, Z. Foldes-Papp, S. Wennmalm, and R. Rigler, "The fluctuating enzyme: A single molecule approach,'" Chem. Phys. **247**, 11–22 (1999)
28. D. Heinova, J. Blahovec, and I. Rosival, "Lactate dehydrogenase isoenzyme patterns in bird, carp and mammalian sera," Eur. J. Clin. Chem. Clin. Biochem. **34**, 91–95 (1996)
29. D. B. Craig and N. J. Dovichi, "E. coli β-galactosidase is heterogeneous with respect to the activity of individual molecules," Can. J. Chem. **76**, 623–626 (1998)
30. K. Fujiyama, H. Takemura, S. Shibayama, K. Kobayashi, J. K Choi, A. Shinmyo, M. Takano, Y. Yamada, and H. Okada, "Structure of the horseradish peroxidase isozyme C genes," Eur. J. Biochem. **173**, 681–687 (1988)
31. K. G. Welinder, "Amino acid sequence studies of horseradish peroxidase. Amino and carboxyl termini, cyanogen bromide and tryptic fragments, the complete sequence, and some structural characteristics of horseradish peroxidase C," Eur. J. Biochem. **96**, 483–502 (1979)
32. D. B. Craig and N. J. Dovichi, "Multiple labeling of proteins," Anal. Chem. **70**, 2493–2494 (1998)

15 The Energy Landscape

H. Frauenfelder and B. H. McMahon

15.1 Why the Energy Landscape is Important

If proteins existed in unique structures, with each atom always in the same place, there would be no need to introduce the concepts of an energy landscape and of conformational substates. Single protein experiments would give the same results as experiments on ensembles and would therefore be unnecessary; this Nobel conference would not have taken place. Proteins would be rigid and could not function. In reality, proteins can assume a very large number of somewhat different conformations [1,2]. In some cases, such as prions and ameloid fibers, the protein conformation may change completely. In other cases, conformational motions activate enzymes, permit access to active sites, or cleave chemical bonds. The energy landscape is the conceptual framework in which these motions and reactions can be described. Here we sketch the experiments that led us to the concepts of substates and landscapes and describe their importance for the interpretation of single-molecule experiments. For surveys, see for instance the papers presented at a conference on landscapes [3]. Some of the data and concepts treated here were discussed from a different point of view [4] at an earlier Nobel Symposium on Structure and Dynamics of Biological Systems.

15.2 Energy Landscape and Conformational Substates

The old textbook picture of proteins is sketched in Fig. 15.1: The primary sequence folds into a unique tertiary structure, shown in beautiful pictures in texts and journals. Experiments prove that this picture is misleading. One of the first experiments was the study of the time dependence of the rebinding of carbon monoxide (CO) to myoglobin (Mb) at temperatures below about 200 K [5,6]. Mb with CO bound at the central heme iron was cooled to low temperatures. The Fe–CO bond was broken by a light flash and the subsequent rebinding, $N(t)$, monitored as a function of time. $N(t)$ denotes the fraction of Mb that have not rebound a CO at the time t after photodissociation. Surprisingly $N(t)$ turned out to be nonexponential in time as shown in Fig. 15.2.

Nonexponential time dependence in physical phenomena has been known since Gauss. Explanations fall into two classes, sketched in Fig. 15.3. In the

The Textbook Picture

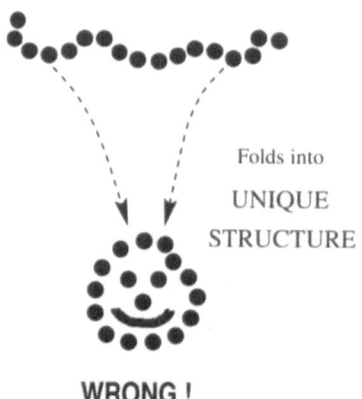

Folds into

UNIQUE
STRUCTURE

WRONG !

Fig. 15.1. The old textbook picture: The primary sequence folds into a unique tertiary structure

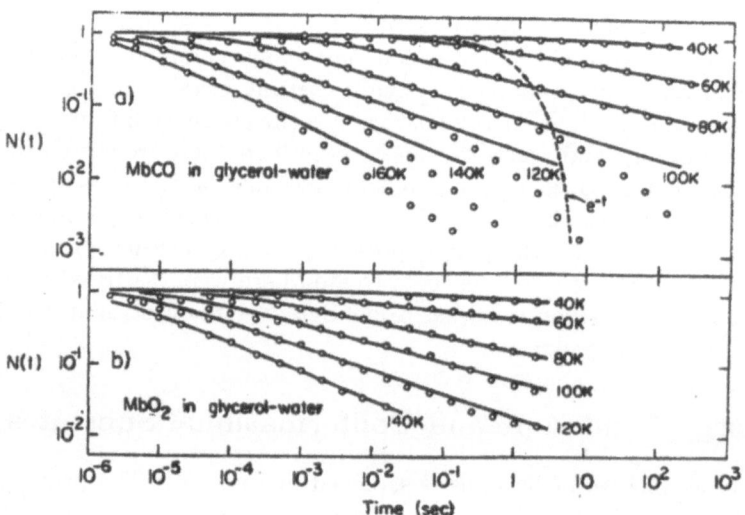

Fig. 15.2. Nonexponential rebinding of CO and O_2 to Mb after photodissociation [6]. In the upper panel, an exponential is shown for comparison

homogeneous case, all Mb molecules are identical, and each rebinds nonexponentially in time, with $\Phi(t)$ described by a rate coefficient, $k(t)$, that decreases with time. In the heterogeneous case, each Mb molecule rebinds with a different rate coefficient so that the overall rebinding is given by

$$N(t) = \int dk f(k) \exp\left(-kt\right). \tag{15.1}$$

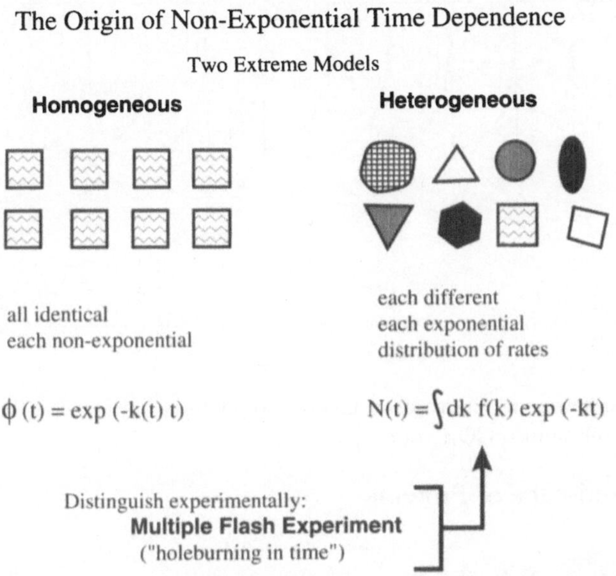

The Origin of Non-Exponential Time Dependence

Two Extreme Models

Homogeneous   **Heterogeneous**

all identical   each different
each non-exponential   each exponential
                                 distribution of rates

$$\phi(t) = \exp(-k(t)\,t)$$ $$N(t) = \int dk\, f(k)\, \exp(-kt)$$

Distinguish experimentally:
Multiple Flash Experiment
("holeburning in time")

Fig. 15.3. A non-exponential time dependence can be caused by each molecule having the same nonexponential time dependence (homogeneous case), or by each molecule having a different rate coefficient (inhomogeneous case)

Here k is the rate coefficient for binding and $f(k)$ the fraction of Mb molecules that rebind CO with a rate coefficient between k and $k + dk$. Single-molecule experiments are perfect for deciding between the two models. However, 25 years ago they did not exist. A poor-man's version, multiple-flash experiments, settled the argument [6,7]. In such an experiment, the sample is photodissociated repeatedly whenever about one half of the Mb molecules have rebound their ligand. In the homogeneous case, repeated flashing pumps more and more Mb into long-lived states and the flashed-off fraction becomes smaller with each flash. In the inhomogeneous case, the repeated flashing selects a fast-rebinding subset ("single molecule") and, after an initial increase because of insufficient laser power (1974!), the flashed-off fraction remains constant. The experiment, shown in Fig. 15.4, proves that the Mb sample is inhomogeneous.

Why are proteins inhomogeneous? When the primary sequence folds, it can lead to a very large number of somewhat different structures or conformations. We call these conformations, represented as valleys in Fig. 15.5, conformational substates. (Substates, because the entire protein can assume a number of different states, each consisting of a large number of substates.) At low temperatures, each protein will remain in a given substate. At high temperatures, each molecule fluctuates from substate to substate. All possible substates are described by the energy landscape.

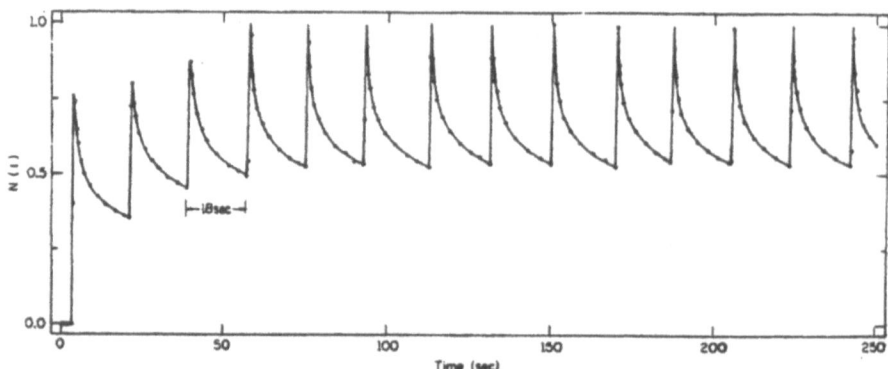

Fig. 15.4. Multiple-flash experiment. Photodissociation is repeated after about half of the Mb molecules have rebound a CO (from [6])

The Energy Landscape of Proteins

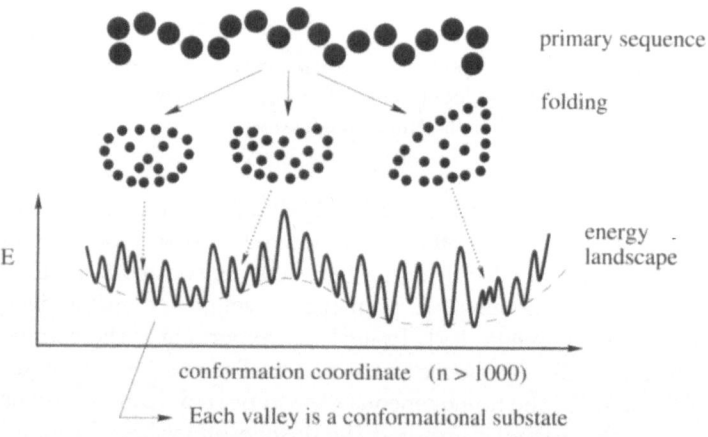

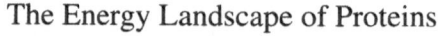

Fig. 15.5. The primary sequence can fold into a large number of somewhat different structures, called conformational substates. To change the structure, or in other words, move from one substate to another, the protein must overcome a barrier

The energy landscape concept requires some explanation. In Fig. 15.5 substates are shown as valleys along a one-dimensional unspecified coordinate. The complete energy landscape is, however, defined as giving the energy (or enthalpy) of a protein as a function of the positions of all atoms, ions, and hydrated waters. For a protein like Mb it is therefore a construct in a space of about 3000 dimensions. A complete knowledge of the energy landscape is equivalent to knowing the energy of all possible structures of the protein. A point in the landscape hyperspace characterizes a particular protein structure. Substates are valleys in this hyperspace. Fig. 15.6 describes this relation.

3000 Dimensional Energy Landscape

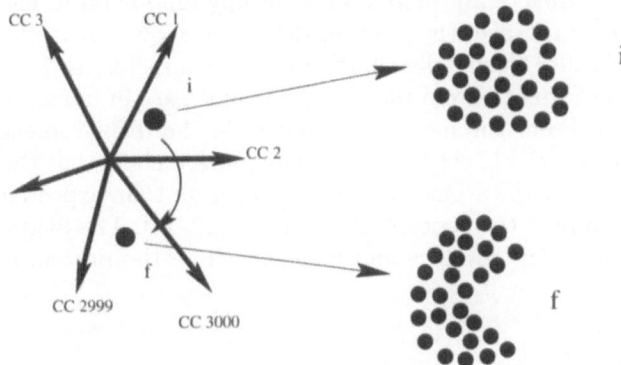

Fig. 15.6. Left: The CC 1 to CC 3000 are the 3000 coordinates that describe the positions of all atoms in a protein with 1000 atoms. A point in this hyperspace characterizes the structure of a protein (right) at a given time. A transition from one point to another (left) corresponds to a change in conformation in which the positions of all atoms may change (right)

At left, some of the few thousand coordinates of the landscape hyperspace are shown schematically. A point in this hyperspace, denoted by i, corresponds to a given structure, shown at right. A second point, f, characterizes another structure, for instance one with a channel wide open. The step $i \rightarrow f$ in the landscape hyperspace describes the transition from the closed to the open structure in the real three-dimensional space. It is difficult, or even impossible, to visualize a landscape in a hyperspace. Only computers can do so [8].

A simple example, however, may clarify the concept. The one-dimensional representation of an energy landscape, Fig. 15.5, implies that the protein must overcome many barriers in order to change from a substate on the left to one on the right and that there is only one path to do so. Already in two dimensions a landscape exhibits more features, as a look at any topographic map shows. Saddles and passes appear. To go from Zermatt to Torino, one can, but does not have to climb over the Matterhorn; many alternative routes are easier. Some are nearly flat and can be traveled by car. In the hyperspace of the energy landscape the number of pathways between any two substates is extremely large and some may involve just one step with a small barrier.

So far, we have used only the non-exponential time dependence of ligand binding to Mb to prove the existence of an energy landscape and conformational substates. Many other techniques support concepts and provide additional information. We sketch two of these techniques. More information can be found in reviews [9,10].

15.2.1   X-ray Diffraction

Substates imply that the atoms of a protein do not occupy unique positions, but are distributed around the average position determined by X-rays. As shown by Debye and by Waller, such a distribution can be characterized by a mean-square displacement for each atom that leads to a decrease in intensity of the diffraction pattern. From the measured intensities, the displacement can be extracted (e.g. Garcia et al., [11]). Experiments indeed show that the mean-square displacements for the atoms of proteins are larger than expected for vibrations and thus support the concept of substates [12–14]. The distributions vary widely for different residues and indicate where the protein is flexible.

15.2.2   Spectral Hole Burning

Further evidence for substates comes from spectral hole burning experiments [15]. Spectral lines in different substates will usually have somewhat different wave numbers. A spectral band observed in a protein sample is consequently inhomogeneous. With lasers holes can be burned into such bands. The width of the overall line compared to the width of the hole gives a lower limit on the number of substates. This number is larger than 10^6. Such hole-burning experiments are also like single-molecule studies: Through the very narrow laser line, a very small set of substates is selected and probed.

15.2.3   Photosynthetic Reaction Center

The discussion so far has been restricted nearly completely to Mb. Other proteins show similar phenomena. The photosynthetic reaction center protein, for instance, allows electron transfer rates to be used to probe the time and temperature dependence of the protein adaptation to internal charge separation [16–18]. At room temperature, most of the relaxations occur as a broad relaxation from 100 ps to 1 μs; at lower temperatures, half of the energy is dissipated before motions of the glassy solvent occur. The landscape that emerges is remarkably similar to the one found in Mb [17].

15.3   The Hierarchy of Substates

A look at various experiments suggests that barrier heights can differ considerably and that substates are organized hierarchically into a number of tiers [19]. Nestled within each of the substates of a given tier are substates with much smaller barriers. The arrangement shown in Fig. 15.7 provides a glimpse into the complexity of the protein energy landscape [20]. An updated view of the energy landscape of myoglobin is displayed in Fig. 15.8 as a tree diagram. CO bound to the heme iron in myoglobin has three different

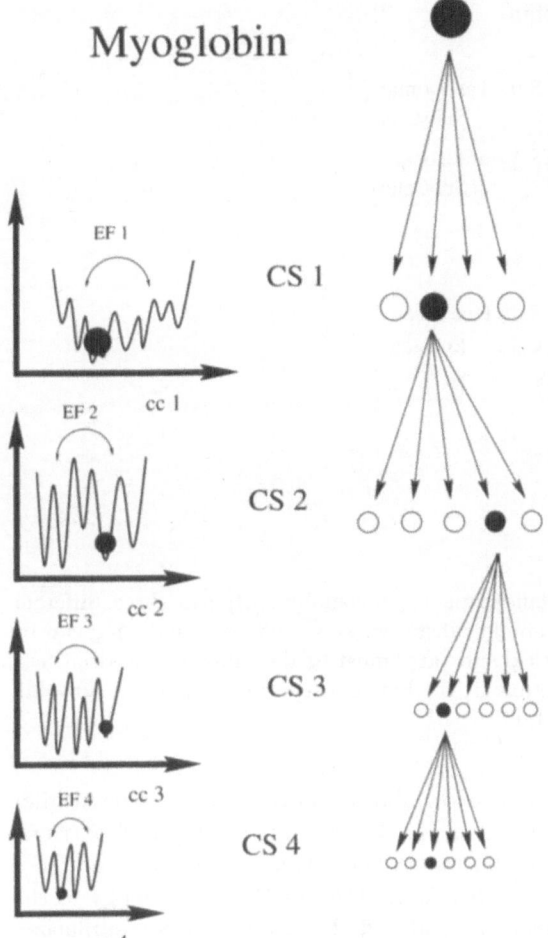

Fig. 15.7. Hierarchy of the energy landscape as envisaged in 1985 [19]. The conformational substates are classified into tiers according to the average barrier height between valleys

stretch bands, with center wave numbers of 1967 cm^{-1} for (A$_0$), 1947 cm^{-1} for (A$_1$), and 1929 cm^{-1} for (A$_3$) [21–23]. These bands correspond to three different substates whose properties can be determined individually and are therefore called taxonomic substates and assigned to tier 0 [24]. Such taxonomic substates also occur in other proteins, for instance in hemoglobin (R → T transition) and in calmodulin (with and without calcium bound).

The substates that give rise to the nonexponential rebinding at low temperatures and the inhomogeneous spectral lines are so numerous that they cannot be described individually; they are consequently called statistical substates. The barriers between the statistical substates are still sizable and

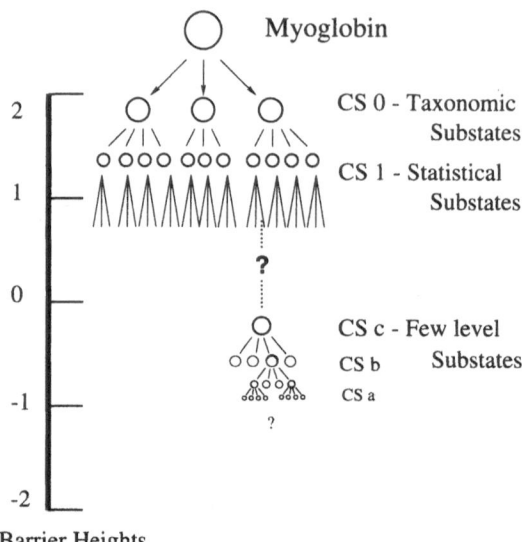

Fig. 15.8. An updated energy landscape for myoglobin displays three different types of substates [20]. The taxonomic substates are so different that they can be characterized individually. Statistical substates must be described by distributions. Few-level substates are small in number; the barriers separating them are so small that tunnel transitions are important

transition rates become very small below about 200 K. Hole burning studies down to the mK range reveal a number of substates with very small barriers, less than 10 kJ/mol [25]. The experiments show that the number of these substates is small and we call them *few-level states* or FLS, in analogy to the famous TLS observed in amorphous solids [26,27]. Figure 15.8 summarizes the situation and shows that the energy landscape in a protein as "simple" as myoglobin is quite complex. Many of the features shown in Fig. 15.8 are not yet fully proven. The energy landscape remains to be further explored and the connection to structure and function constitutes a major challenge.

15.4   Dynamics: Relaxations and Fluctuations

The energy landscape provides the framework to describe and understand motions in proteins. Three categories of motions can be distinguished: vibrations, fluctuations, and relaxations. In vibrations, the protein remains in one substate and the motion is nearly harmonic along each normal mode. Vibrations can be studied best at very short times or low temperatures. Fluctuations are nonlinear equilibrium transitions between substates. As the temperature is lowered, fluctuations become increasingly slower. The hierarchy of substates implies that fluctuations in different tiers possess different

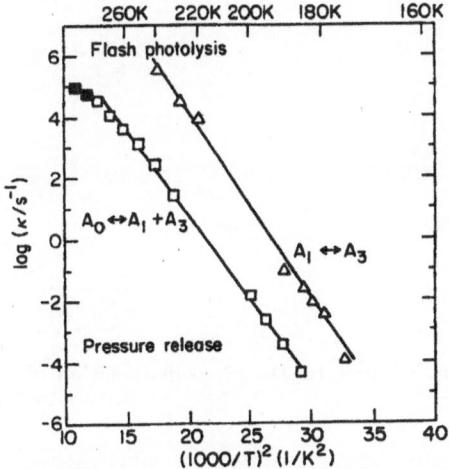

Fig. 15.9. Rate coefficients for the fluctuations among the taxonomic substates in myoglobin (from [23])

characteristic rate distributions. In Fig. 15.7, equilibrium fluctuations are denoted by EF 1 to EF 4. Fluctuations in tier 0 are easiest to study because the taxonomic substates can be unambiguously identified. Figure 15.9 shows the average fluctuation rate coefficients in tier 0 as a function of $(1000/T)^2$. The structural origin of the transitions between A_0 and A_1 is understood. In A_1, the distal histidine, His64 (E7) is uncharged and inside the heme cavity; in A_0, it is protonated and points toward the solvent [28]. The transition consequently involves motions of a large part of the protein and of a proton.

If a protein is brought to a non-equilibrium state by a reaction or a perturbation such as a change in temperature or pressure, it relaxes to equilibrium by transitions to low-lying substates. One extreme relaxation process is folding [29,30]. A one-dimensional picture of the energy landscape such as in Fig. 15.5 creates the impression that a relaxation process must always proceed through a large number of steps. However, as discussed earlier there are many pathways from one conformation to another and a relaxation process may find a path with only a few steps.

15.5   Protein Quakes

Because covalent bond energies are comparable to the entire protein's stabilization energy, it is likely that nearly all protein reactions are accompanied by relaxation of the structure of the protein [31]. Such a process can be called a protein quake because a local *high-energy* event leads to a change in structure and a dissipation of the released energy [19]. A cartoon of a quake is displayed in Fig. 15.10. Myoglobin with CO bound and deoxy myoglobin have different structures. When CO is photodissociated, the protein changes its structure in

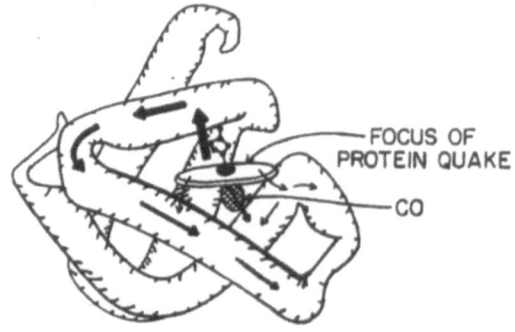

Fig. 15.10. A protein quake is caused for instance by the photodissociation of a bound CO from myoglobin (from [19])

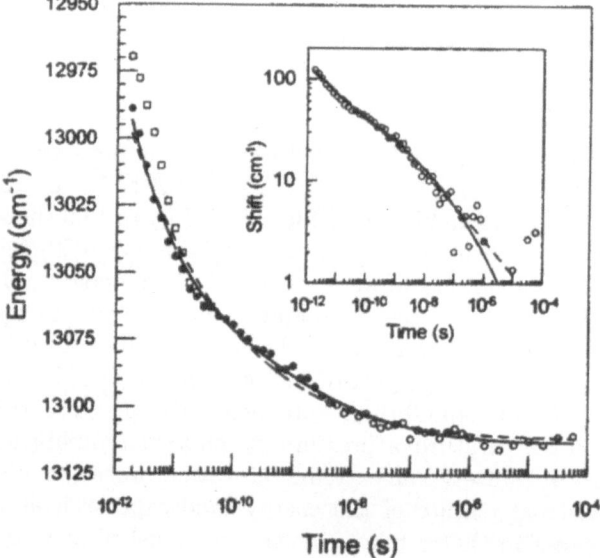

Fig. 15.11. A protein quake shifts a spectral band in myoglobin (from [34]). The wave number of band III is shown as a function of the time after photodissociation

a quake-like process. Studies of this relaxation permit conclusions concerning the energy landscape in the different tiers [32,33]. A beautiful example of the effect of a quake, shown in Fig. 15.11, comes from the work of Anfinrud and collaborators who studied the shift in a spectral band (band III) in myoglobin after photodissociation of CO [34]. Similar phenomena should occur in any protein reaction.

15.6   Glass-Like Properties of Proteins

Proteins and glasses appear at first to have little in common. Proteins are macromolecules with well-defined functions; glasses are arrested liquids. They share, however, one property that appears to be general for complex systems, the energy landscape. The organization of the landscape is different, but general features are similar [35]. We compare here some of these features.

15.6.1   Disorder

Glasses are frozen liquids; they are disordered. Proteins are aperiodic systems; each protein can exist in a large number of different conformations. Because the different substates all may have functional importance it is, however, misleading to call the protein disordered. One could equally well call a Beethoven sonata disordered.

15.6.2   Nonexponential Time Dependence

Relaxation phenomena in glasses show a nonexponential time dependence. Fig. 15.2 proves that protein processes can also have a nonexponential time dependence. This behavior has been observed not just in low-temperature binding, but also in relaxation phenomena, as demonstrated by Fig. 15.11. Single-molecule techniques will be important to study such processes and to distinguish between homogeneous and inhomogeneous behavior [36,37].

15.6.3   Non-Arrhenius Temperature Dependence

In a glass, the temperature dependence of rate coefficients deviates from an Arrhenius expression and can, for instance, be described by the Ferry law [38],

$$k_r = k_0 \exp\left[-\left(\frac{E}{RT}\right)^2\right],$$

(15.2)

where E quantifies the roughness of the energy landscape. Where protein relaxations have been measured over an extended range in time and temperature, they also show deviations from the Arrhenius law, but can be described by (2) [23,39]. Figure 15.9 shows that transitions between the taxonomic substates in myoglobin follow (2) over more than eight orders of magnitude in time.

15.6.4   Slaved Glass Transitions

Glasses are formed when, on cooling, a liquid becomes a structurally disordered solid. The temperature at which the viscosity reaches 10^{13} poise is called the glass temperature T_g. Viscosity and the relaxation rate coefficient k_r are related by [40],

$$k_r = C/\eta. \tag{15.3}$$

Here C is a coefficient that depends on the particular relaxation being studied. T_g can thus also be defined as the temperature where k_r is of the order of 0.01 s^{-1}. Myoglobin, embedded in a 3/1 v/v glycerol–water solvent displays behavior that is similar to glasses [35]. Below a *slaved glass transition* temperature T_{sg}, conformational motions appear to be frozen; above T_{sg} they are fluid. T_{sg} depends on the surroundings of the protein. In a solid environment, many protein motions are still arrested at temperatures as high as 350 K [6,41]. In a low-viscosity solvent, in contrast, T_{sg} can be well below 180 K. In general, T_{sg} is close to T_g of the solvent and it is therefore called a slaved glass transition [24].

15.6.5   Specific Heat

The specific heat, C_p, appears to behave differently in glasses and proteins. In glasses, C_p drops appreciably within a few tens of K below T_g [40]. In proteins, however, the range over which C_p drops is much larger [42]. The hierarchy of substates provides a natural explanation. Proteins do not just have one T_{sg}. Each tier has its own T_{sg} and even within a given tier, T_{sg} is distributed. Figure 15.9 provides an example. If we define T_{sg} by $\log \tau_R = -2$, Fig. 15.9 shows that T_{sg} for the faster fluctuation is 180 K, for the slower one 200 K.

15.6.6   Low-Temperature Specific Heat

In glasses, C_p is linear in T below about 1 K [43,44]. Proteins show the same behavior [45–47]. The linear dependence in glasses is ascribed to two-level states (TLS) [26,27]. In glasses, the TLS are not yet fully understood. In proteins, hole-burning experiments demonstrate that the number of substates is small, but usually greater than two [25]. We therefore call them few-level substates (FLS). A structural understanding of these substates may be easier to find in proteins than in glasses. FLS could form an exciting field for single-molecule experiments.

15.6.7   Nonergodicity

The properties of a glass depend on the path by which it was formed, but the proof of nonergodicity is not always easy. In proteins, the nonergodicity can be proven directly. Consider the taxonomic substates in myoglobin. They are characterized by their CO stretch frequencies. In an ergodic system, the ratio of these substates should only depend on temperature T and pressure

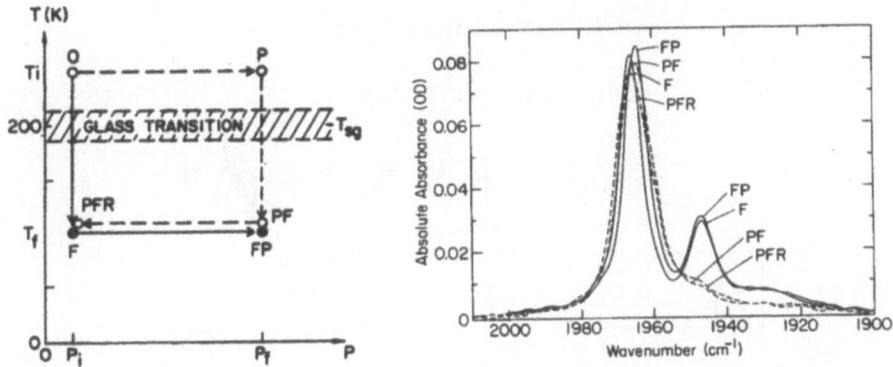

Fig. 15.12. Proteins are metastable below the slaved glass transition temperature T_{sg}. Left: Different pathways through T_{sg}. Right: The CO stretch bands differ depending on the path taken in the $T - P$ plane (from [39])

P regardless of the path taken to a selected $T - P$ point. To prove nonergodicity below the slaved glass temperature, a Mb sample was taken from an initial state ($T_i = 225$ K, $P_i = 0.1$ MPa) above $T_{sg} \approx 200$ K to a final state ($T_f = 100$ K, $P_f = 200$ MPa) below T_{sg} by different pathways, as shown in Fig. 15.12a [39]. The result, shown in Fig. 15.12b is unambiguous: different pathways lead to different ratios of the taxonomic substates below T_{sg}. Thus Mb below T_{sg} is metastable and its structure depends on its history; it is nonergodic.

In summary, proteins and glasses share crucial properties, not because of a structural similarity, but because of a similar energy landscape.

15.7   Energy and Entropy

The entropy plays a central role in proteins [48]. The unfolded protein has a very large number, n, of substates, of the order of 10^{100} [49]. For simplicity we assume that the entropy is proportional to the energy, $S/k_B = \log n \sim E$, as shown in Fig. 15.13a. The real energy landscape then is not as sketched in Fig. 15.5; the number of substates or valleys increases exponentially with E. The projection of the *real* energy landscape on one conformational coordinate is sketched in Fig. 15.13b. The probability of a protein being in a substate with high energy is therefore not negligible. In myoglobin the difference E between the unfolded and the folded state is about 175 kJ/mol [50]. Without entropy, the probability of Mb being unfolded at 350 K would be about 10^{-26}. The difference in the free energy, $G = E - TS$, however, is only 56 kJ/mol [50]; the probability of being unfolded is therefore about 10^{-10}. Figure 15.13c sketches the probability $P(E)$ of finding a substate with energy between E and $E + \delta E$. The rate coefficients for most protein motions are of the order of 10^6 s^{-1} or larger. Substates present with a probability of about 10^{-6} or

Energy Landscape

enthalpy -- entropy

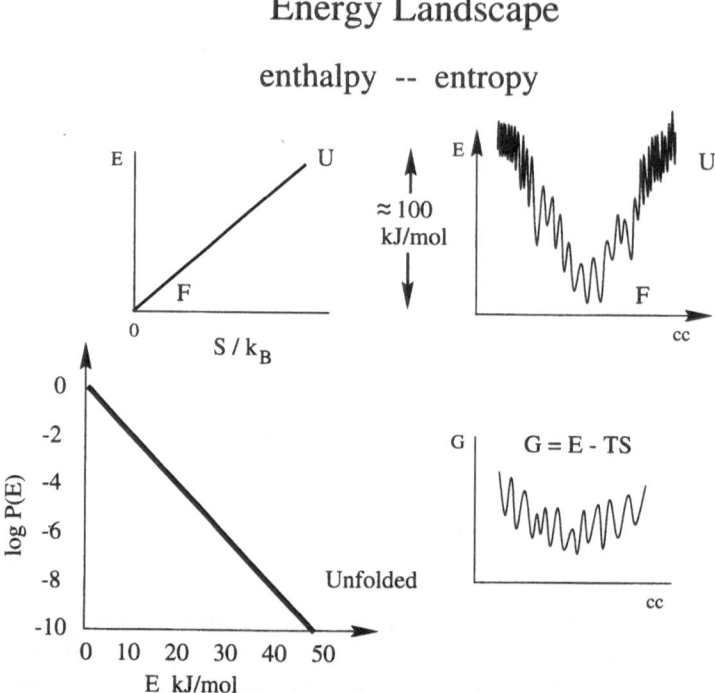

Fig. 15.13. a We assume that the entropy is proportional to the energy. **b** A projection of the energy landscape onto a single conformational coordinate looks different from the landscape in Fig. 15.5. The number of substates increases exponentially with energy. The folded state, **F**, is stabilized by ~ 100 kJ/mol of enthalpy relative to the unfolded state, **U**. **c** The exponential increase in the number of substates implies that substates high in enthalpy are populated with probabilities that are relevant for the dynamics of the protein. **d** A plot of free energy vs. conformational coordinate is much shallower than the plot **b** of enthalpy vs. CC

less, high on the energy surface, thus represent the working protein. X-ray diffraction cannot easily find and characterize structures that are present with less than 10% occupancy. The working proteins consequently can have structures that are far different from what the standard X-ray pictures show.

15.8   Protein Reactions

Protein reactions and the protein energy landscape are intimately linked. Consider as an example the entrance and exit of a CO or O_2 molecule into or out of myoglobin. Even larger molecules such as isocyanides enter myoglobin and bind to the heme iron [51,52]. The average structure of myoglobin shows no opening; the transit must therefore involve fluctuations [53]. Such a reaction is shown schematically in Fig. 15.14 as a two-state process, in-

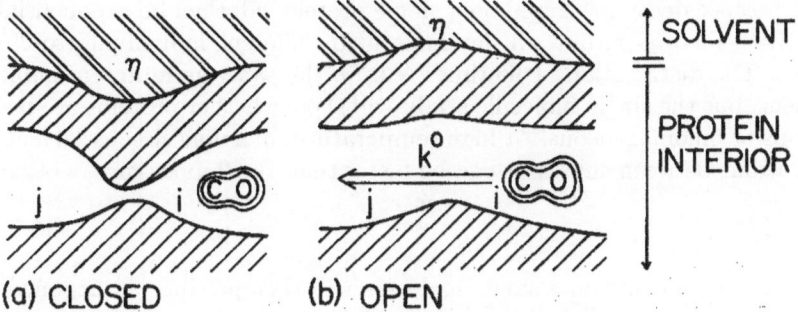

Fig. 15.14. The motion of ligands such as CO or O_2 into and inside proteins may involve gated control. Residues that block the passage may fluctuate so as to create an opening. The gating can be described in terms of the energy landscape by transitions between substates characterizing open and closed channels (from [54])

volving opened and closed gates [54]. The sketch applies not just to ligand entry; many catalytic processes can be described similarly. The fluctuations are transitions involving high-lying substates. Such processes are not simply motion of the CO over a fixed barrier, because the transition rates depend on the external viscosity [54–56]. The protein, slaved to the solvent, controls such fluctuating barriers. For a theoretical treatment of motions over fluctuating barriers, see [57,58].

15.9   Single-Molecule Experiments

Single-molecule experiments, the subject of the present conference, have become a powerful tool to explore the dynamics of proteins [59,60]. Most single-molecule experiments involve signal averaging by repeating the same observation on the same molecule until, for instance, photobleaching sets in. The energy landscape is important for planning and interpreting such experiments.

15.9.1   Conformational Dynamics

Single-molecule experiments can explore conformational motions. Assume that different substates are characterized by different markers, for instance different fluorescence lifetimes τ, and that transitions (fluctuations) between different substates are given by rate coefficients k_f. Now consider two extreme cases. At low temperatures or in a solid matrix, the transitions between substates are slow so that

$$k_f \ll 1/\tau. \tag{15.4}$$

The fluorescence decay measured on a single protein will then be exponential in time, with lifetime τ, but will be different in different individual protein molecules. The distribution of lifetimes will be the same as in an ensemble experiment, but the single-molecule study will show whether proteins are homogeneous or inhomogeneous. At high temperatures in a non-viscous solvent, the transitions between substates can be faster than the fluorescence, so that

$$k_f \gg 1/\tau. \tag{15.5}$$

Repeated measurements on a single molecule will then produce an exponential time dependence, as would be obtained by an ensemble experiment.

15.9.2 Reactions

Consider now an experiment in which a reaction with rate coefficient k_r is observed. If k_r has the same value in all substates, single-molecule experiments give the same information as ensemble experiments, but with smaller statistics. We therefore assume that different substates possess different k_r. The outcome of a single-molecule study then depends on the ratio k_r/k_f, where k_f is the rate coefficient for fluctuations between the substates. Two extreme situations are easy to describe. If the fluctuations are much faster than the reaction, or

$$k_f \gg k_r, \tag{15.6}$$

the fluctuations will average the process under observation and the single-molecule experiment is equivalent to a standard ensemble observation. If the fluctuations are much slower than the reaction, or

$$k_f \ll k_r, \tag{15.7}$$

each run determines the properties of the molecule in a single substate. Different runs will give different answers and a complete set of observations produces the distribution of k_r. Most fluctuations in proteins at physiological temperatures are faster than, say, 10^6 s^{-1}. It will therefore be difficult with the present techniques to observe the distributions of fast reactions. Experiments may require low temperatures or reducing the rate of fluctuations in some other way, for instance by immobilizing single molecules.

15.10   A Note on Double Exponentials

When chemists measure an observable as a function of the hydrogen ion concentration, H, they plot the observable not as a function of H, but of pH, where $pH \equiv \log H$. The reason is, of course, that H extends over many orders of magnitude. In contrast, when experimenters measure an observable, $N(t)$,

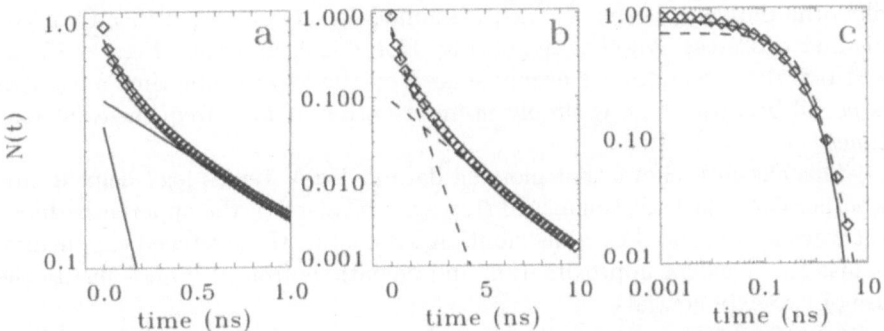

Fig. 15.15. A stretched exponential decay $N(t)$ (9) plotted as diamonds in three different ways. **a** $N(t)$ vs. t plotted and fit to a two-exponential process from $t = 0$ to 1.0 ns ($k_1 = 7.7$ ns$^{-1}, k_2 = 1.1$ ns^{-1}). **b** $N(t)$ vs. t plotted and fit to a two-exponential process from $t = 0$ to 10 ns. ($k_1 = 1.5$ ns$^{-1}, k_2 = 0.4$ ns^{-1}). **c** The same data as **a** and **b**, but plotted as log $N(t)$ vs. log t, along with the fits derived from **a** and **b**. When plotted in this way, the relevant deviations from the fits are visible

as a function of t, they routinely plot log $N(t)$ versus t, even though the data may extend over many orders of magnitude in time. If log $N(t)$ versus t is a straight line, they are happy and extract the rate coefficient from the slope of log $N(t)$. If, however, log $N(t)$ deviates from a straight line, they usually fit the data to two exponentials,

$$N(t) = c_1 \exp\left[-(k_1 t)\right] + c_2 \exp\left[-(k_2 t)\right]. \tag{15.8}$$

Such a fit, however, is not unique because exponentials are not orthogonal functions.

The non-uniqueness can be understood with Fig. 15.15, which shows a stretched exponential curve,

$$N(t) = N(0) \exp\left[-(3.00t)^{0.55}\right], \tag{15.9}$$

plotted as diamonds in three different ways. If the data are taken only to 1 ns (Fig. 15.15a) a fit to (8) gives $k_1 = 7.7$ ns$^{-1}, k_2 = 1.1$ ns^{-1} If, however, data are taken to 10 ns (Fig. 15.15b) the result is $k_1 = 1.53$ ns$^{-1}, k_2 = 0.4$ ns^{-1}. Both the individual exponentials and the sum of the two are plotted; in each case, the sum fits the data reasonably well. Because the resulting fit parameters are completely different, it is clear that the expansion into two exponentials is not unique. (Of course, if a decay process really occurs by two exponentials, and the rate coefficients differ by at least a factor ten, the two-exponential fit will be robust.)

The data are redrawn in Fig. 15.15c as log $N(t)$ versus log t (or $pN(t)$ versus pt!), along with the fits to the data in Figs. 15.15a and b. It is clear that both fits to double exponentials are inadequate. In practice, we have found that unless an inflection point is visible in such a log $N(t)$ versus log t

plot, the data are better fit with a unimodal distribution, rather than two distinct processes. Another reason Fig. 15.15c is clearer than Figures 15.15a and b is that the data are evenly spaced on the logarithmic time axis. This is useful because there is simply no information at high frequencies at long times.

This example shows that plotting data as $\log N$ versus $\log t$ helps decide whether two clearly distinguished states are involved or the apparently different rates k are caused by a distribution. Of course, the stretched exponential is just a convenient approximation and deviations from it do not justify the use of two exponentials.

An alternative approach to the analysis of a nonexponential time dependence is to determine the underlying rate distribution by using the maximum entropy method [61].

15.11   Conclusions

The existence of conformational substates described by an energy landscape is important in understanding protein dynamics. Its existence and organization may be a prerequisite to analyzing single-molecule experiments. Conversely, single-molecule experiments may help to unravel puzzles that are difficult to solve in standard approaches.

Acknowledgments. HF thanks Rudolf Rigler and the organizing committee for the invitation to attend the Nobel conference. We both thank our many collaborators and colleagues for many discussions that have helped to shape our (still incomplete) understanding of the energy landscape in proteins. This research was supported by the Department of Energy, under contract W-7405-ENG-36.

References

1. K. U. Linderstrøm-Lang and J. A. Schellman, Enzymes **1**, 443 (1959)
2. I. M. Klotz, Arch. Biochem. Biophys. **116**, 92 (1966)
3. H. Frauenfelder, A. R. Bishop, et al., *Landscape Paradigms in Physics and Biology-Concepts, Structures, and Dynamics* (North-Holland, Amsterdam, 1997)
4. H. Frauenfelder, P. J. Steinbach, and R. D. Young, Chemica Scripta **29A**, 145–150 (Nobel Symposium 71, Structure and Dynamics of Biological Systems B) (1989)
5. R. H. Austin et al., Phys. Rev. Lett. **32**, 403–405 (1974)
6. R. H. Austin et al., Biochem. **14**, 5355 (1975)
7. H. Frauenfelder, Methods of Enzymology **54**, 506–532 (1978)
8. A. E. Garcia, R. Blumenfeld, G. Hummer, and J. A. Krumhansl, Physica D **107**, 225–239 (1997)

9. H. Frauenfelder, G. U. Nienhaus, and R. D. Young, *Disorder Effects on Relaxational Processes*, Richter and Blumen (eds.), pp. 591–614 (Springer, Berlin, 1994)

10. G. U. Nienhaus and R. D. Young, Encyclopedia of Applied Physics **15**, 163–184 (1996)

11. A. E. Garcia, J. A. Krumhansl, and H. Frauenfelder, Proteins **29**, 153–160 (1997)

12. H. Frauenfelder, G. A. Petsko, and D. Tsernoglou, Nature **280**, 331 (1979)

13. G. A. Petsko and D. Ringe, Ann. Rev. Biophys. Bioeng. **17**, 451 (1984)

14. P. A. Rejto and S. T. Freer, Progr. Biophys. Molec. Biol. **66**, 167–196 (1996)

15. J. Friedrich, Methods in Enzymology **246**, 226–259 (1995)

16. D. Kleinfeld, N. Okamura, and G. Feher, Biochem. **23**, 5780–5786 (1984)

17. B. H. McMahon, J. D. Müller, C. A. Wraight, and G. U. Nienhaus, Biophys. J. **74**, 2567–2587 (1998)

18. J. M. Peloquin et al., Biochem. **33**, 8089–8100 (1994)

19. A. Ansari et al., Proc. Natl. Acad. Sci. USA **82**, 5000–5004 (1985)

20. H. Frauenfelder, S. G. Sligar, and P. G. Wolynes, Science **254**, 1598–1603 (1991)

21. M. W. Makinen, R. A. Houtchens, and W. S. Caughey, Proc. Natl. Acad. Sci. USA **76**, 6042–6046 (1979)

22. J. O. Alben et al., Proc. Natl. Acad. Sci. USA **79**, 3744–3748 (1982)

23. J. B. Johnson et al., Biophys. J. **71**, 1563–1573 (1996)

24. A. Ansari et al., Biophys. Chem. **26**, 337–355 (1987)

25. D. Thorn Leeson et al., J. Phys. Chem. **101**, 6331–6340 (1997)

26. P. W. Anderson, B. I. Halperin, and C. M. Varma, Phil. Mag. **25**, 1 (1972)

27. W. A. Phillips, J. Low Temp. Phys. **7**, 351 (1972)

28. F. Yang and G. N. Phillips Jr., J. Mol. Biol. **256**, 762–774 (1996)

29. J. D. Bryngelson and P. G. Wolynes, J. Phys. Chem. **93**, 6902–6915 (1989)

30. J. N. Onuchic, Z. Luthey-Schulten, and P. G. Wolynes, Ann. Rev. Phys. Chem. **48**, 545–600 (1997)

31. N. Agmon and J. J. Hopfield, J. Chem. Phys. **79**, 2042–2053 (1983)

32. G. U. Nienhaus et al., Proc. Natl. Acad. Sci. USA **89**, 2902–2906 (1992)

33. J. Huang, A. Ridsdale, J. Q. Wang, and J. M. Friedman, Biochem. **36**, 14353–14365 (1997)

34. T. A. Jackson, M. Lim, and P. A. Anfinrud, Chem. Phys. **180**, 131–140 (1994)

35. I. E. T. Iben et al., Phys. Rev. Lett. **62**, 1916–1919 (1989)

36. L. Edman, D. C. Mets, and R. Rigler, Proc. Natl. Acad. Sci. USA **93**, 6710–6715 (1996)

37. Y. W. Jia et al., Proc. Natl. Acad. Sci. USA **94**, 7932–7936 (1997)

38. P. G. Wolynes, J. Res. Natl. Inst. Stand. Technol. **102**, 187–194 (1997)

39. H. Frauenfelder et al., J. Phys. Chem. **94**, 1024–1037 (1990)

40. S. Brawer, *Relaxation in Viscous Liquids and Glasses* (American Ceramic Society, Columbus, Ohio 1985)

41. S. J. Hagen, J. Hofrichter, and W. A. Eaton, J. Phys. Chem. **100**, 12008–12021 (1996)

42. G. Sartor, E. Mayer, and G. P. Johari, Biophys. J. **66**, 249–258 (1994)

43. R. Zallen, *The Physics of Amorphous Solids* (Wiley, New York, 1983)

44. W. A. Phillips, Rep. Prog. Phys. **50**, 1657 (1987)

45. V. I. Goldanskii, I. F. Krupyanskii, and V. N. Flerov, Dokl. Akad. Nauk SSSR **272**, 23 (1983)

46. G. P. Singh et al., Z. Phys. B **55**, 23 (1984)
47. A. Schulte and R. Murray, Phys. Rev. A. **36**, 1722 (1987)
48. P. G. Wolynes, J. N. Onuchic, and H. Frauenfelder, to be published
49. G. I. Makhatadze and P. L. Privalov, Protein Science **5**, 507–510 (1996)
50. C. Tanford, Adv. Protein Chem. **23**, 121 (1968)
51. J. S. Olson et al., J. Am. Chem. Soc. **105**, 1522–1527 (1983)
52. D. Ringe et al., Biochem. **23**, 2–4 (1984)
53. D. A. Case and M. Karplus, J. Mol. Biol. **132**, 343–368 (1979)
54. D. Beece et al., Biochem. **19**, 5147–5157 (1980)
55. T. Kleinert et al., Biochem. **37**, 717–733 (1998)
56. A. Ansari et al., Biochem. **33**, 5128–5145 (1994)
57. J. A. McCammon and S. H. Northrup, Nature **293**, 316 (1981)
58. J. Wang and P. G. Wolynes, Chem. Phys. Lett. **212**, 427–433 (1993)
59. S. S. Xie and J. K. Trautman, Ann. Rev. Phys. Chem. **49**, 441–480 (1998)
60. S. Nie and R. N. Zare, Ann. Rev. Biophys. Biomol. Struct. **26**, 567–596 (1997)
61. P. J. Steinbach et al., Biophys. J. **61**, 235–245 (1992)

16 Coherent Intramolecular Dynamics in Small Enzyme Populations

A. S. Mikhailov[1], P. Stange[1], and B. Hess[2]

Complex internal dynamics of conformational transformations in single enzyme molecules is essential for their catalytic function [1,2]. Such experimental methods as time-resolved X-ray spectroscopy of protein crystals or single-molecule fluorescence microscopy [3–6] allow observation of individual enzymic turnover cycles [7,8]. Enzymic reactions in living biological cells are confined to very small spatial volumes and this can lead to qualitative changes in the kinetics of biochemical reactions [9,10]. It has recently been suggested that classical concepts of chemical kinetics must be revised when processes inside living cells are considered [11]. New approaches are needed to take into account interactions between protein machines of the living cell and treat the cell as a whole [12]. Investigations of complete enzymic reactions in biologically relevant nanoenviroments are now possible [13]. The studies of collective dynamics of enzyme molecules in small volumes are thus important from the perspective of molecular biology applications.

Another aspect of this research is that experiments with small enzymic populations can provide valuable information about intramolecular processes that accompany turnover cycles in single enzyme molecules. If one can synchronize turnover cycles of enzymes, the collective behavior of the entire molecular population would repeat that of a single molecule. Indeed, experiments with the photosensitive cytochrome P-450 dependent monooxygenase system have shown [14–16] that the turnover cycles of individual enzymes can be externally synchronized by applying periodic light flashes. A different type of external synchronization has been discussed in the membrane transport system of the enzyme Na,K-ATPase using external electric fields [17]. In a series of theoretical investigations we have found that spontaneous mutual synchronization of individual turnover cycles of enzymes can take place in allosteric [18–21] and nonallosteric [22] enzymic reactions in small spatial volumes. The common conclusion of such experimental and theoretical studies is that *coherent* intramolecular dynamics may emerge as a consequence of external forcing or interactions between the molecules. This microscopic coherence in ensembles of protein macromolecules is a new property absent in macroscopic chemical kinetics.

We have recently reviewed our research on mutual synchronization in allosterically regulated reactions in several publications [23–25]. The aim of the present article is to illustrate the basic mechanism of mutual synchronization for nonallosteric self-regulated enzymic reaction cycles.

Metabolic pathways in living cells consist of a large number of reactions catalyzed by enzymes. A particular reaction is embedded in such a pathway, taking product molecules of the previous reaction step and supplying the substrate for the next step. Additionally, links between different metabolic cycles can be formed through common use of some reactants. The structure of metabolic pathways and interactions between them shows great variety. Some reaction chains, such as substrate cycles [26], contain closed loops where a large fraction of the product molecules are converted back to the substrate molecules of the same reaction. As an example, we consider below a hypothetical nonallosteric enzymic reaction chain with such a closed loop. The kinetics of this reaction is investigated in a small volume, where diffusional mixing and transport times of substrate and products are much shorter than the turnover time of a single catalytic cycle. Under such conditions each of the chemical reactions in the chain is coupled to others via fast diffusing substrate and product molecules.

The next section is devoted to a general discussion of kinetics of self-regulated macromolecular reactions in small volumes. In Sect. 2 we formulate a stochastic model of the reaction considered. The data of numerical simulations of this model are presented and discussed in Sect. 3. To interpret these numerical results, a mean-field analysis is then performed in Sect. 4.

16.1   Molecular Networks in Microvolumes

In volumes of micrometer and submicrometer sizes, containing a few hundred or a thousand enzyme molecules, all characteristic kinetic times related to diffusional transport of substrates and intermediate products through the volume to their enzymic targets can be much shorter than the time scales of slow intramolecular processes associated with conformational transformations in enzymes [10]. The two principal kinetic time scales are the mixing time t_{mixing} and the transit time $t_{transit}$ for regulatory molecules, whereas the characteristic time associated with intramolecular processes in an enzyme is its mean turnover time τ.

The mixing time is defined as the time needed by a regulatory molecule to forget its initial location inside the volume. For a volume of linear size L and diffusion constant D, it is estimated as $t_{mixing} = L^2/D$. To introduce the transit time, we first suppose that the volume contains only one enzyme target. A regulatory molecule is released somewhere inside the volume and performs diffusional wandering inside it. If it approaches an enzyme and touches an atomic target group on its surface, binding takes place with high probability and the regulatory molecule becomes docked at the enzyme sur-

face. The characteristic time needed to find a single immobile atomic target
of radius R in a three-dimensional volume of linear size L can be estimated
as L^3/DR (see [10,27,28]). If we have N identical immobile targets that are
randomly distributed inside the volume, the characteristic time for docking
to any one of them is the transit time $t_{transit} = L^3/NDR$.

If the condition $t_{transit} > t_{mixing}$ is satisfied, a regulatory molecule will
diffusively explore the entire volume before docking to a target. By the moment when docking occurs, any memory of the initial location of this regulatory molecule will have been lost. Hence, a regulatory molecule will have
the same probability to find any enzymic target, irrespective of its spatial
position in the volume. Substituting the above expressions for t_{mixing} and
$t_{transit}$, one can see that this condition essentially requires that the number
N of enzymic targets in the volume be relatively small, i.e. $N < N_{cr}$ where
$N_{cr} = L/R$. If we take for numerical estimates the values $L = 1\ \mu m$ and
$R = 1$ nm, this yields $N_{cr} = L/R = 1000$, which corresponds to the maximum volume enzymic concentration of about $10^{-6} M$ [10]. If the diffusion
constant of small regulatory molecules is $D = 10^{-5}$ cm^2/s, the mixing time
of such molecules is $t_{mixing} \sim 1$ ms. When the number of enzymic targets is
sufficiently small ($N < N_{cr}$), the transit time for the same molecules is even
shorter. These times should be further compared with the turnover times of
enzymes.

The actual time needed for an enzyme to complete its catalytic cycle is
influenced by fluctuations. Hence, the mean turnover time $\bar\tau$ is not enough
to specify this process and the mean dispersion $\Delta\tau$ of turnover times is also
important. The mean turnover time $\bar\tau$ is usually estimated by measuring the
maximum turnover rate under the condition of the substrate saturation. Except for some very rapidly operating enzymes, the typical mean turnover
times lie in the interval from 10 to 100 ms. For some slow enzymes, the
turnover times may even reach seconds (e.g. $\bar\tau = 1.53$ s for the above mentioned cytochrome P-450 dependent monooxygenase system). The dispersion
$\Delta\tau$ cannot be so easily determined and is mostly unknown. It is one of the
aims of single-molecule spectroscopy to provide measurements of this property. The statistical analysis of experimental data obtained under external
optical synchronization of the photosensitive cytochrome P-450 dependent
monooxygenase system reveals that in this slow enzyme the relative statistical dispersion of turnover times is about 20% [16].

An important conclusion is that characteristic turnover times of enzymes
may be *much longer* than mixing and transit times of regulatory molecules
in microvolumes. When the conditions

$$\tau \gg t_{transit} > t_{mixing} \tag{16.1}$$

are satisfied, the enzymic reaction in a microvolume will proceed in a special
kinetic regime. It can be viewed as a collective evolution of a population of
cyclic protein machines (i.e. enzyme molecules). The intramolecular dynamics

of individual enzymes becomes coupled through regulatory molecules that are produced by machines, rapidly diffuse through the reaction volume, and influence the intramolecular dynamics of other enzymes. We have suggested that such biochemical reaction systems should be called *molecular networks* [9,10,23,29]. Under certain conditions, collective intramolecular dynamics of enzymes in a molecular network may spontaneously become coherent.

16.2   Stochastic Model of an Enzymic Reaction

As an example, we shall consider a hypothetical reaction chain including a closed loop:

$$S + E \xrightarrow{\alpha} ES \rightarrow P + E$$
$$P \xrightarrow{\kappa} S, \; P \xrightarrow{\gamma_1} C \tag{16.2}$$
$$A \xrightarrow{\zeta} S, \; S \xrightarrow{\gamma} B$$

Binding of a substrate molecule S initiates the catalytic turnover cycle of enzyme E, during which a product molecule appears. A free product molecule P is transformed through a different chemical reaction back into a substrate molecule. Moreover, substrate and product molecules are consumed by other reactions, transforming them into molecules B and C that do not further participate in the set of reactions considered. Substrate is supplied through chemical conversion of molecules A. The concentration of these molecules A is maintained constant. The classical kinetic equations of such a reaction chain do not show self-sustained oscillations. Note that the enzyme is not allosteric and its catalytic activity is not regulated in the system considered.

In molecular networks the dynamical properties of each single enzyme are important. The actual intramolecular dynamics of such macromolecules is extremely complex and is not sufficiently known. To model these intramolecular processes large simplifications are needed. Following our previous publications [19–21] (see also [16]) the dynamics of a single enzyme molecule during the catalytic turnover cycle is modeled below as diffusive motion along a certain reaction coordinate. The enzymic cycle also includes such stochastic events as binding and dissociation of substrate and product molecules (Fig. 16.1).

It is convenient to define for each enzyme i a binary state variable u_i, such that $u_i = 0$, if the enzyme is in its free state and ready to bind a substrate molecule. The formation of an enzyme–substrate complex with rate α is then described as a transition into the state with $u_i = 1$. This transition initiates the turnover cycle, which consists of the catalytic conversion of the substrate into the product and subsequent return of the enzyme into its free state. This process is modeled as diffusional drift through an energy landscape along the reaction coordinate ϕ_i. The coordinate $\phi_i = 0$ corresponds to the beginning

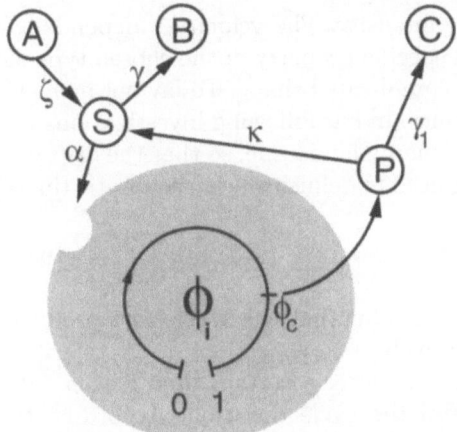

Fig. 16.1. Schematic representation of an enzymic turnover cycle. The cycle is initiated by binding of a substrate molecule at rate α and modeled as diffusive motion along the reaction coordinate ϕ, which increases from $\phi = 0$ to $\phi = 1$. At stage $\phi = \phi_c$ in the cycle a product molecule is released that can later be converted into molecules of type C at rate γ_1 and can be converted back into substrate molecules at rate κ. The substrate molecules are permanently supplied by converting molecules of type A into substrate at rate ζ. Additionally, substrate can be converted into molecules of type B at rate γ

of the cycle. The cycle ends when $\phi_i = 1$ and the enzyme returns to its free state with $u_i = 0$. The release of the product molecule takes place at point $\phi_i = \phi_c$ in the cycle. Thus the point ϕ_c on the reaction coordinate separates two different processes. In the coordinate interval $0 < \phi_i < \phi_c$ the enzyme–substrate complex exists, whereas later in the interval $\phi_c < \phi_i < 1$ the enzyme undergoes conformational relaxation back to its free state. There, it can bind again a substrate molecule to start a new cycle.

Introducing the probability distribution $p(\phi_i, t)$ over the reaction coordinate ϕ_i, we assume that this distribution satisfies the diffusion equation

$$\frac{\partial p}{\partial t} = -v \frac{\partial p}{\partial \phi_i} + \sigma \frac{\partial^2 p}{\partial \phi_i^2}. \tag{16.3}$$

The first term in this equation describes drift and the second term takes into account thermal fluctuations within the cycle. This diffusion equation is equivalent to the following stochastic Langevin equation:

$$\frac{d\phi_i}{dt} = v + \eta_i(t), \tag{16.4}$$

where v is the drift velocity and $\eta_i(t)$ is white Gaussian noise with the correlation function

$$\langle \eta_i(t)\eta_j(t')\rangle = 2\sigma \delta_{ij}\delta(t - t'), \tag{16.5}$$

The parameter σ determines the noise intensity. The velocity v depends on the free energy profile. This profile is a specific property of the chosen type of enzyme and may generally have a very complicated shape. Today not much is known about this property of real enzymes. In our following investigations we assume a constant negative slope of the energy landscape, so that the drift velocity v is constant. The drift velocity is used to define two characteristic times

$$\tau_c = \frac{\phi_c}{v} \text{ and } \tau_0 = \frac{1}{v} \tag{16.6}$$

needed after cycle initiation in the absence of fluctuations to release a product molecule and to finish the cycle, respectively.

The cycle starts at $\phi = 0$ and is finished after a certain time τ at $\phi = 1$. This is the turnover time that specifies the cycle duration. According to Eq. (16.4), motion inside the cycle is a random process and therefore the turnover time fluctuates from one realization of this process to another. The mean turnover time is estimated to be (see [20])

$$\langle \tau \rangle = \frac{1}{v} - \frac{\sigma}{v^2} \left[1 - e^{-\frac{v}{\sigma}} \right] \tag{16.7}$$

and its statistical dispersion $\Delta \tau = \sqrt{\langle \tau^2 \rangle - \langle \tau \rangle^2}$ is given by (see [20])

$$\Delta \tau^2 = \frac{2\sigma}{v^3} - \frac{5\sigma^2}{v^4} + \frac{4\sigma}{v^3} e^{-\frac{v}{\sigma}} + \frac{4\sigma^2}{v^4} e^{-\frac{v}{\sigma}} + \frac{\sigma^2}{v^4} e^{-\frac{2v}{\sigma}}. \tag{16.8}$$

The relative statistical dispersion is defined as

$$\xi = \frac{\Delta \tau}{\langle \tau \rangle} \tag{16.9}$$

For small noise intensities σ, the approximations $\langle \tau \rangle \approx 1/v$, $\Delta \tau \approx \sqrt{2\sigma/v^3}$, and $\xi \approx \sqrt{2\sigma/v}$ hold.

The dispersion ξ of turnover times is an important statistical property of the intramolecular dynamics that is more amenable to direct measurement than the intensity σ of the intramolecular noise. The two parameters are connected by a simple relationship (see Eqs. (16.8) and (16.9)). When results of our numerical simulations are presented below we typically give only the value of the dispersion ξ corresponding to a particular simulation. In all cases considered, the intramolecular noise is always relatively small and its intensity can be estimated to be $\sigma = v\xi^2/2$.

The first step of the enzymic reaction (16.2) is the formation of an enzyme substrate complex at rate constant α. When the enzyme passes the phase state ϕ_c a product molecule is released and increases the number of free enzymes by 1. Subsequent relaxations of the enzymic conformation brings the enzyme i back to the free state where $u_i = 0$. The number of free products decreases at rate constant κ due to the conversion to substrate molecules. To prevent accumulation of products, we assume that the product molecules decay at rate γ_1. The "decay" is actually represented by the conversion of products into another chemical species C along the reaction pathway; these do not further participate in the reaction. Typically we assume that $\kappa \leq \gamma_1 \ll \tau_0^{-1}$, so that the lifetime of free product molecules is short compared to the turnover time. Besides the decrease of substrate molecules at a rate constant α through formation of the enzyme–substrate complex, a second reaction at rate γ proceeds by transforming the substrate into molecules of type B, which do not further participate in the reaction. The substrate is supplied through conversion of the chemical species A at a rate constant ζ. The detailed algorithm of this stochastic model is given in [22].

16.3   Synchronization of Turnover Cycles

Now we present results of numerical investigations of the stochastic model. The total number of enzymes is $N = 1000$ in all our simulations. As the initial condition, we assume that the enzymic states are randomly distributed over their cycle phase. Simulations of this model show, depending on the parameters $\alpha, \gamma_1, \gamma, \xi, \phi_c, \kappa$, and ζ, the existence of two qualitatively different types of behavior.

At small values of α and large values of the relative statistical dispersion ξ of the turnover time, the numbers of product, substrate, and free enzymes fluctuate randomly around a certain mean value. This is seen in Fig. 16.2a where the number $s(t)$ of substrate molecules (black) and the number $n(t)$ of free enzymes (gray) in the reaction volume are shown as a function of time. If the dispersion ξ of molecular turnover times is relatively small, the system can however exhibit rapid oscillations (spiking) in the number of substrate, product, and free enzymes. This behavior is displayed in Fig. 16.2b, where again (using the same notation as in Fig. 16.2a) the number of substrate and free enzymes are shown.

We see in Fig. 16.2b that a certain phase shift between the number of free enzymes and substrate is present in the spiking regime. Whenever the substrate reaches its minimum number, the amount of free enzymes has its maximum. For different parameters the spiking frequency can be higher than in Fig. 16.2b. Depending on the value of the parameter ϕ_c, which characterizes the moment inside the cycle when a product molecule is released, spiking double the frequency (Fig. 16.2c, $\phi_c = 0.55$) or three times the frequency in Fig. 16.2b (Fig. 16.2d, $\phi_c = 0.36$) is found.

Periodic spiking in the number of free enzyme molecules indicates the presence of coherence in the intramolecular dynamics of individual enzyme molecules. Indeed, this implies that the moments at which the cycles of different enzymes are initiated are strongly correlated and synchronous. The duration and other parameters of the individual molecular cycle, such as ϕ_c,

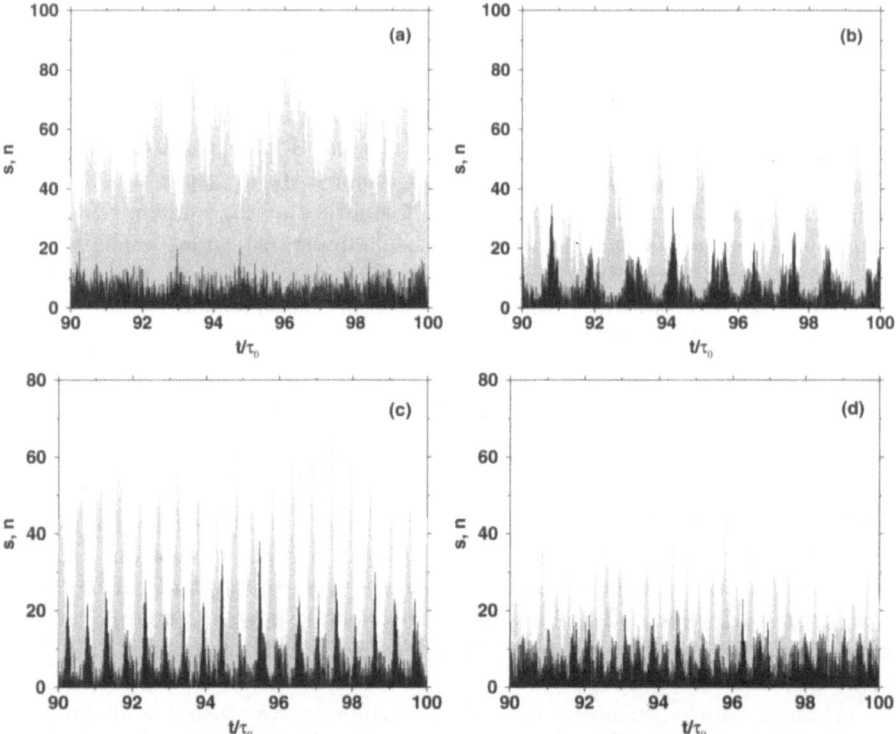

Fig. 16.2. Time dependence of the number of substrate molecules (black) and free enzymes (gray) in the stochastic model of reaction (16.2) for **a** no spiking at $\alpha = 3\tau_0^{-1}$, $\phi_c = 0.2$, and $\xi = 0.4$; **b** spiking with $\omega \approx 2\pi\tau_0^{-1}$ at $\alpha = 10\tau_0^{-1}$, $\phi_c = 0.2$, and $\xi = 0.15$; **c** spiking with $\omega \approx 4\pi\tau_0^{-1}$ at $\alpha = 10\tau_0^{-1}$, $\phi_c = 0.55$, and $\xi = 0.02$; and **d** spiking with $\omega \approx 6\pi\tau_0^{-1}$ at $\alpha = 10\tau_0^{-1}$, $\phi_c = 0.72$, and $\xi = 0.01$. Other parameters are $N = 1000$, $\gamma = 25\tau_0^{-1}$, $\gamma_1 = 5\tau_0^{-1}$, and $\zeta = 200\tau_0^{-1}$

play an important role in this kinetic regime by directly determining the spiking period. Spiking periods that are half or one third the turnover time indicate that the entire population of enzyme molecules breaks down into two or three coherently operating groups whose turnover cycles are shifted by half or a third of the period.

The original model (16.2) can be simplified without losing its principal properties. We assume below that the decay of the product is absent ($\gamma_1 = 0$). Moreover, the rate constant κ of the back reaction is assumed to be much greater than the inverse of the turnover time τ_0^{-1}. The release of a product molecule by enzyme i at $\phi_i = \phi_c$ then immediately leads to an increase in the number of free substrate molecules by 1. In this case the number $n(t)$ of free products is negligibly small (it goes to zero for $\kappa \to \infty$) and it suffices to follow only the number $s(t)$ of substrate molecules in the reaction volume. Under these assumptions the reaction scheme (16.2) can be effectively written

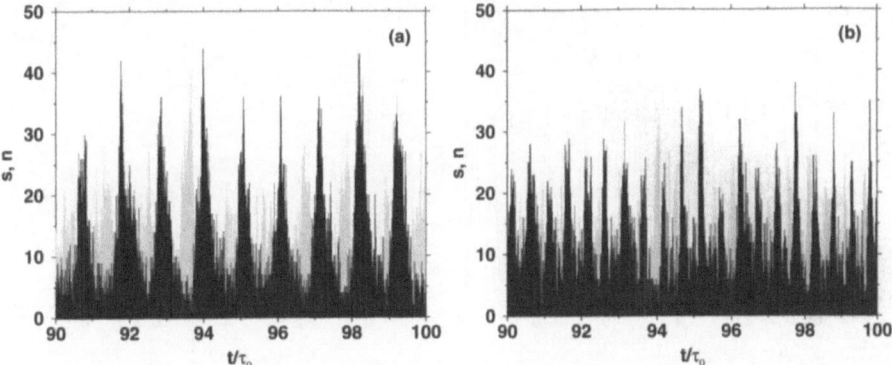

Fig. 16.3. Time dependence of the number of substrate molecules (black) and free enzymes (gray) in the stochastic model of reaction (16.10) for **a** spiking with $\omega \approx 2\pi\tau_0^{-1}$ at $\phi_c = 0.2$, and **b** spiking with $\omega \approx 4\pi\tau_0^{-1}$ at $\phi_c = 0.55$. Other parameters are $N = 1000$, $\gamma = 15$, $\zeta = 200\tau_0^{-1}$, $\alpha = 10\tau_0^{-1}$, and $\xi = 0.02$

as

$$S + E \xrightarrow{\alpha} ES \rightarrow S + E$$
$$A \xrightarrow{\zeta} S, \; S \xrightarrow{\gamma} B . \tag{16.10}$$

According to this scheme, substrate molecules become bound to the enzyme, spend some time in the enzyme–substrate complex, and are then released as product molecules that are instantaneously converted back to the substrate. The enzyme returns after a while to its free state and can again bind the substrate. The substrate molecules are supplied at rate ζ and decay at rate γ. The complete algorithm of the simplified model is described in [22].

Results of numerical simulations of this simplified model are shown in Fig. 16.3. We see that both spiking and random fluctuations around a certain mean value are possible, similar to what has been found in the original model (Fig. 16.2). Figure 16.3a displays the time-dependent number of free enzymes (gray) and substrate molecules (black) in the spiking regime at $\phi_c = 0.2$ and $\xi = 0.02$. This spiking has a period close to the turnover time τ. By varying parameter ϕ_c, spiking with shorter periods can also be achieved. For example, when $\phi_c = 0.55$ and $\xi = 0.02$ the spiking period is approximately $0.5\tau_0$ (Fig. 16.3b). Our simulations have shown that spiking with even greater period is also observed in the model, but it is possible only at extremely low relative statistical dispersions ξ of the turnover time. Therefore, we investigate only spiking regimes with periods close to τ_0 and $0.5\tau_0$, which are more robust against fluctuations. Increasing the intensity σ of intramolecular noise, which determines the dispersion ξ of turnover times, leads to the breakdown of spiking and the onset of the incoherent kinetic regime, similar to that shown in Fig. 16.2a.

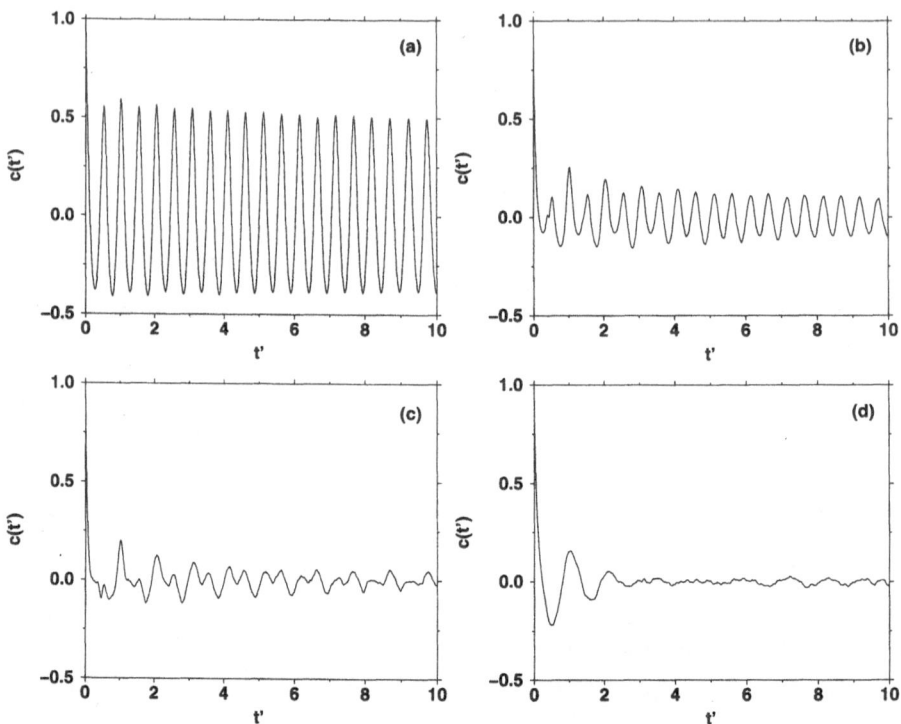

Fig. 16.4. Correlation function $c(t')$ at **a** $\xi = 0.02$, **b** $\xi = 0.035$, **c** $\xi = 0.05$, and **d** $\xi = 0.19$. Other parameters are the same as in Fig. 16.3b

To analyze the time dependence of the number of substrate molecules in the reaction volume we use the autocorrelation function

$$c(t') = \frac{\langle y(t)y(t+t')\rangle}{\langle y^2(t)\rangle} \ , \ y = s(t) - \langle s(t)\rangle. \tag{16.11}$$

Figure 16.4 displays this autocorrelation function of the time-dependent number of substrate molecules at four different values of the relative statistical dispersion ξ of the turnover time, with the same other reaction parameters as in Fig. 16.3b.

When the system exhibits spiking, the correlation function shows weakly damped periodic oscillations with period close to $\tau_0/2$ (Fig. 16.4a with $\xi = 0.02$). The amplitude of these oscillations decreases and the damping rate increases if ξ is increased slightly (Fig. 16.4b with $\xi = 0.035$). In Fig. 16.4c with $\xi = 0.05$, the oscillations of the correlation function are more strongly damped and contain several frequencies. At $\xi = 0.02$ (Fig. 16.4d), the correlation function exhibits heavily damped oscillations.

16.4   The Mean-Field Approximation

The mean-field approximation in chemical kinetics consists of neglecting fluctuations in concentrations of reactants that are due to the atomistic stochastic nature of reaction processes. In the limit $N \to \infty$ our system can then be completely described by the density distribution $\tilde{n}(\phi, t)$ such that $\tilde{n}(\phi, t)\Delta\phi$ gives the mean number of enzymes at time t inside the cycle in the phase states between ϕ and $\phi + \Delta\phi$. We also introduce the mean number n of free enzymes that are ready to bind the substrate. Moreover, $s(t)$ is the mean number of free substrate molecules at time t.

The evolution equation for $\tilde{n}(\phi, t)$ is

$$\frac{\partial \tilde{n}}{\partial t} = -v\frac{\partial \tilde{n}}{\partial \phi} + \sigma\frac{\partial^2 \tilde{n}}{\partial \phi^2} \tag{16.12}$$

Thus, evolution of the density function $\tilde{n}(\phi, t)$ represents a drift with constant velocity v accompanied by diffusion along the reaction coordinate ϕ. The effective diffusion is caused by intramolecular fluctuations with intensity σ.

The reaction coordinate varies from $\phi = 0$ to $\phi = 1$. Additionally, there is a special point $\phi = \phi_c$ inside the cycle. It is convenient to divide the whole variation interval of the reaction coordinate into two segments $[0, \phi_c]$ and $[\phi_c, 1]$ and separately formulate boundary conditions at the ends of these two segments.

The flux through the left boundary $\phi = 0$ of the segment $[0, \phi_c]$ is given by the number of enzymes per unit time that bind a substrate and thus start their cycles:

$$\left[v\tilde{n} - \sigma\frac{\partial \tilde{n}}{\partial \phi}\right]_{\phi=0} = \alpha s n. \tag{16.13}$$

At point $\phi = \phi_c$ inside the cycle a product molecule is irreversible released (and quickly transformed into the substrate), and the enzyme–substrate complex disappears. Since this occurs with any enzyme when it reaches the state $\phi = \phi_c$, an absorbing boundary condition should be placed at the right end of the segment $[0, \phi_c]$, i.e. we have

$$\tilde{n}|_{\phi=\phi_c} = 0 . \tag{16.14}$$

The segment $[\phi_c, 1]$ corresponds to enzymes that are in the process of conformational relaxation to their free states. The flux through its left boundary $\phi = \phi_c$ should be equal to the flux of enzymes that reach at the considered moment the right boundary of the segment $[0, \phi_c]$. Thus, we obtain

$$\left[v\tilde{n} - \sigma\frac{\partial \tilde{n}}{\partial \phi}\right]_{\phi=\phi_c+0} = \sigma\frac{\partial \tilde{n}}{\partial \phi}|_{\phi=\phi_c} . \tag{16.15}$$

Here we have used the notation $\phi = \phi_c + 0$ to indicate that the respective flux should be taken at the left boundary of the segment $[\phi_c, 1]$. Moreover, we have taken into account that the absorbing boundary condition (16.14) holds on the right boundary of segment $[0, \phi_c]$ and therefore $\tilde{n} = 0$ at $\phi = \phi_c$.

Finally, any enzyme that reaches $\phi = 1$ is finishing its cycle and becoming free. Therefore, Eq. (16.12) should have an absorbing boundary at the right end $\phi = 1$ of the segment $[\phi_c, 1]$, i.e.

$$\tilde{n}|_{\phi=1} = 0. \tag{16.16}$$

The mean-field evolution equation for the mean number n of free enzymes is

$$\frac{dn}{dt} = -\alpha s n + \left[v\tilde{n} - \sigma \frac{\partial \tilde{n}}{\partial \phi} \right]_{\phi=1} . \tag{16.17}$$

The first term on the right-hand side of this equation takes into account that the number n of free enzymes decreases when they bind a substrate molecule. The second term describes the increase of n through enzymes finishing their cycles and returning to the free state.

Finally, the mean-field evolution equation for the mean number s of substrate molecules in the reaction volume is

$$\frac{ds}{dt} = -\alpha s n + \left[v\tilde{n} - \sigma \frac{\partial \tilde{n}}{\partial \phi} \right]_{\phi=\phi_c} - \gamma s + \zeta. \tag{16.18}$$

The first term on the right side describes the decrease in the number of substrate molecules by binding to free enzymes, and the second term takes into account the release of new substrate molecules by enzymes at point $\phi = \phi_c$ inside the cycle. The third and fourth terms describe decay and supply of substrate molecules.

Note that the combination

$$N = n(t) + \int_0^1 \tilde{n}(\phi, t) d\phi \tag{16.19}$$

is conserved, since it represents the total number of enzyme molecules in the system, given by the sum of enzymes in the free state plus enzymes inside the catalytic cycle.

Equations (16.12–16.19) constitute the mean-field description of the considered reaction. These equations are formulated neglecting fluctuations in the numbers of substrate molecules and enzyme molecules in different cycle states, i.e. by replacing the respective exact fluctuating variables by their time-dependent ensemble averages. The effects of intramolecular noise are however taken into account in the equations as leading to an effective diffusion along the reaction coordinate. Below we numerically integrate these equations and compare their predictions with the behavior described by the full stochastic model.

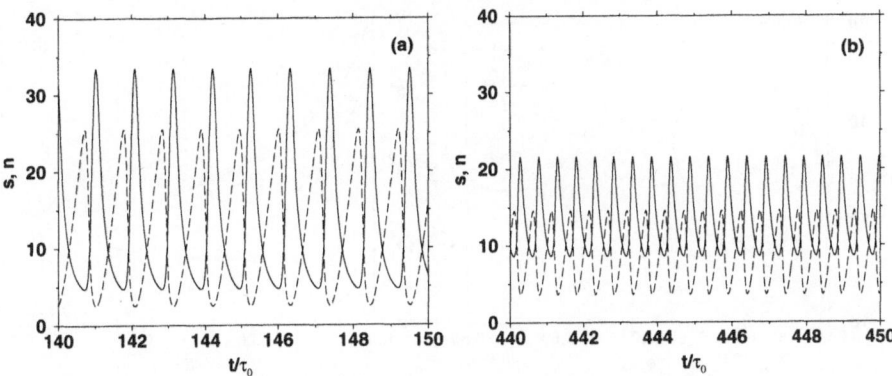

Fig. 16.5. Time dependence of the number of substrate molecules (*solid line*) and free enzymes (*dashed line*) obtained by numerical integration of the mean-field model for the same parameters as in Figs. 16.3a and b, respectively

Figure 16.5a displays the time dependence of the substrate (solid line) and free enzyme (dashed line) concentrations in the mean-field approximation (16.12–16.19) with the same reaction parameters as in Fig. 16.3a. We see that rapid oscillations with period $T \approx \tau_0$ are observed at $\phi_c = 0.2$. The comparison with the respective stochastic simulation in Fig. 16.3a reveals good agreement: both the oscillation frequency and the amplitude of oscillations are equal in the two approaches (although, as should have been expected, fluctuations are absent in the mean-field description). If we take $\phi_c = 0.55$, spiking with period $T \approx 0.5\tau_0$ is found at low intensities of intramolecular noise in the mean-field description (Fig. 16.5b, $\xi = 0.02$) in good agreement with the result of the corresponding stochastic simulation (Fig. 16.3b).

In the mean-field approximation the onset of spiking corresponds to a bifurcation of the steady state (the fixed point) of Eqs. (16.12-16.19). This Hopf bifurcation transforms the stable fixed point into a stable limit cycle. Thus, by analyzing the stability of this fixed point with respect to small perturbations, boundaries of spiking regimes in the parameter space can be determined.

The steady state is given by a stationary solution of the diffusion equation (16.12) with boundary conditions (16.13) to (16.16), and of equations (16.17) and (16.18). The stability of this stationary solution can be analyzed by investigating the evolution of small perturbations, introduced as $s(t) = s_0 + \delta s(t)$, $n(t) := n_0 + \delta n(t)$, $\tilde{n}(\phi, t) = \tilde{n}_0(\phi) + \delta\tilde{n}(\phi, t)$. This analysis has been performed in our recent publication [22].

Figure 16.6a displays bifurcation boundaries obtained in the parameter plane (ϕ_c, α). The steady state is unstable and spiking with periods $T_1 \approx \tau_0$ and $T_2 \approx 0.5\tau_0$ develops inside the regions whose boundaries are indicated by the solid and dashed lines, respectively. Thus, if we keep ϕ_c constant and increase the binding probability rate constant α, the steady state becomes

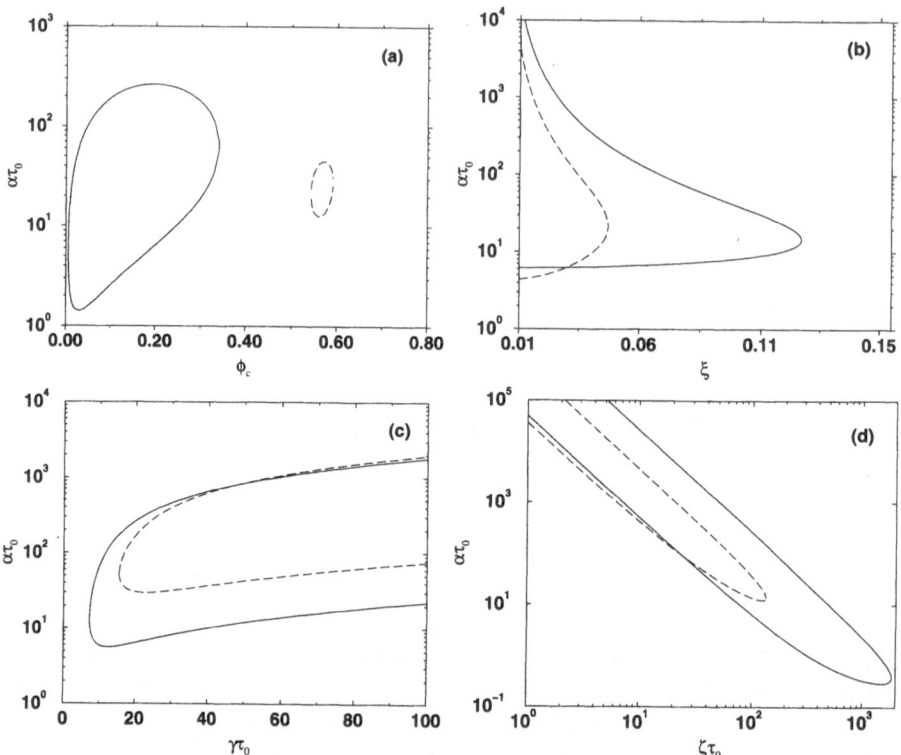

Fig. 16.6. The mean-field bifurcation diagrams for a population of $N = 1000$ enzymes **a** in the parameter plane (α, ϕ_c), **b** in the parameter plane (α, ξ), **c** in the parameter plane (α, γ), and **d** in the parameter plane (α, ζ). When the parameters are fixed, they take values $\gamma = 20\tau_0^{-1}$, $\zeta = 100\tau_0^{-1}$, $\xi = 0.04$. Solid curves show the boundaries of instabilities with respect to synchronous oscillations of the entire population, dashed curves display the respective boundaries for the regimes with two coherent enzymic groups. In diagrams (b,c,d) the solid curves correspond to $\phi_c = 0.2$ and the dashed curves correspond to $\phi_c = 0.55$

unstable and oscillations develop when the lower boundary is crossed. However, if we further increase α and cross the upper boundary, the oscillations disappear and the stable steady state is reestablished. Hence, oscillations are expected only in a certain window of the parameter α.

Figure 16.6b shows the bifurcation boundaries in the parameter plane (ξ, α) for $\phi_c = 0.2$ (solid line) and $\phi_c = 0.55$ (dashed line). Oscillations with periods $T_1 \approx \tau_0$ and $T_2 \approx 0.5\tau_0$ take place on the left side of these two curves, respectively. The bifurcation diagram in the parameter plane (γ, α) is displayed in Fig. 16.6c. In the last bifurcation diagram (Fig. 16.6d) the dependence of the synchronization threshold on the rate ζ of the substrate supply is shown. Note that the region of oscillations with period $T_2 \approx 0.5\tau_0$ (dashed curve) extends over three orders of magnitude of the supply rate

ζ, and for oscillations with period $T_1 \approx \tau_0$ (solid curve) it even covers four orders of magnitude of this parameter. At small values of ζ the oscillations are found at large substrate binding rate constants α.

16.5   Conclusions

We have presented a theoretical investigation of coherent intramolecular dynamics in a model reaction chain including a reaction step catalyzed by an enzyme. The analysis is performed under the assumption of a small spatial reaction volume, where the characteristic diffusive transport and mixing times are much shorter than the mean turnover time of the enzymes. The synchronization of enzymic molecular turnover cycles and rapid spiking in the product concentration are found in this system. This conclusion, obtained by direct numerical simulations of the stochastic reaction model, is confirmed by the analysis of the mean-field approximation for the reaction considered. Using this approximation, the bifurcation diagram is derived, and regions in the parameter space of the model where coherent collective enzyme dynamics takes place are determined.

Depending on the microscopic properties of an individual molecular turnover cycle, the same reaction chain can form two coherently operating enzymic groups whose phases are shifted by half the period or show complete synchronization. The synchronization of turnover cycles and coherent intramolecular dynamics were found even at relatively high intensities of intramolecular fluctuations. It should be emphasized that the enzymic reaction chain considered does not show slow kinetic oscillations under incoherent conditions.

Similar behavior was found in our previous studies [20,21] for allosterically activated and inhibited enzyme reactions. Thus, the results of our investigations suggest that the spontaneous emergence of coherent intramolecular dynamics is a robust phenomenon that can be expected both for allosteric and nonallosteric enzymes in the reactions proceeding in biologically relevant microvolumes.

Acknowledgments. The authors acknowledge financial support from Peter und Traudl Engelhorn Stiftung zur Förderung der Biotechnologie und Gentechnik (Germany).

References

1. L. A. Blumenfeld and A. N. Tikhonov, *Biophysical Thermodynamics of Intracellular Processes* (Springer, Berlin, 1994)
2. S. Subbiah, *Protein Motions* (Springer, Berlin, 1996)
3. G. A. Petsko, Nature **371**, 740 (1994)

4. I. Schlichting, J. Berendzen, G. N. Phillips, and R. M. Sweet, Nature **371**, 808 (1994)
5. M. Eigen and R. Rigler, Proc. Nat. Acad. Sci. USA **91**, 5740 (1994)
6. S. Weiss, Science **283**, 5408 (1999)
7. H. P. Lu, X. Luying, and X. S. Xie, Science **282**, 1877 (1998)
8. A. Ishijima, H. Kojima, T. Funatsu, M. Tokunaga, H. Higuchi, M. Tanaka, and T. Yanagida, Cell **92**, 161 (1998)
9. B. Hess, and A. S. Mikhailov, Science **264**, 223 (1994)
10. B. Hess and A. S. Mikhailov, J. Theor. Biol. **176**, 181 (1995)
11. B. Alberts, Cell **92**, 291 (1998)
12. R. F. Service, Science **284**, 80 (1999)
13. D. T. Chiu, F. W. Clyde, F. Ryttsen, et al., Science **283**, 1892 (1999)
14. H. Gruler and D. Müller-Enoch, Eur. Biophys. J. **19**, 217 (1991)
15. W. Häberle, H. Gruler, P. Dutkowski, and D. Müller-Enoch, Z. Naturforsch. **45c**, 237 (1990)
16. M. Schienbein and H. Gruler, Phys. Rev. E **56**, 7116 (1997)
17. I. Derenyi and D. R. Astumian, Phys. Rev. L **80**, 4602 (1998)
18. B. Hess and A. S. Mikhailov, Biophys. Chem. **58**, 365 (1996)
19. A. S. Mikhailov and B. Hess, J. Phys. Chem. **100**, 19059 (1996)
20. P. Stange, A. S. Mikhailov, and B. Hess, J. Phys. Chem. B, **102**, 6273 (1998)
21. P. Stange, A. S. Mikhailov, and B. Hess, J. Phys. Chem. B **103**, 6111 (1999)
22. P, Stange, A. S. Mikhailov, and B. Hess, J. Phys. Chem. B, **104**, 1844 (2000)
23. P. Stange, D. Zanette, A. S. Mikhailov, and B. Hess, Biophys. Chem. **79**, 233 (1999)
24. A. S. Mikhailov, P. Stange, and B. Hess, "Self-organizing networks of molecular machines in allosterically regulated enzymic reactions," in: *Statistical Mechanics of Biocomplexity*, D. Reguerra, J. M. G. Vilar, and J. M. Rubi (eds.), p. 72 (Springer, Berlin, 1999)
25. P. Stange, A. S. Mikhailov, and B. Hess, "Coherent intramolecular dynamics in populations of allosteric enzymes," in: *Transport and Structure*, S. C. Müller, J. Parisi, and W. Zimmerman (eds.), p. 231 (Springer, Berlin, 1999)
26. L. Stryer, *Biochemistry*, 3rd ed. (Freeman: New York, 1995)
27. G. Adam and M. Delbrück, "Reduction of dimensionality in biological diffusion processes," in: *Structural Chemistry and Molecular Biology*, A. Rich and N. Davidson (eds.), p. 198 (Freeman, San Francisco, 1968)
28. J. B. Wittenberg and B. A. Wittenberg, Ann. Rev. Biophys. Chem. **19**, 217 (1990)
29. B. Hess and A. S. Mikhailov, Ber. Bunsen-Ges. Chem. **98**, 1198 (1994)

17 Single-Molecule Dynamics in Biosystems

T. Yanagida

Biomolecules assemble to form molecular machines such as molecular motors, cell signal processors, DNA transcription processors and protein synthesizers to fulfill their functions. Their collaboration allows the activity of biological systems. The reactions and behaviors of molecular machines must be flexible in order to respond to their surroundings. This flexibility is essential for biological organisms. The underlying mechanism of molecular machines is not as simple as one would predict from analogy to man-made machines. Since molecular machines are only nanometers in size and have a flexible structure, they are very prone to thermal agitation. Molecular machines operate under a strong influence of thermal noise, with a high efficiency of energy conversion [1]. This is in sharp contrast to man-made machines that operate precisely and rapidly at energies much higher than thermal noise. The aim of our research is to approach the essential engineering principle of the adaptive biological systems by uncovering the unique operations of molecular machines. For this aim, we have developed several new technologies for single-molecule imaging and manipulation of biomolecules. These new techniques allow the dynamic properties of individual molecules in molecular machines, which were previously hidden in averaged ensemble measurements, to be unveiled [2,3,4]. I will survey the applications of single-molecule detection (SMD) techniques to several biological molecular machines and briefly discuss the unique mechanism of motion underlying molecular motors, the system on which SMD has been most successfully used.

17.1 Single-Molecule Imaging

Biomolecules and even their assemblies are on the order of nanometers in size, so it is impossible to observe them using optical microscopy. To overcome this problem, biomolecule can be fluorescently labeled and visualized using fluorescence microscopy. However, single fluorophores had only been imaged in nonaqueous solution and on a dry surface [5]. Biomolecules, however, are active only in aqueous solution. In 1995, we demonstrated that single fluorophors could be seen in aqueous solution [6]. The key point to overcome in

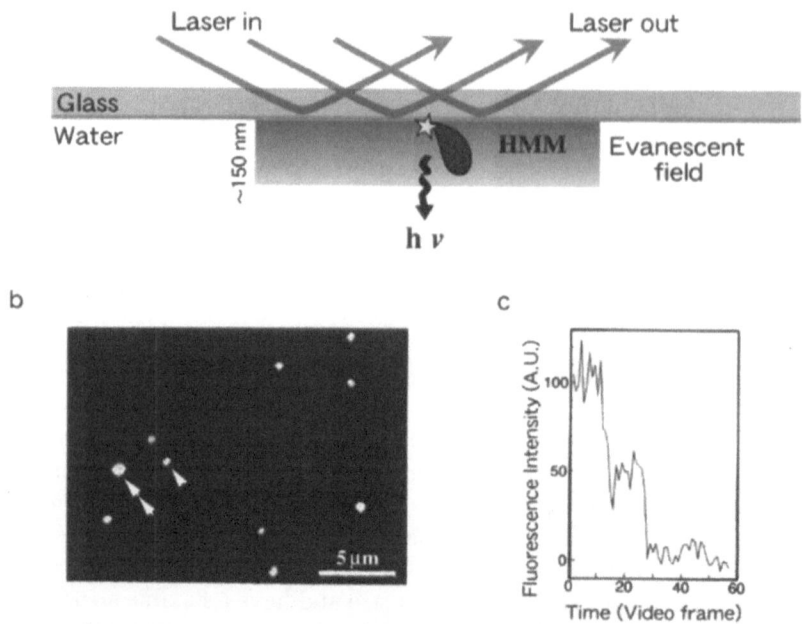

Fig. 17.1. Single-molecule imaging [6]: **a** Schematic drawing of the total internal reflection fluorescence microscopy for visualization of single fluorescent molecules in aqueous solution. **b** Fluorescence images of single fluorophores (Cy3) attached to proteins (myosin) bound to a glass surface. Larger spots are due to proteins that bind two fluorophores. **c** Quantized photobleaching of fluorescent molecules (two molecules) observed at the video rate

order to visualize single fluorophores in aqueous solution was how to reject the background noise due to Raman scattering from the solution, incident light breaking through filters, and luminescence arising from objective lens, immersion oil, and dust. To overcome this problem, we used the evanescent field produced when a laser beam was totally reflected by the interface between the water and glass (Fig. 17.1a). The evanescent field is localized near the glass surface and the penetration depth is ~150 nm. Therefore, illumination is limited only to fluorophores bound to the glass surface and their vicinity, so that the background can be greatly reduced. Furthermore, by appropriately choosing optical elements, the background noise could be reduced 2000-fold compared to that of conventional fluorescence microscopy. Thus, single fluorophores could be clearly observed in aqueous solution (Fig. 17.1b).

Molecular Motors: We have used this technique to observe sliding motion of single kinesin molecules on a microtubule [7]. Kinesin is a molecular motor that transports organelles along a microtubule in cells. Fluorescently labeled

a   b

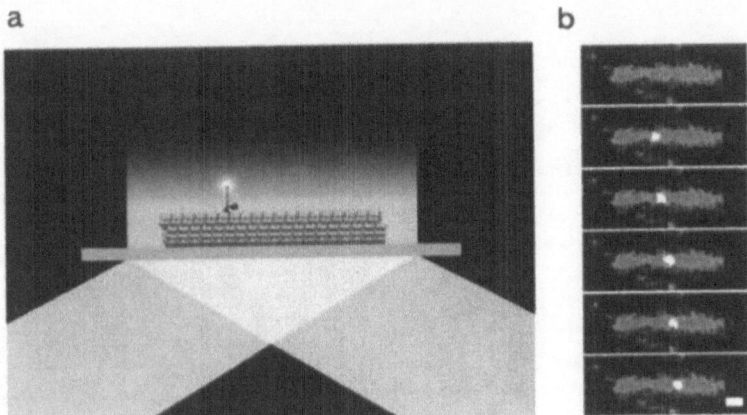

Fig. 17.2. Sliding movement of a single molecular motor, kinesin, along a microtubule [7]. Top panel: fluorescently labeled kinesin was visualized in TIRFM, while
it was moving along a microtubule placed on the glass surface. Lower panels: time
records of the fluorescence image show that the kinesin slides on the microtubule

kinesin molecules were added to microtubules, which had been previously
adsorbed onto the surface of a glass, in the presence of ATP (Fig. 17.2a).
Figure 17.2b shows series of fluorescence images of a single kinesin molecule
moving along a microtubule. The labeling of kinesin was at its tail and did
not damage the molecule. ATP synthase (F_1F_0-ATPase) produces ATP from
ADP and Pi in the F_1 portion by using proton- or sodium-motive force in
the F_0 portion. The process is reversible. When hydrolyzing ATP in F_1, the
enzyme pumps protons in F_0 in the opposite direction. ATPase and proton
flow are coupled by the rotation of a shaft which links F_1 and F_0. Noji et al.
[8] have demonstrated that the rotation of the shaft of F1 can be visualized
by tracing the position of fluorescently labeled actin filaments attached to
the top of the shaft. Sambongi et al. [9] have recently shown that an actin
filament connected to a c subunit oligomer of F_0 is able to rotate by using
the energy of ATP hydrolysis (Fig. 17.3).

Enzymatic (ATPase) Reaction: Motions of molecular motors and operations of most of other molecular machines are fueled by the chemical energy
released from ATP hydrolysis. The enzyme, ATPase, catalyzes ATP hydrolysis. Kinesin and myosin are both motors and ATPase. In order to uncover
the operations of these molecular machines, it is important to observe the
individual cycles of ATP hydrolysis by single ATPase molecules. This has
been achieved by using the single-molecule imaging technique described above
(TIRFM) and the fluorescent ATP analog, Cy3-ATP (Fig. 17.4a,b) [6,10].
Cy3-ATP is hydrolyzed in the same way as ATP [11]. Recording the fluorescence of Cy3-ATP at the position of a single myosin molecule revealed

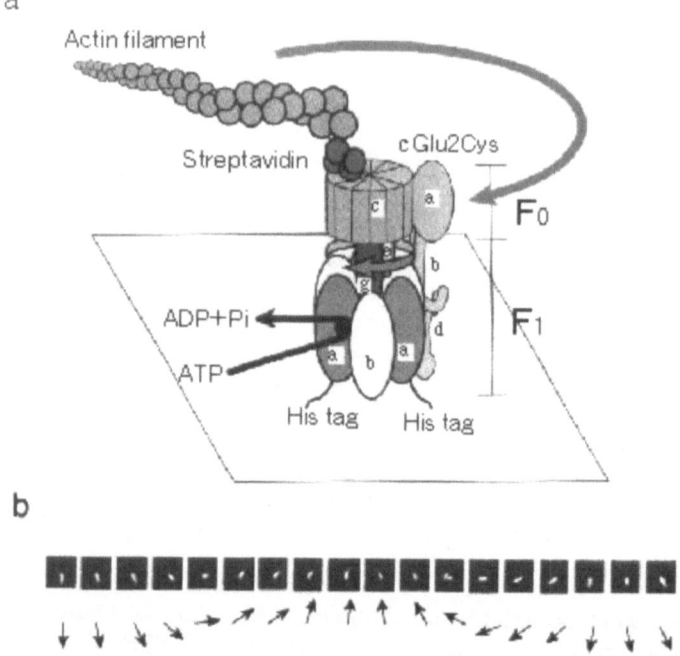

Fig. 17.3. Rotary motion of an F_1F_0 motor [9]: **a** Architecture of the F_1F_0 motor. A side view of the motor is shown in the top panel. The hydrolysis of ATP at $(\alpha\beta)_3$ causes the rotary motion of the γ subunit relative to $(\alpha\beta)_3$. **b** Rotary motion of the F_1F_0 motor was detected by visualizing the motion of a fluorescently labeled actin filament attached to the γ subunit, while $(\alpha\beta)_3$ is immobilized. Bottom panel shows the time record of the image of the actin filament

the turnovers of ATP during its hydrolysis by a single myosin molecule. Figure 17.4c shows individual cycles of ATP hydrolysis by a single myosin molecule, monitored by measuring the fluorescence intensity of bound Cy3-nucleotides.

DNA Transcription: The initial steps of gene expression include the binding of RNA polymerase (RNAP) to DNA, the search for a promoter in the DNA sequence, where transcription starts, and the synthesis of RNA based on the information coded in the DNA. These steps are central regulatory mechanisms of gene expression and have been extensively investigated [12]. We have observed single fluorescently labeled RNAP molecules interacting with a single molecule of DNA suspended in solution using optical traps (Fig. 17.5a) [13]. Figure 17.5b shows series of fluorescence images of a single molecule of RNAP undergoing linear sliding along DNA. This observation provides direct evidence that a sliding motion is the mechanism used to search for the

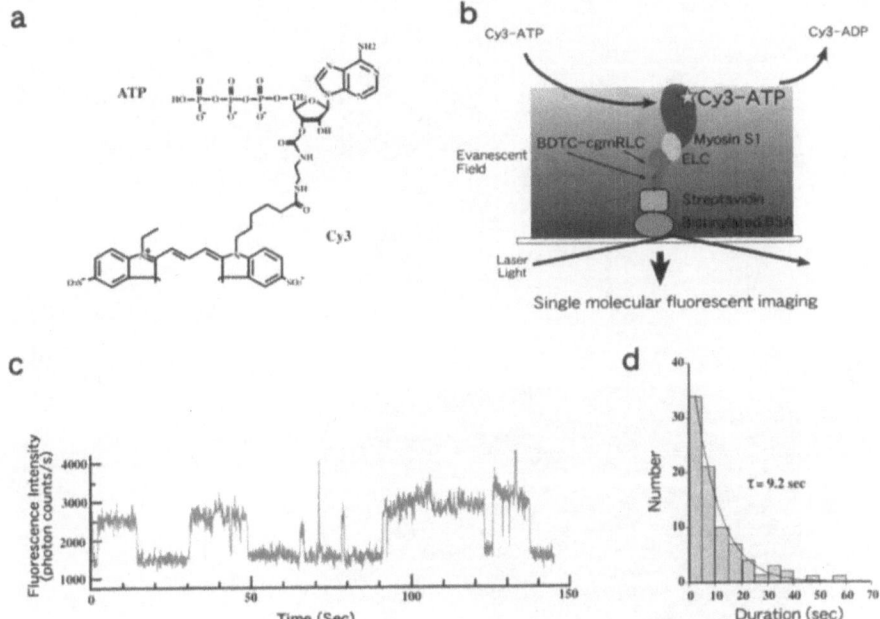

Fig. 17.4. Single-molecule imaging of individual ATP turnovers by single myosin molecules [6,10]. **a** Structure of Cy3-ATP. **b** Schematic diagram of how the ATPase used by a single myosin molecule has been measured. A single myosin head was fixed on the glass surface and the turnover of ATP was monitored by the fluorescence of labeled ATP or ADP. The fluorescence could be visualized when fluorescent ATP or ADP was bound to myosin and no fluorescent spot was detected when it was in solution because of its rapid Brownian motion. **c** Time records of the fluorescence intensity from Cy3-ATP or -ADP bound to the myosin head. **d** Histogram of the lifetime of Cy3-ATP or -ADP bound to myosin heads. The dissociation rate (1/average lifetime) was determined to be 0.06/s, which was consistent with that detected using the suspension

promoter. When RNAP binds to the promoter of the DNA, transcription starts. Harada et al. are extending these experiments to visualize the mechanism for DNA transcription, including termination processes, for which many unanswered questions remain.

Protein Dynamics: Proteins, the main constituents of molecular machines, behave dynamically. The dynamic behavior is crucial for their function. Fluorescence spectroscopy can be used to monitor the structural changes of proteins [4]. The fluorescence spectrum is sensitive to the microenvironment of the fluorescent probe. In addition to spectral fluctuation, tetramethyrhodamine attached to the most reactive cysteine on the myosin molecule shows

a

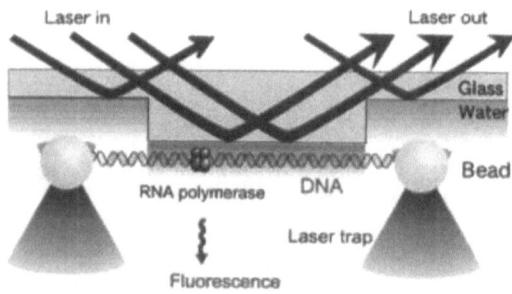

b

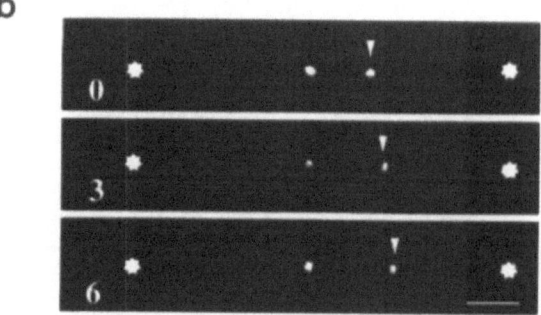

Fig. 17.5. DNA transcription processes by RNA polymerase [13]: **a** Schematic diagram of a measurement system. Fluorescently labeled RNA polymerase (RNAP) was visualized on a single DNA molecule trapped by a laser through beads at both ends. **b** RNAP (indicated by an arrow) is moving along a DNA molecule which is suspended between two optically-trapped beads

spontaneous fluctuations on a time scale of seconds (Fig. 17.6) [14]. This may be due to slow structural changes of the myosin.

Fluorescence resonance energy transfer (FRET) is a more direct method to determine the structure of proteins. When donor and acceptor fluorophores are located close to each other (on the order of nanometers), the donor fluorescence is quenched and the excited energy of the donor is transferred to the acceptor without radiation, resulting in acceptor fluorescence. Thus, we can determine the distance between the donor and acceptor fluorophores attached to two different sites of a protein by monitoring the color of the fluorescence (Fig. 17.7a). Figure 17.7b shows fluorescence spectra of donor and acceptor fluorophores attached to a myosin molecule [15]. The fluorescence intensities of the donor and the acceptor varied in a flip-flop fashion, consistent with FRET. The FRET also changed slowly on a time scale of seconds.

These results suggest that myosin can take several metastable states (cf. [16,30]) and undergo slow transitions among them. This idea challenges the widely accepted view that proteins have unique conformations [17].

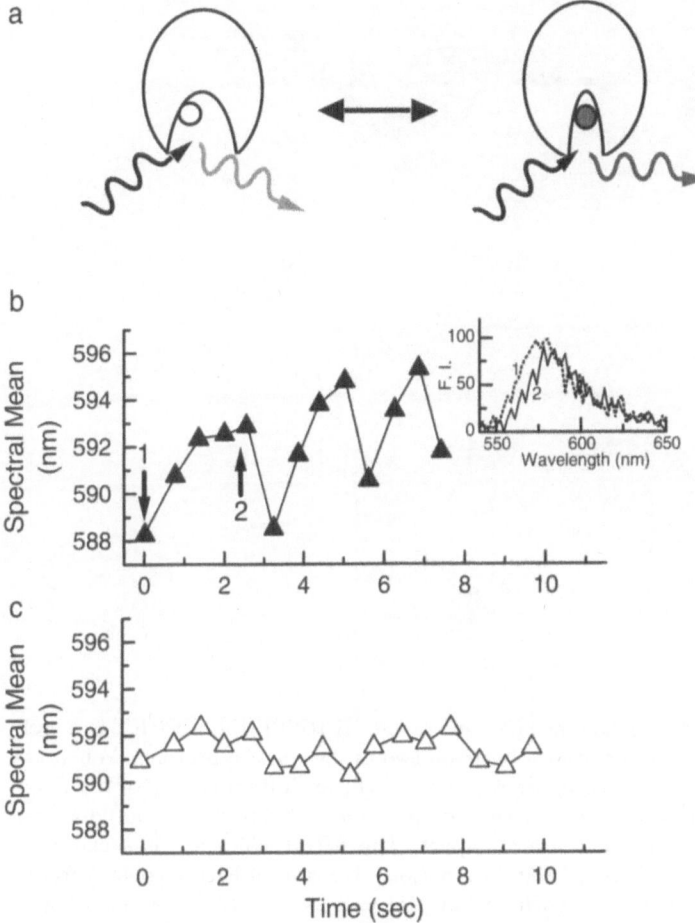

Fig. 17.6. Protein dynamics revealed by single-molecule fluorescence spectroscopy [14]: **a** Schematic diagram. **b** The fluorescence spectrum of tetramethylrhodamine attached to myosin S1 fluctuates on a time scale of seconds. The spectral mean was plotted as a function of time. The inset shows the two fluorescence spectra, which give different spectral means. **c** The time course of the spectral mean of a fluorphore bound to a glass surface

Protein Folding: Protein folding is an intriguing problem. Does every molecule fold using the same sequence of events or is the sequence variable? This topic has attracted the attention of many investigators. In vivo, proteins fold with the aid of a molecular machine called a chaperone [18]. SMD will help us to understand the mechanism of this machine [19].

Cell Signaling: Cell signaling is one of the major target areas for investigation using single-molecule imaging. Cells are complex but have well con-

a

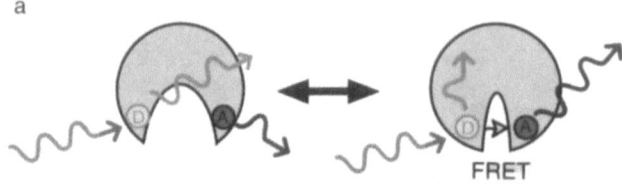

b

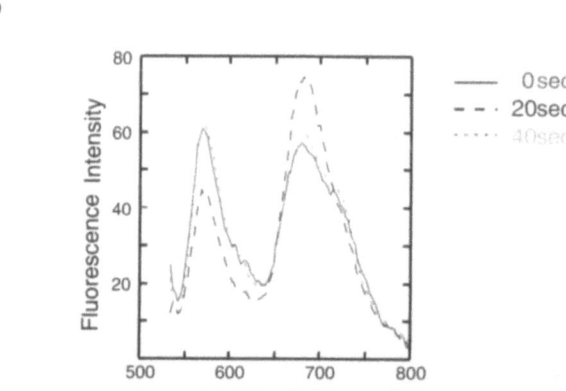

Fig. 17.7. Fluorescence resonance energy transfer (FRET) [15]. **a** FRET is a technique that determines the distance between two probes on the proteins. When two probes called donor D and acceptor A are close, the excitation energy of the donor is transferred to the acceptor and the acceptor emits fluorescence, but does not when the donor and acceptor are far apart. The FRET efficiency is sensitive to protein conformation and to protein interactions. **b** Series of FRET spectra from a single myosin molecule labeled with Cy3 at SH1 and Cy5 at SH2. The increase (decrease) in the donor fluorescence at 570 nm and decrease (increase) in the acceptor fluorescence at 670 nm are coupled, indicating changes in FRET efficiency

trolled systems consisting of many kinds of molecular machines. Cells work very flexibly and autonomously, responding to external stimuli. The problem of how signals are transmitted and processed in cells is a central theme for life sciences.

Cell signaling is triggered by external signals, and the first event in this process occurs on the cell membrane. Recently, we demonstrated binding of single molecules of fluorescently labeled epidermal growth factor (EGF) to its receptor (EGFR). Following binding, dimerization of EGF-EGFR complexes can be directly visualized on the apical surface of living cells using TIRFM (Fig. 17.8) [20]. The autofluorescence of these cells was small enough to observe single fluorophores, and the evanescent field produced in the interface

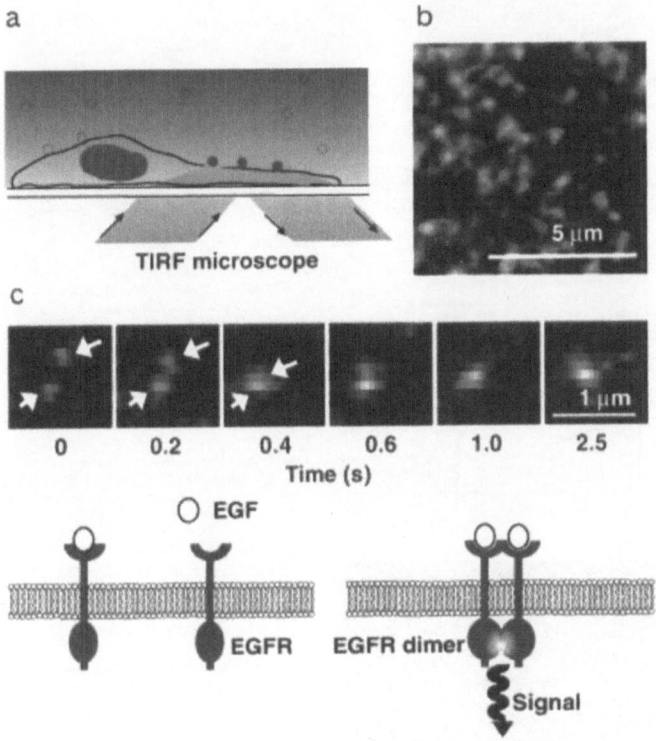

Fig. 17.8. Cell signaling [20]: **a** Single fluorescently-labeled epidermal growth factors (EGF) were visualized on the living cell surface by TIRFM. **b** Fluorescence image of the surface 1 min after addition of fluorescently labeled EGF to culture medium. A single EGF molecule can be seen as a spot. **c** Dimerization of EGF and EGF receptor complexes. A time sequence of images shows dimerization of the EGF/EGF receptor, triggering transmission of signal

between the cell surface and the medium allows direct observation of single fluorophores on the apical surface.

Ion Channel: Ion channels precisely regulate the ionic flow across cell membranes and generate ionic gradients that are responsible for nerve and muscle excitability. Single ion-channel current recording [21] has shed light on the kinetics and pharmacological properties of many kinds of ion channels. Their detailed mechanisms, however, are still being elucidated. The ability to monitor the conformation and chemical state of ion channels, combined with measurement of single-channel ion current, would advance the investigation of this central area of research. We have combined the planar lipid bilayer and single-molecule imaging technique, and thus simultaneously observed the ion current and fluorescence image of single ion channels (Fig. 17.9) [22]. Expanding this technique, it should be possible to directly observe the

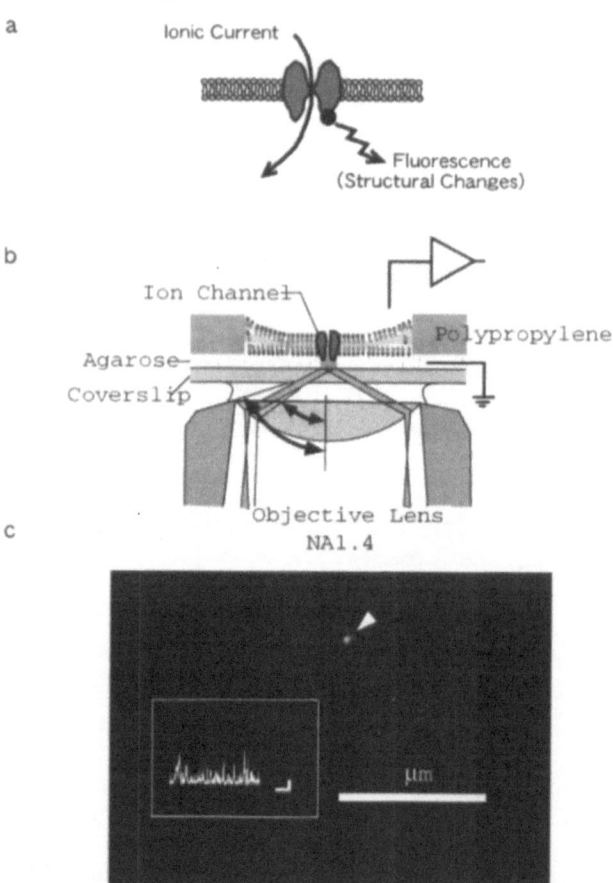

Fig. 17.9. Simultaneous electrical and optical measurement of single ion channels [22]: **a,b** Schematic diagram for measurement. Fluorescently labeled ion channels are incorporated into a membrane formed on the agarose-coated glass and observed by an objective type of TIRFM. Ion current from a single channel is measured electrodes simultaneously. **c** Image of a single ion channel and a current recording from it (insertion)

interaction between a single ion channel and its regulator proteins, and the subsequent change in channel activity.

Single-Molecule Nano-manipulation

Biomolecules, and even single molecules, can be caught by glass-microneedles [23,24] or by beads trapped by optical tweezers [25,26]. A microneedle and a bead trapped by laser are similar to a spring: when force is applied on the spring from the outside, the spring expands in proportion to the applied force.

Utilizing this relationship, the force and displacement due to single molecular motors can be measured. The displacement of a microneedle and a bead has been determined with subnanometer accuracy by a paired or quadrant photo-diode (much less than diffraction-limited optical measurements) [24–26]. This accuracy of the displacement corresponds to sub-piconewton accuracy in the force measurement.

Figure 17.10 shows measurement of forces and displacements of single myosin molecules by microneedle nanometry [24]. Individual displacements and forces with amplitudes of 5 to 20 nm and 3 to 5 pN, respectively, were observed. Similar mechanical events were observed by optical trapping nanometry [26].

Individual mechanical events of single kinesin molecules were measured by manipulating kinesin molecules attached to a bead with an optical trap (Fig. 17.11a). Thus, it was found that a kinesin molecule moves along a microtubule with regular steps of 8 nm [25], reflecting the periodicity of the tubulin $\alpha\beta$ heterodimers in a microtubule (Fig. 17.11b). Based on this finding, a hand-over-hand mechanism has been proposed (Fig. 17.11c upper) [27]. Recently, we have found that the 8-nm steps consist of two substeps 4 nm in size [28] and a single one-headed kinesin mutant can also continuously move although the movement is not smooth, and fluctuates between the forward and backward directions [29]. These results suggest that as for myosin, Brownian motion appears to play an essential role in the mechanism for movement.

17.2   Simultaneous Observation of Individual Mechanical and Chemical Events by a Single Motor

In order to elucidate the mechanism of mechanochemical energy conversion by an actomyosin motor, it is crucial to examine how the mechanical events are coupled to the ATP turnover. We have combined optical trapping nanometry with a single-molecule imaging technique to simultaneously observe individual mechanical events and ATP turnover of a single myosin molecule during force generation (Fig. 17.12) [30]. Each overall displacement corresponded to a single ATP turnover. The timing of displacements, however, did not always coincide with that of nucleotide dissociation (probably, ADP). Sometimes (30% of total events) the mechanical event was delayed for several hundred milliseconds after ADP dissociation. Although photobleaching of the fluorescent ATP analog cannot be completely ruled out, this result suggests that myosin has a hysteresis or memory state, which stores chemical energy from ATP hydrolysis.

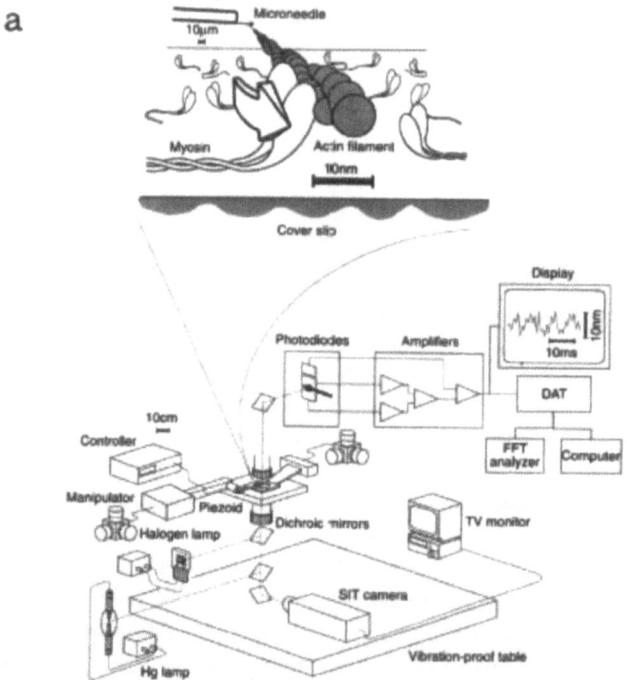

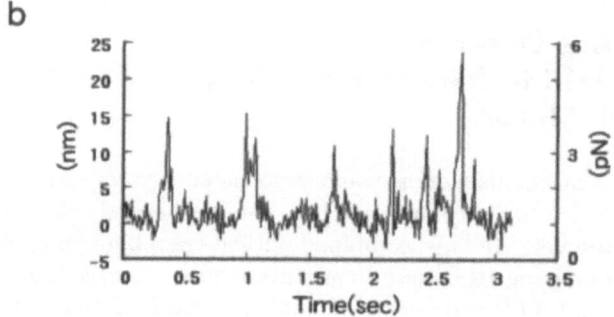

Fig. 17.10. Micromanipulation of a single actin filament by glass microneedle and subpiconewton force measurement system in vitro [24]: single actin filaments labeled with fluorescent phalloidin were observed using an inverted fluorescence microscope. One end of an actin filament was caught by a glass microneedle mounted on a mechanical manipulator and a piezo-actuator, while the other end was brought into contact with the myosin-coated glass surface in the presence of ATP. The force was determined by measuring the displacement of the needle resulting from the actomyosin interaction. The image of the needle was projected onto a pair of photodiodes, and the displacement of the needle was determined from the differential photocurrents. Thus, the displacement could be measured at resolutions of 0.1 nm, corresponding to 0.1 pN and 0.2 ms, respectively

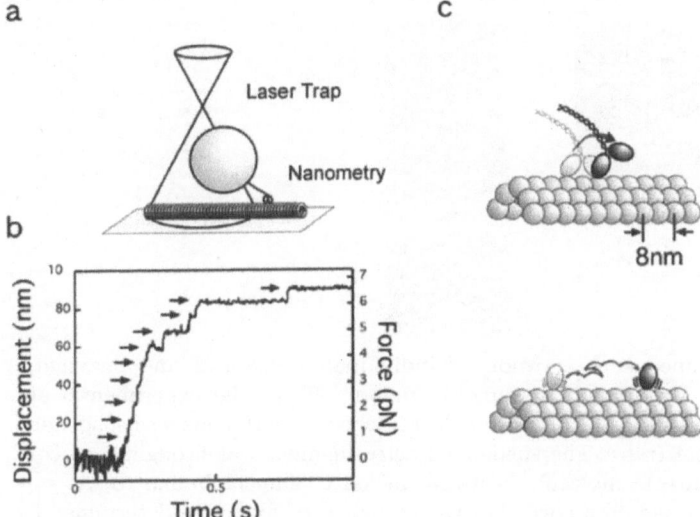

Fig. 17.11. Mechanical measurement of a single kinesin molecule by optical trapping nanometry [28]: **a** A single kinesin molecule was attached to a bead trapped by a laser. Displacement of the kinesin was measured by monitoring the displacement of the bead. **b** Displacement records that the step size is 8 nm. **c** Model. (Upper) A single two-headed kinesin molecule moving in one direction in a hand-over-hand fashion. (Lower) Biased thermal ratchet model (see text)

17.3   Myosin Walks Along an Actin Filament by Brownian Motion

Movement of the myosin motor results from the relative sliding between the head portion of a myosin molecule and the actin filament coupled to the ATP hydrolysis. Recently, many studies have shown that the neck region of the myosin head undergoes conformational changes. The angle of the myosin head changes relative to the neck region, depending on the form of bound nucleotide (see [31] for review). Based on these findings, a "swinging lever-arm" model has been proposed and is currently accepted (Fig. 17.13) [32]. According to this model, the neck region acts as a lever arm and swings to pull the actin filament. Furthermore, it is postulated that the conformational change in the neck region is tightly coupled to the biochemical cycle of ATP hydrolysis cycle in a one-to-one fashion. However, conformational changes in the neck region observed during generation of force are much smaller than those predicted by this model [33]. In order to understand how myosin works, it is crucial to measure the mechanical properties of a single molecule of myosin with high resolution.

We have developed a new assay for directly manipulating a single myosin head (subfragment-1) and measuring displacements using a scanning probe

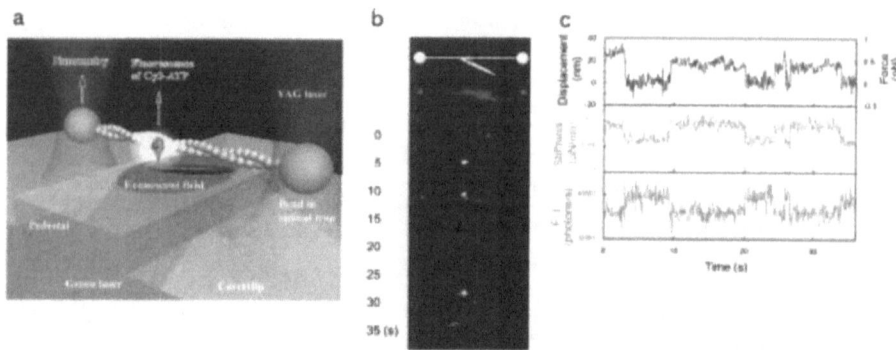

Fig. 17.12. Simultaneous observation of individual mechanical and enzymatic (ATPase) reactions by a single actomyosin motor [30]. **a** The experimental apparatus. A single acting filament with beads attached to both ends was suspended in solution by optical traps. The suspended actin filament was brought into contact with a single myosin molecule in a myosin-rod cofilament bound to the surface of a pedestal formed on a coverslip. Displacement or force was determined by measuring bead displacements with nanometer accuracy. Using TIRFM, individual ATPase turnovers were monitored (see Fig. 17.4). **b** Individual Cy3-ATP turnovers produced by a single myosin molecule during force generation. (Top) Fluorescence image of a suspended actin filament interacting with a myosin-rod cofilament bound to a pedestal surface and a corresponding schematic drawing. Asterisks indicate the positions of the beads as observed under bright field illumination. Both filaments labeled with Cy5. (Bottom) The lower panels indicate sequential images of association–(hydrolysis)–dissociation events of Cy3-ATP with the same myosin head within the filament shown above. **c** The time courses of individual mechanical (displacement and stiffness due to actomyosin interaction) and Cy3-ATP turnover events. Each overall mechanical event (displacement and stiffness, upper and middle traces, respectively) corresponds to a single ATP turnover (lower trace)

[34]. A single myosin head, which was fluorescently labeled at its tail end and visualized by TIRFM, was attached to the tip of a scanning probe mounted on a fine glass needle. The captured myosin head was brought into contact with an actin bundle adsorbed on a glass surface. The displacement and force due to actin–myosin interaction in the presence of ATP were determined by measuring the displacement of the needle with nanometer accuracy (Fig. 17.14a). The displacement did not take place abruptly but instead developed in stepwise fashion (Fig. 17.14b). The number of substeps in the rising phase of a displacement varied from one to five, and the dwell time between substeps varied stochastically. However, the step size was approximately 5.5 nm, coinciding with the periodicity of actin monomers in one strand of an actin filament.

When the load exerted on the myosin head was increased from approximately zero to 3 pN, which was >50% of the maximum force that the myosin head can produce, the number of substeps decreased and the dwell time in-

a
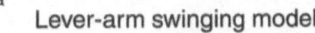
Lever-arm swinging model

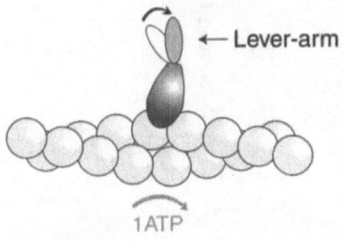

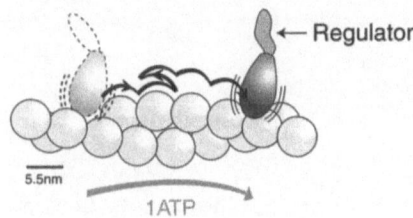

Fig. 17.13. Models of actomyosin motor. Two models have been proposed to explain the mechanism of myosin movement. In the lever-arm swinging model (upper), the myosin molecule moves in a single step. This step occurs by the swinging motion of the lever-arm relative to the head portion of myosin, coupled with the ATPase reaction. In the biased Brownian ratchet model (lower), movement of myosin is driven by Brownian motion, and ATP hydrolysis biases the direction of motion

creased, but the step size was constant. Some of substeps (10% of the total number) were backwards, and the size was the same as that of a forward step (∼5.5 nm) [35].

Simultaneous measurement of mechanical and ATP turnover events has shown that each (overall) displacement corresponds to a single ATP turnover (Fig. 17.12). Therefore, each substep is not coupled to ATP hydrolysis, but instead multiple substeps take place during a single ATP turnover (Fig. 17.13b).

17.4   Model of the Actomyosin Molecular Motors

Our results demonstrate that myosin can perform multiple mechanical cycles during the hydrolysis of a single ATP molecule, and the number of mechanical cycles varies depending on the load. Since the chemical energy from ATP hydrolysis is 20 k_BT, the energy for each mechanical cycle (substep) is only a few k_BT, i.e., close to the average thermal noise energy. Therefore, the biological motor constitutes a very skillful molecular machine that can operate with very high efficiency (>50%), even when the input energy level is close to the average energy of thermal noise. The behavior thus contrasts with a

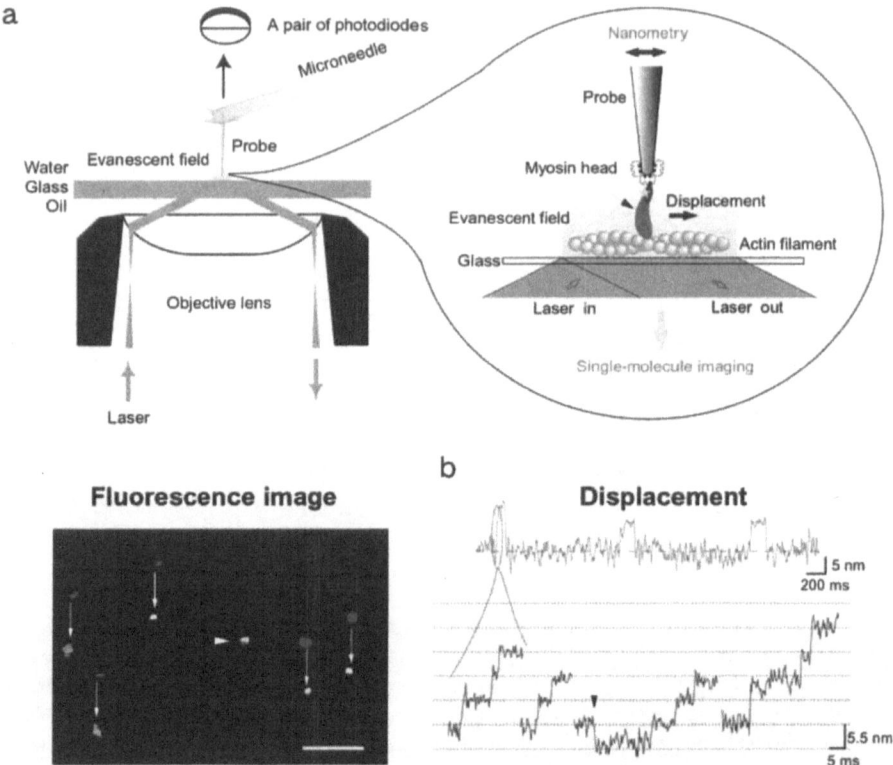

Fig. 17.14. Sliding movement of a single molecular motor, myosin subfragment 1 (S1) [34]: **a** Fluorescently-labeled S1 molecules on the surface of the glass were illuminated by an evanescent field generated by the total internal reflection of laser light. A single S1 molecule was captured by a scanning probe through the biotin–avidin system. The bottom panel shows an image of single S1 molecules when the stage was moved. The S1 molecule captured by the probe remained in the same position, while others moved with the stage, indicating that the single S1 molecule has been captured by the probe. **b** The S1 molecule captured by the scanning probe was allowed to interact with actin bundles in the presence of ATP and the displacement of S1 was measured by monitoring the displacement of the probe. The rising phase of the displacement records in a long time range has been expanded in a short time range. A 5.5-nm step could be clearly observed. The arrows indicate backward steps, which implies that the sliding movement is derived from thermal motion

manmade machines, which are designed to consume large amounts of energy to overcome noise and perform properly.

Myosin may not overcome but rather use Brownian motion to effectively move large distances with small expenditures of energy. Brownian motion, however, is random, so it must be biased in the forward direction by chemical energy from ATP hydrolysis. A. F. Huxley proposed a model in which

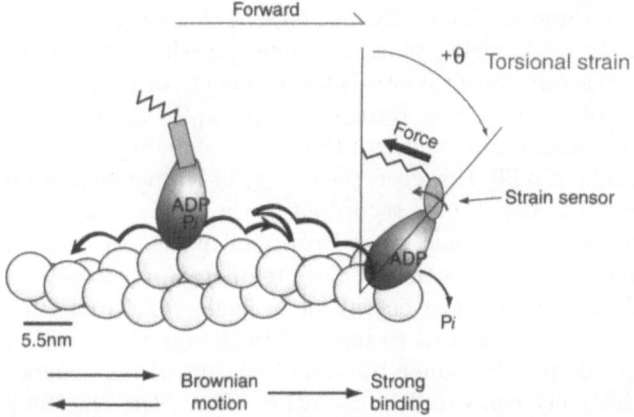

Fig. 17.15. Model of the actomyosin motor. The neck region acts as a strain sensor. When the myosin head steps forwards along an actin filament (which has helical structure) by Brownian motion, the myosin head rotates around the actin filament (26° per step) and thus the neck region is torsionally strained. This model assumes that the torsional strain causes conformational changes in the neck region and accelerates the release of bound Pi from the myosin head to make the bond between actin and the myosin head rigid, and to stop the movement of the myosin head at the positive position. When the myosin head moves backwards, the neck region would be conversely strained. The converse strain is assumed not to cause conformational changes in the neck region, and thus the myosin head returns to the original position. As a result, the myosin head moves forward

the affinity of a myosin head for actin is asymmetric, so that the myosin head, which undergoes Brownian motion, more favorably binds to actin in the forward direction [36]. In this model, ATP hydrolysis links to a single displacement of myosin, and our data demonstrate that a myosin head produces multiple steps during a single ATP turnover. To explain our results, we propose a strain-based Brownian ratchet model (Fig. 17.15). We assume that when the myosin head moves in the forward direction by Brownian motion, its neck region is torsionally strained due to the helical structure of the actin filament. The torsional strain causes conformational changes and accelerates the release of Pi from the head, which is strongly bound to actin. However, when the myosin head moves in the backward direction, the neck region is strained in the opposite direction, and the conformational change which results in Pi release does not take place. Thus, the myosin head diffuses back to its original position and the net displacement is zero. As a result, the myosin head moves forward. In this model, the origin of motion is Brownian motion, and the myosin head moves forward due to random Brownian motion by using chemical energy from ATP hydrolysis.

The strain sensor also switches the direction of movement. If the strain sensor is designed to respond to a strain in the opposite direction, the myosin will move in the opposite direction. Actually, a motor in the myosin family, called myosin VI, with the neck region modified, moves in the opposite direction. Furthermore, the strain sensor controls the efficiency of energy conversion and the velocity of motion. The average energy $(k_B T)$ of Brownian motion of a myosin is only about one-twentieth the chemical energy from hydrolysis of a single molecule of ATP. However, the energy of Brownian motion of myosin is not constant, but distributes according to the Boltzmann equation, $\exp(-E/k_B T)$, where E is the energy of Brownian motion of myosin, k_B is the Boltzmann constant, and T is the absolute temperature, so it could be, e.g., 10 $k_B T$, one-half the chemical energy of a single ATP molecule. Therefore, if the strain sensor is designed to respond to the force of a spring extended with an energy 10 $k_B T$, Brownian motion of myosin with an energy of >10 $k_B T$ will be selectively converted into motion. Thus, the efficiency of chemomechanical energy conversion might be $>50\%$. The probability that the myosin gains larger thermal energy from a heat bath is smaller according to the Boltzmann equation, so the frequency of events is smaller.

This model could explain how myosin can work under the strong influence of thermal agitation, with high efficiency of energy conversion.

Acknowledgment. I thank the members of the Biomotron Project and Single-Molecule Project of ERATO JST, and Dept. of Phys. Med. School of Osaka Univ. for valuable discussion.

References

1. T. Yanagida, Y. Harada, and A. Ishijima, "Nano-manipulation of actomyosin molecular motors in vitro: A new working principle," Trends in Biochemical Science **18**, 319–324 (1993)
2. Y. Ishii and T. Yangida, "Single molecule detection in life science," Single Molecules **1**, 5–13 (2000)
3. A. D. Mehta, J. A. Spudich, D. A. Smith, and R. M. Simmons, "Single-molecule biomechanics with optical methods," Science **283**, 1689–1695 (1999)
4. S. Weiss, "Fluorescence spectroscopy of single biomolecules," Science **283**, 1676–1683
5. W. E. Moerner and M. Orrit, "Illuminating single molecules in condensed matter,' Science **283**, 1670–1676 (1999)
6. T. Funatsu, A. Harada, M. Tokunaga, M. and T. Yanagida, "Imaging of Single Fluorescent Molecules and individual ATP turnovers by single myosin molecules in aqueous solution," Nature **374**, 555–559 (1995)
7. R. D. Vale, T. Funatsu, D. W. Pierce, L. Romberg, Y. Harada, and T. Yanagida, "Direct observation of single kinesin molecules moving along microtubules," Nature **380**, 451–453 (1996)
8. H. Noji, R. Yasuda, M. Yoshida, M., and K. Kinosita, "Direct observation of the rotation of F1-ATPase," Nature **386**, 299–302 (1997)

9. A. Sambongi, Y. Iko, M. Tanabe, H. Omote, A. Kihara, I. Ueda, T. Yanagida, Y. Wada, and M. Futai, "Mechanical rotation of the c subunit oligomer in ATP synthase (F_1F_0): Direct observation," Science **286**, 1722–1724 (1999)

10. M. Tokunaga, K. Kitamura, A. H. Iwane, and T. Yanagida, "Single molecule imaging of fluorophores and enzymatic reactions achieved by objective-type total internal reflection fluorescence microscopy," Biochem, Biophys. Res. Comm. **235**, 47–53 (1997)

11. J. F. Eccleston, K. Oiwa, M. A. Ferenczi, M. Anson, J. E. T., A. Corrie, H. Yamada, D. Nakayama, and R. Trentham, "Ribose-linked sulfoindocyanine conjugates of ATP: Cy3-EDA-ATP and Cy5-EDA-ATP," Biophys. J. **70**, A159 (1996)

12. M. D. Wang, M. J. Schnitzer, H. Yin, R. Landick, J. Gelles, and S. M. Block, "Force and velocity measured for single molecules of RNA polymerase," Science **282**, 902–907 (1998)

13. Y. Harada, T. Funatsu, K. Murakami, Y. Nonoyama, A. Ishihama, and T. Yanagida, "Single molecule imaging of RNA Polymerase–DNA interactions in real time," Biophys. J. **76**, 709–715 (1999)

14. T. Wazawa, Y. Ishii, T. Funatsu, and T. Yanagida, "Spectral fluctuation of a single fluorophore conjugated to a protein molecule," Biophys. J. **78**, 1561–1569 (2000)

15. Y. Ishii, T. Yoshida, T. Funatsu, T. Wazawa, and T. Yanagida, "Fluorescence resonance energy transfer between single fluorophores attached to a coiled-coil protein in aqueous solution," Chem. Phys. **247**, 163–173 (1999)

16. H. P. Lu, L. Xun, and X. S. Xie, "Single-molecule enzymatic dynamics," Science **282**, 1877–1881 (1998); H. Yin, M. D. Wang, K. Svoboda, R. Landick, S. M. Block, and J. Gelles, "Transcription against an applied force," Science **270**, 1653–1657 (1995)

17. H. Frauenfelder, S. G. Sligar, and P. G. Wolynes, "The energy landscapes and motions of proteins," Science **254**, 1598–1603 (1991)

18. B. Bukau and A. L. Horwich, "The Hsp70 and Hsp60 chaperone machines," Cell **9**, 351–366 (1998)

19. R. Yamasaki, M. Hoshino, T. Wazawa, Y. Ishii, T. Yanagida, Y. Kawata, T. Higurashi, K. Sasaki, J. Nagai, and Y. Goto, "Single molecular observation of the interaction of GroEL with substrate proteins," J. Mol. Bio., in press (2000)

20. Y. Sako, S. Minoguchi, and T. Yanagida, "Single molecule imaging of EGFR signal transduction on the living cell surface," Nature Cell Biol. 168–172 (2000)

21. B. Sakmann and E. Neher (eds.), *Single-Channel Recording* (Plenum Press, 1995)

22. T. Ide and T. Yanagida, "An artificial lipid bilayer formed on an agarose-coated glass for simultaneous electrical and optical measurement of single ion-channels," Biochem. Biophys. Res. Comm. **265**, 595–599 (1999)

23. A. Kishino and T. Yanagida, "Force measurements by micromanipulation of a single actin filament by glass needles," Nature **334**, 7476 (1988)

24. A. Ishijima, T. Doi, K. Sakurada, and T. Yanagida, "Sub-piconewton force fluctuations of actomysin in vitro," Nature **352**, 301–306 (1991)

25. K. Svoboda, C. F. Schmmidt, B. J. Schnapp, and S. M. Block, "Direct observation of kinesin stepping by optical trapping interferometry," Nature **365**, 721–727 (1993)

26. J. T. Finer, R. M. Simmons, and J. A. Spudich, Single myosin molecule mechanics: Piconewton forces and nanometer steps," Nature **368**, 113–119 (1994)

27. S. M. Block and K. Svoboda, "Analysis of high resolution recordings of motor movement," Biophy. J. **68**, 230–241 (1995)
28. M. Nishiyama et al., Biophy. J. (supplement) (2000)
29. Y. Inoue et al., Biophy. J. (supplement) (2000)
30. A. Ishijima, H. Kojima, T. Funatsu, M. Tokunaga, H. Higuchi, and T. Yanagida, "Simultaneous measurement of chemical and mechanical reaction," Cell **92**, 161–171 (1998)
31. T. Yanagida, K. Kitamura, H. Tanaka, A. H. Iwane, and S. Esaki, "Single molecule analysis of the actomyosin motor," Current Opinion in Cell Biology, in press
32. J. A. Spudich, "How molecular motors work," Nature **372**, 515–518 (1994)
33. J. E. Corrie, B. D. Brandmeier, R. E. Ferguson, D. R. Trentham, J. Kendrick-Jones, S. C. Hopkins, U. A. van der Heide, Y. E. Goldman, C. Sabido-David, R. E.,Dale, S. Criddle, and M. Irving, "Dynamic measurement of myosin light-chain-domain tilt and twist in muscle contraction," Nature **400**, 425–430 (1999)
34. K. Kitamura, M. Tokunaga, A. H. Iwane, and T. Yanagida, "A single myosin head moves along an actin filament with regular steps of 5.3 nanometres," Nature **397**, 129–134 (1999)
35. K. Kitamura, H. A. Iwane, and T. Yanagida, A single myosin head moves along an actin filament with regular 5.5 nm steps under loaded condition," Biophy. J. (supplement) 1383 (2000)
36. A. F. Huxley, "Muscle structure and theories of contraction," Prog. Biophys. Biophys. Chem. **7**, 255–318 (1957)

18 Single-Molecule Dynamics Associated with Protein Folding and Deformations of Light-Harvesting Complexes

D. S. Talaga, Y. Jia, M. A. Bopp, A. Sytnik, W. A. DeGrado, R. J. Cogdell, and R. M. Hochstrasser*

Experimental understanding of biological and chemical systems is based primarily on measurements of many molecules and therefore kinetic measurements generally display the evolution of the mean of that ensemble. However, since heterogeneity of structure and mechanism is required to describe complex systems such as proteins and other biological assemblies, this useful paradigm can break down. Recent developments in single molecule fluorescence detection have allowed the study of single biological molecules and single biological assemblies under physiological conditions [1–28]. By following the trajectory of individual members of an equilibrium ensemble as they evolve in time, the fluctuation rates, reaction rate constants, and distributions of other properties can be evaluated.

We present two applications of single molecule spectroscopy that illustrate how ensemble averaging masks important dynamic properties of fluctuating systems. In the first, mobile elliptical structural deformations are observed in single assemblies of the light harvesting complex, LH2. These mobile structural deformations are averaged in bulk measurements resulting in the erroneous conclusion that LH2 is a circular absorber. In the second application trajectories of individual members of a folding ensemble of coiled coil GCN4-P1 peptides allow us to determine distributions of properties not available from bulk studies.

18.1 Structural Deformations of Surface-Immobilized Single Light-Harvesting Complexes [27]

The crystal structure of the LH2 complex from photosynthetic bacteria *Rhodopseudomonas acidophila* is notable for its high symmetry arrangement of the nine $\alpha\alpha\beta\beta$-dipeptides which form the scaffold holding the associated bacteriochlorophyll (Bchl) cofactors [29]. The LH2 Bchls are arranged into

* *Corresponding author:* Robin M. Hochstrasser, Chemistry Department, University of Pennsylvania, Philadelphia, PA, 19104, Telephone: (215) 898-8410, FAX: (215) 898-0590, E-mail: hochstra@sas.upenn.edu

two rings that have approximate 9-fold rotation symmetry. The B800 ring contains nine monomeric Bchls located between the β-apoproteins. The B850 ring consists of nine pairs of Bchls each associated with one $\alpha\alpha\beta\beta$-dipeptide. In the LH2 complex the B800 ring absorbs light and transfers the excitation energy to B850 in less than a picosecond [30]. The excitation properties of macromolecular systems depend on the interplay between the nuclear motions that tend to localize excitations, and the delocalizing effect of the interaction between the molecules [31]. Therefore the nature of the excitation and energy transfer in the LH2 complex must depend not only on the static or average structure but also on the structural fluctuations that can occur in bacterial membrane. Single-molecule methods are well suited for the investigation of the microseconds-to-seconds structural dynamics [32]. Previously we [25] and others [33] applied single-molecule confocal microscopy for the photophysical and photochemical characterization of the LH2 complexes.

The LH2 complexes were immobilized on a mica surface to model the protein–membrane interactions occurring in bacterial cells. The mica has some negative charges on its atomically flat surface but it can also have hydrophobic interactions with the protein. The N-terminal regions of LH2 are largely negatively charged because of 27 glutamates and should avoid the mica surface. As a result, the LH2 is expected to be bound to the mica via C-termini. In that case the B850 ring will be closest to the surface.

The fluorescence images and trajectories were recorded with a confocal microscope [27] on samples deoxygenated by an enzymatic system. Because the transition dipoles for B850 and B800 both lie in the xy-plane of the LH2 complex [34], both B850 and B800 should behave as essentially circular oscillators with the same absorption cross section for all linear polarizations in the xy-plane (x, y, z are the molecule-fixed coordinates, with z the cylindrical axis of the assembly, and X, Y, Z are the laboratory-fixed coordinates with X, Y as the focal plane of the microscope). If LH2 had these ideal circular characteristics, then its absorption cross section in circularly polarized light would depend only on the tilt angle θ between the z-axis and the direction of light propagation, Z:

$$A_{\text{CIRC}} = 1/2(1 + \cos^2 \theta) . \tag{18.1}$$

For linear polarization the ideal absorption cross section also depends on the direction in the XY-plane about which tilt occurs:

$$A_\chi = \{\cos^2 \chi + \sin^2 \chi \cos^2 \theta\} , \tag{18.2}$$

where $(\phi - \alpha) = \chi$ is the angle in the XY-plane between the linear polarization axis α and the axis of tilt ϕ, both determined with respect to X. The total angular parts of the fluorescence signal intensities for circular and linearly polarized excitation, in the ideal case, are:

$$S_{\text{CIRC}} = 1/2(1 + \cos^2 \theta)^2 , \tag{18.3}$$

$$S_\chi = (\cos^2 \chi + \sin^2 \chi \cos^2 \theta)(1 + \cos^2 \theta) . \tag{18.4}$$

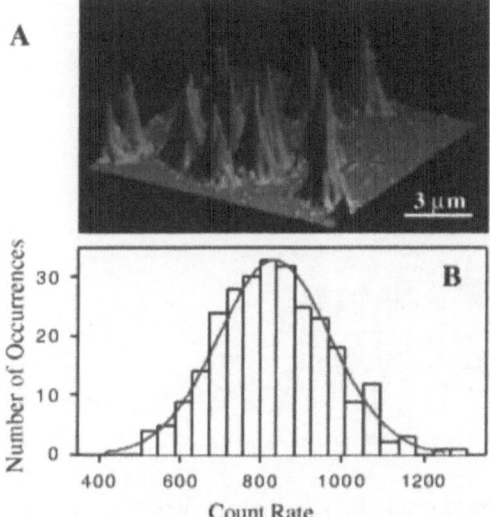

Fig. 18.1. The fluorescence images of single mica-bound LH2 complexes excited with circularly polarized light at 794 nm (**a**) and the distribution of the fluorescence count rates of the 273 single LH2s fitted with a Gaussian function with a mean = 833, and sd = 140 counts (**b**). The buffer is 50 mM TrisHHCl, pH 7.8/0.1% lauryldimethylamine oxide

The high sensitivity of S_{CIRC} and S_χ to the tilt angle is evident from (3) and (4).

When we excited the single LH2 complexes via the B800 ring with circularly polarized light, the fluorescence images (Fig. 18.1) had a narrow distribution of count rates (Fig. 18.1b). The width of this distribution contains a contribution of ~20% from the fluorescence signal blinking, which is different for each single assembly. As shown by (3), the fluorescence signals of single mica immobilized LH2 complexes excited with circularly polarized light should not depend on the azimuthal angles. Therefore, the variance in the tilt angle must be significantly less than the width of the fluorescence count rate distribution.

In the next set of experiments the 794-nm excitation light was switched between two orthogonal polarization directions X and Y at 16 Hz. The corresponding trajectories for fluorescence from B850 are labeled I_x and I_y. The total signal is $I_T = I_x + I_y$. Figure 18.2a shows one fluorescence polarization trajectory having $\langle I_x \rangle = \langle I_y \rangle$ and constant $\langle I_T \rangle$. However, it was more common to observe different signals for X and Y polarization (Fig. 18.2b). The fluctuations in I_x and I_y suggest that for the majority of the single LH2 complexes either the principal absorption axes rotate in the XY-plane or the z-axis precesses about the Z-axis, keeping the tilt angle constant. The total emission signal does not change appreciably during these motions.

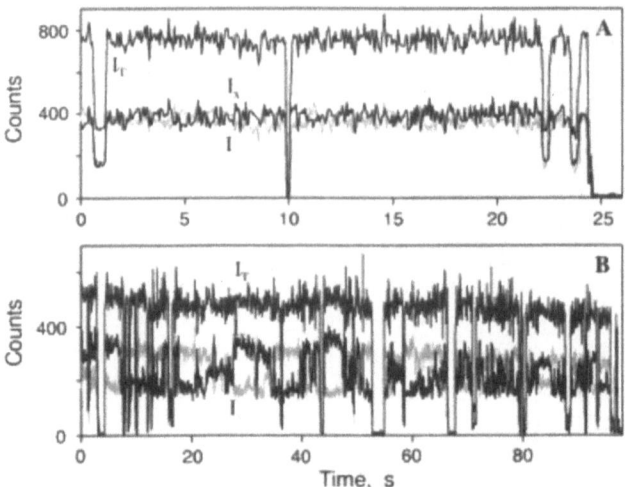

Fig. 18.2. The fluorescence trajectories of single LH2 complexes detected with the excitation polarization switching between 0 and pi/2 at 794 nm. **a** LH2 assembly with overlapping I_x and I_y trajectories. **b** LH2 having substantial polarization changes

In contrast to the results for B800, when the single LH2 assemblies were excited with linearly polarized light at 850 nm (B850), the majority of I_x and I_y trajectories were very similar. The majority of LH2 complexes excited via B850 had polarization ratio values close to unity indicating an apparently more circular oscillator response.

To obtain a more complete picture of the linear polarization properties we swept the 794-nm excitation over 155°. During the sweep the detected fluorescence of a single LH2 complex (S) should have the general form

$$S = A\cos^2(\alpha + \delta) + B \,, \tag{18.5}$$

where A is an amplitude, δ is a relative phase, B is an offset, and $\alpha = 0$ to 155° is the sweep angle. According to (5), only B would contribute to the total fluorescence if the single LH2 assemblies were lying flat and B800 and B850 were optically ideal. Figure 18.3 shows the phase and count rate trajectories of typical single LH2 complexes, determined from the detected fluorescence trajectories using (5). Figure 18.3a shows an example with a relatively constant δ during the whole measurement. The δ trajectory of single LH2 complex shown on Fig. 18.3c undergoes numerous transitions while the total count rate remains essentially constant. Figure 18.4 shows the time-resolved probability histogram of the phase of this assembly. Figure 18.3b and d shows the distributions of δ for the single LH2 complexes presented in Fig. 18.3a and c. The data of Figs. 18.3 and 18.4 are typical of what was seen for a large number of single LH2 complexes undergoing frequent changes in δ, often in jumps of $\sim$40°.

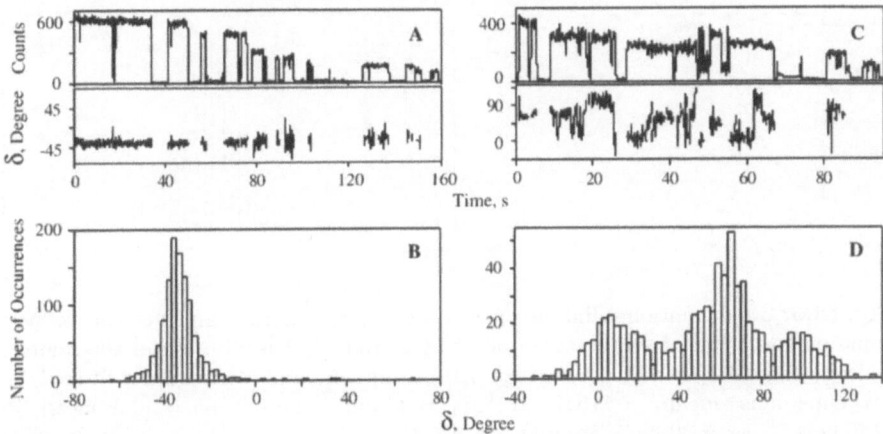

Fig. 18.3. The polarization asymmetry of single LH2 complexes measured by sweeping the excitation polarization over 155°. **a** The dT and count rate trajectory of single LH2 complex with a stable dx value during the whole measurement. **b** The distribution of the dT values determined for single LH2 complex shown in **a**. **c** The dT and count rate trajectory of single LH2 assembly undergoing numerous changes in da. **d** The distribution of the dt values determined from the dv trajectory presented in **c**

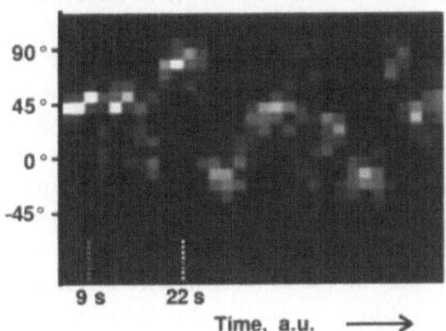

Fig. 18.4. The time-resolved probability histogram of the phase of single LH2 presented in Fig. 18.3 (**a,b**). The number of occurrences varies from 0 (black) to 15 (white)

If the polarization and its dynamics are due to tilting of cylindrical LH2, there should be a correlation between the apparent ellipticity and the total signal. A statistical analysis revealed a correlation coefficient of about 0.28, suggesting that tilting is not the main contribution to the polarization anisotropy magnitudes or changes.

It was also found that free and mica-bound LH2 complexes have overlapping fluorescence spectra (Fig. 18.5a). The fluorescence peak position of the single LH2 complexes fluctuates by more than 100 cm^{-1}, and the width

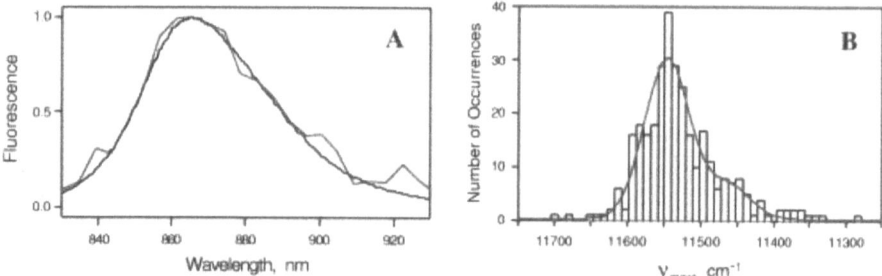

Fig. 18.5. a Conventional fluorescence spectra of free LH2 complexes (black) and emission spectrum of single mica-bound LH2 (red). **b** Distribution of the fluorescence peak positions, nD_{max}, of single LH2 complexes and least-squares fit (red) to two Gaussians (mean$_1$ = 11540 cm^{-1}, s$_1$ = 30 cm^{-1}, amplitude = 30.0; mean$_2$ = 11460 cm^{-1}, s$_2$ = 30 cm^{-1}, amplitude = 6.7)

by over 200 cm^{-1} during the measurement. The distribution of *fluorescence peaks* (Fig. 18.5b) is peaked at 11540 cm^{-1} with the suggestion of a shoulder near 11460 cm^{-1}. We have suggested that the shoulder might belong to the LH2 complexes that emit from the lowest exciton level of the B850 band of states [29,35]. This transition is forbidden for the circular structure but allowed for structurally perturbed molecules displaying elliptical absorption properties.

The main results of these single-molecule investigations came from the polarization switching and sweeping experiments. The circular oscillator model does not reasonably explain the data. The fact that changes in the fluorescence polarization signals were not often accompanied by changes in the total emission suggested that the fluctuations in the polarization signals of single LH2s were not caused by variations in tilt angle, but instead by variations in the ϕ angle of a tilted molecule. However, significant changes in δ were seen when the 794-nm excitation polarization was swept. In an ideal circular model this result would have required all LH2 complexes to be tilted within a narrow distribution of angles. Moreover, a given single LH2 would need to undergo significant (0–140°) precession in ϕ angle without any significant changes in the θ angle. In a structural model in which immobilized LH2 complexes are dynamic, elliptic absorbers and emitters seems much more reasonable. The polarization changes seen in the experiments are therefore attributed to fluctuations of the electronic ellipticity and of the directions of the principal axes of the absorption ellipse. An LH2 an elliptical absorber would have two nearly degenerate excitations with unequal transition dipoles whose vectors are fixed in the molecular frame. Any structural distortion that reduces the rotational symmetry of the B800 and/or B850 electronic states to C_2, but maintains a vertical symmetry plane would yield an electronically elliptical absorber and emitter. For example, an in-place electric field would cause this effect. However, further work is needed to fully determine the nature of the electronic states responsible for these observations of florescence polarization anisotropy.

The polarization sweeping measurement of δ locates the principal axes of the ellipse in the XY plane. Significant changes in δ for a single molecule are most reasonably explained by distortions occurring at different locations around the ring structures rather than by a precession around the z axis of the whole LH2 assembly. Interestingly, the phase fluctuations of approximately $2\pi/9 = 40°$ are very commonly observed in the data, as if a distortion at one location often shifts around the structure by $2\pi/9$ steps. This picture is not unreasonable, because a distortion of the structure near the interface between two $\alpha\beta$-dipeptides might couple more effectively and more easily transfer to a neighboring interface than to some random location around the ring. It is important to note that for the 155° excitation polarization sweeping, eight phase peaks separated by $\pi/9$ are not observed, as would be expected if sequential distortions were occurring in a random stochastic manner around the circle of Bchls.

The different polarization behavior observed on excitation of the B800 and B850 rings suggests either a larger effect of the distortion on the electronic properties, or a larger distortion of B800 compared with B850. The B850 absorbance may be less sensitive to distortions due to its larger exciton bandwidth, which implies more effective motional averaging of the distortion [35]. In addition, the B800 Bchls are held in the LH2 more peripherally than are those of B850. The B850 Bchls are each constrained by histidines of the α-, β-peptides, whereas the B800 Bchls are located between β-peptides and coordinated by the N-terminal formylmethionines of α-peptides [29]. The electronic distortion might arise from a small displacement of one or two of the B800 Bchls, and/or it could involve a partial dissociation of the assembly at the interface between two dipeptides.

18.2   Single-Molecule Protein Folding Trajectories [28,36]

GCN4 was studied at the single-molecule level using fluorescence energy transfer between donor and acceptor dyes labeled at the N-termini of the crosslinked monomers. Distributions of fluctuating molecular structures in terms of the distance between the donor and acceptor, as inferred from the energy transfer efficiency, were measured at a series of positions in the folding equilibrium both on a modified surface and while freely diffusing in solution. Distributions of the energy transfer efficiency were obtained under both conditions, permitting us to examine the folded and unfolded states and the influence of interactions with the modified surface. The time scales associated with the fluctuations that give rise to these distributions were also examined.

A peptide derived from the yeast transcription factor, GCN4, was used in this study [28,36]. The DNA binding domain of this peptide includes a

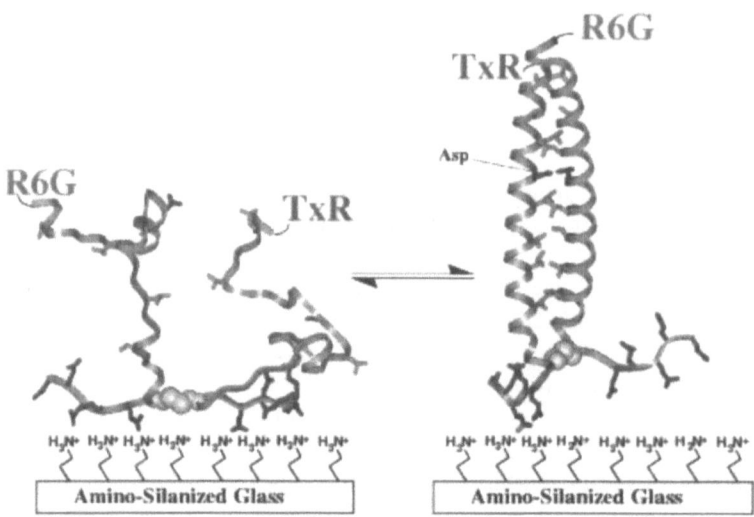

Fig. 18.6. Schematic representation of the folding of GCN4-Pf. The right panel shows the crystal structure of folded GCN4-P1 with a hypothetical unfolded structure at the left. The peptide adheres to the positively charged surface by electrostatic interaction with the negatively charged glutamic acids at the C terminus of the peptide. Conformational fluctuations cause changes in the donor–acceptor distance, resulting in an anticorrelated modulation in the donor and acceptor fluorescence intensities

sequence that forms a short segment of a two-stranded coiled coil [37,38], as shown in Fig. 18.6. Coiled coils provide a very simple model system for the study of the folding of water-soluble proteins [39–41]. A peptide spanning the coiled coil of GCN4 (GCN4-P1) has been shown to form a cooperatively folded helical dimer [39,41–46]. This peptide is an excellent system for studying protein folding because it is quite simple, and yet contains a well-packed helix/helix interface, as found in globular proteins. It has been shown to exist in a two-state equilibrium between unstructured monomers and fully alpha-helical dimers [39]. The alpha-helical secondary structure and the double-helical folded structure apparently form concomitantly [40–44,47]. Introduction of a covalent disulfide tether between the two-peptide chains simplifies the folding reaction and thermally stabilizes GCN4, yet the peptide continues to fold in an apparent two-state equilibrium [48]. One purpose of this study is to investigate the microscopic features of a macroscopically observed kinetic model. GCN4-P1 exhibits two-state folding kinetics when in bulk solution [40,44].

The GCN4-P1 variant employed in this study, designated GCN4-Pf [28], has the sequence GGRMKQLEDK[10]VEELLSKDYH[20]LENEVARLKK[30]LVGERGGCGE[40]EEEE (Fig. 18.6). Five glutamic acid residues were appended to the C-terminus, providing a flexible appendage to allow oriented electrostatic adsorption of the peptides onto a positively charged surface for single-molecule studies. Texas Red-X (TxR) was used as the energy acceptor and 5-carboxyrhodamine 6G (R6G) as the energy donor attached to the N-termini.

Single-molecule fluorescence intensity fluctuations can arise from a variety of photophysical sources, such as dynamic shifts in the fluorescence spectrum [49,50], transient non-fluorescent states of the system [19], including triplet states [51,52], and irreversible photobleaching [32]. Angular motions of the transition dipoles of the probes R6G and TxR can contribute to the fluctuations in the present example. The relative signal intensities from donor and acceptor depend not only on the angles involved in the dipole–dipole interaction, but also on the transition dipole colatitudes, $\theta\theta_A$ and $\theta\theta_D$. The fluctuations in $\theta\theta_A$ and $\theta\theta_D$ are expected to be more correlated when GCN4-Pf is folded than when it is unfolded.

The corrected and optimally filtered trajectories I_D, I_A are used with the quantum yields for unsensitized donor and acceptor fluorescence and energy transfer $\Theta_D, \Theta_A, \Theta_{ET}$ and the donor and acceptor extinction coefficients to determine the quantum yield for energy transfer:

$$\Theta_{ET} = (1 + (R/R_0)^6)^{-1} = \frac{I_A\Theta_D - I_D\varepsilon_{A/D}\Theta_A}{\Theta_D(I_A + I_D\Theta_A)} , \qquad (18.6)$$

where $\varepsilon_{A/D}$ is the ratio of the donor and acceptor extinction coefficients at the excitation wavelength. R_0 is the Förster distance [53] between chromophores that gives a quantum yield for energy transfer of 50%.

The goal of this work was to measure conformational fluctuations of GCN4-Pf in the folded and unfolded states, and the dynamic equilibrium between these two conformational ensembles. Under 532-nm excitation mainly the R6G is excited, but with no urea the TxR channel shows significantly more intensity, consistent with efficient energy transfer in a folded state. At 7.4 M urea the R6G channel shows significantly more emission, indicating less effective energy transfer.

We observe a number of different types of trajectories for GCN4-Pf. The acceptor signal dominates most trajectories until photobleaching occurs, at which time the donor signal jumps to its non-perturbed level, indicating that the photobleached acceptor does not act as an energy acceptor. When the donor bleaches, first we see a reduction of the acceptor signal to the level of its direct excitation. Figure 18.7 shows a single-molecule trajectory on an expanded time scale that more clearly shows anticorrelated fluctuations in the donor and acceptor signals.

We were able to measure the donor and acceptor fluorescence intensity autocorrelation and cross-correlation functions. The cross-correlation function is negative, but its magnitude at time zero is less than the geometric mean of the amplitudes of the autocorrelation functions, indicating that energy transfer is not the only mechanism for modulating the signal. Uncorre-

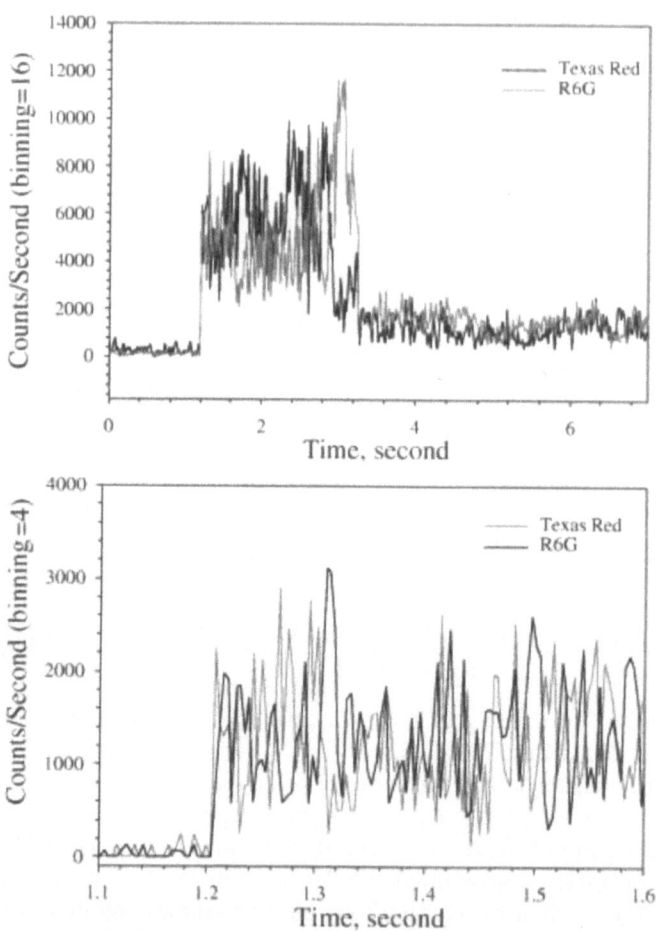

Fig. 18.7. High time-resolution detection of donor (*solid line*) and acceptor (*dashed line*) fluorescence signals from a single GCN4-Pf molecule at pH 6.1

lated fluctuations of the colatitudes will increase the magnitude of the auto-correlation functions, whereas their correlated fluctuations will contribute to a positive cross-correlation function, effectively canceling some of the negative cross correlation. Nevertheless, it is clear that the signals are exhibiting dynamic modulation of the energy transfer distance.

For a given concentration of urea, the two autocorrelation functions and the cross-correlation function were quite non-exponential, indicating the occurrence of a range of types of structural fluctuations. This suggests that the distributions should be dependent on the time gate used in the experiment, and experiments have shown this to be the case. Different portions of the distributions coalesce on different time scales. The broad feature in the 7.4 M distribution of Θ_{ET} centered at 55% does not narrow on a 25-ms gating time

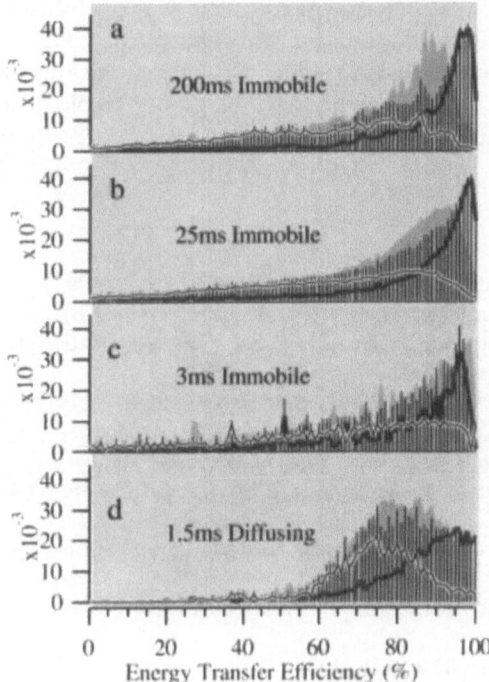

Fig. 18.8. This figure illustrates the distributions obtained under various conditions of denaturant vs. observation time. The filled curve represents the distribution obtained at 3 M urea. The sticks represent a linear combination of the 0M folded distribution and the 7M unfolded distribution, shown as solid black curve and the white highlighted curve. Distributions obtained from 200 ms (**a**), 25 ms (**b**), and 3 ms (**c**) averaging times while immobilized on the modified surface. Part (**d**) shows the distributions obtained for freely diffusing GCN4 averaged for 1.5 ms

scale. These results show that there is a distribution of time scales for fluctuations in the energy transfer efficiency and a correlation between kinetics and structure. Figure 18.8 compares the results we obtain for various observation times on the modified surface with the results for an observation time of 1.5 ms freely diffusing in solution. The broad feature at $\Theta_{ET} \simeq 55\%$ is absent in the freely diffusing distribution. Therefore we attribute this portion of the unfolded ensemble to slowly interconverting frustrated peptides interacting with the surface. Apart from the broad feature in the 7 M distribution on the surface, we see that the distributions observed on the surface and freely diffusing in solution are quite similar, leading us to conclude that the main peak in the distribution, which is substantially broader than shot noise, is representative of the freely fluctuating peptide.

324   D. S. Talaga et al.

Acknowledgments. This research was supported by the National Institutes of Health and the National Science Foundation, Program Project Grant NIH-GM48130, with instrumentation developed under National Institutes of Health Grant RR03148 (R.M.H.), and by the Biotechnology and Biological Sciences Research Council, U.K. (R.J.C.).

References

1. M. Eigen and R. Rigler, Proc. Natl. Acad. Sci. USA **91**, 5740–5747 (1994)
2. L. Edman, Ü. Mets, and R. Rigler, Exp. Tech. Phys. **41**, 157–163 (1995)
3. L. Edman, Ü. Mets, and R. Rigler, Proc. Natl. Acad. Sci. USA **93**, 6710–6715 (1996)
4. S. Nie, D. T. Chiu, and R. N. Zare, Science **266**, 1018–1021 (1994)
5. S. Nie, D. T. Chiu, and R. N. Zare, Anal. Chem. **67**, 2849–2857 (1995)
6. D. T. Chiu and R. N. Zare, J. Am. Chem. Soc. **118**, 6512–6513 (1996)
7. T. Funatsu, Y. Harada, M. Tokunaga, K. Saito, and T. Yanagida, Nature **374**, 555–559 (1995)
8. R. D. Vale, T. Funatsu, D. W. Pierce, L. Romberg, Y. Harada, and T. Yanagida, Nature **380**, 451–453 (1996)
9. K. Kitamura, M. Tokunaga, A. H. Iwane, and T. Yanagida, Nature **397**, 129–134. (1999)
10. Q. F. Xue and E. S. Yeung, Nature **373**, 681 (1995)
11. W. Tan and E. S. Yeung, Anal. Chem. **69**, 4242–4248 (1997)
12. I. Sase, H. Miyata, J. E. T. Corrie, J. S. Craik, and K. Kinosita, Jr., Biophys. J. **69**, 323–328 (1995)
13. I. Sase, H. Miyata, S. Ishiwata, and K. Kinosita, Jr., Proc. Natl. Acad. Sci. USA **94**, 5646–5650 (1997)
14. K. Kinosita, Jr., R. Yasuda, H. Noji, S. Ishiwata, and M. Yoshida, Cell **93**, 21–24 (1998)
15. X. S. Xie, Acc. Chem. Res. **29**, 598–606 (1996)
16. H. P. Lu, L. Y. Xun, and X. S. Xie, Science **282**, 1877–1882 (1998)
17. R. M. Dickson, D. J. Norris, Y.-L. Tzeng, and W. E. Moerner, Science **274**, 966–969 (1996)
18. R. M. Dickson, A. B. Cubitt, R. Y. Tsien, and W. E. Moerner, Nature **388**, 355–358 (1997)
19. T. Ha, Th. Enderle, D. S. Chemla, P. R. Selvin, and S. Weiss, Phys. Rev. Lett. **77**, 3979–3982 (1996)
20. T. Ha, Th. Enderle, D. F. Ogletree, D. S. Chemla, P. R. Selvin, and S. Weiss, Proc. Natl. Acad. Sci. USA **93**, 6264–6268 (1996)
21. Th. Schmidt, G. J. Schütz, W. Baumgartner, H. J. Gruber, and H. Schindler, Proc. Natl. Acad. Sci. USA **93**, 2926–2929 (1996)
22. G. J. Schütz, W. Trabesinger, and Th. Schmidt, Biophys. J. **74**, 2223–2226 (1998)
23. P. M. Goodwin, W. P. Ambrose, and R. A. Keller, Acc. Chem. Res. **29**, 607–613 (1996)
24. Y. Jia, A. Sytnik, L. Li, S. Vladimirov, B. S. Cooperman, and R. M. Hochstrasser, Proc. Natl. Acad. Sci. USA **94**, 7932–7936 (1997)

25. M. A. Bopp, Y. Jia, L. Li, R. J. Cogdell, and R. M. Hochstrasser, Proc. Natl. Acad. Sci. USA **94**, 10630–10635 (1997)
26. A. Sytnik, S. Vladimirov, Y. Jia, L. Li, B. S. Cooperman, and R. M. Hochstrasser, J. Mol. Biol. **285**, 49–54 (1999)
27. M. Bopp, A. Sytnik, T. Howard, R. J. Cogdell, and R. M. Hochstrasser, Proc. Natl. Acad. Sci. USA **96**, 11271–11276 (1999)
28. Y. Jia, D. S. Talaga, W. L. Lau, H. S. M. Lu, W. F. DeGrado, and R. M. Hochstrasser, Chem. Phys. **247**, 69–83 (1999)
29. G. McDermott, S. M. Prince, A. A. Freer, A. M. Hawthornthwaite-Lawless, M. Z. Papiz, R. J. Cogdell, and N. W. Isaacs, Nature (London) **374**, 517–521 (1995)
30. V. Sundström, T. Pullerits, and R. J. Van Grondelle, Phys. Chem. B **103**, 2327–2346 (1999)
31. A. S. Davydov, *Theory of Molecular Excitons* (Plenum, New York, 1971)
32. X. S. Xie and J. K. Trautman, Ann. Rev. Phys. Chem. **49**, 441–480 (1998)
33. A. M. van OIjen, M. Ketelaars, J. Kohler, T. J. Aartsma, and J. J. Schmidt, Phys. Chem. B **102**, 9363–9366 (1998)
34. A. A. Freer, S. M. Prince, K. Sauer, M. Papiz, A. M. Hawthornthwite-Lawless, G. McDermott, R. J. Cogdell, and N. W. Isaacs, Structure **4**, 44–462
35. R. Kumble and R. M. Hochstrasser, J. Chem. Phys. **109**, 855–865 (1998)
36. D. Talaga, W. Lau, Y. Jia, H. Roder, W. Degrado, and R. M. Hochstrasser, Proc. Natl. Acad. Sci. USA **97**, 13021–13026 (2000)
37. E. K. O'Shea, J. D. Klemm, P. D. Kim, and T. Alber, Science **254**, 539 (1991)
38. T. E. Ellenberger, C. J. Brandl, K. Struhl, and S. C. Harrison, Cell **71**, 122–1237 (1992)
39. K. J. Lumb, C. M. Carr, and P. S. Kim, Biochemistry **33**, 7361 (1994)
40. T. R. Sosnick, S. Jackson, R. R. Wilk, S. W. Englander, and W. F. Degrado, Proteins **24**, 427–432 (1996)
41. H. Wendt, L. Leder, H. Harma, I. Jelesarov, A. Baici, and H. R. Bosshard, Biochemistry **36**, 204–213 (1997)
42. H. Wendt, C. Berger, A. Baici, R. M. Thomas, and H. R. Bosshard, Biochemistry **34**, 4097–4107 (1995)
43. H. Wendt, A. Baici, and H. R. Bosshard, J. Amer. Chem. Soc. **116**, 6973–6974 (1994)
44. J. A. Zitzewitz, O. Bilsel, J. Luo, B. E. Jones, and C. R. Matthews, Biochemistry **34**, 12812–12819 (1995)
45. I. Jelesarov, E. Durr, R. M. Thomas, and H. R. Bosshard, Biochemistry **34**, 7539–7550 (1998)
46. E. Durr, I. Jelesarov, and H. R. Bosshard, Biochemstry **38**, 870–880 (1999)
47. S. Ozeki, T. Kato, M. E. Holtzer, and A. Holtzer, Biopolymers **31**, 957–966 (1991)
48. E. K. O'Shea, R. Rutkowski, and P. S. Kim, Science **243**, 538–542 (1989)
49. J. K. Trautman, J. J. Macklin, L. E. Brus, and E. Betzig, Nature **369**, 40 (1994)
50. H. P. Lu and X.S. Xie, Nature **385**, 143 (1997)
51. T. Basche, S. Kummer, and C. Brauchle, Nature: **373**, 132 (1995)
52. T. Ha, T. Enderle, D. S. Chemla, P. R. Selvin, and S. Weiss, Chem. Phys. Lett. **271**, 1 (1997)
53. L. Streyer, Ann. Rev. Biochem. **47**, 819–846 (1978)

19 The Study of Single Biomolecules with Fluorescence Methods

T. Ha, X. Zhuang, H. Babcock, H. Kim, J. W. Orr, J. R. Williamson, L. Bartley, R. Russell, D. Herschlag, and S. Chu

The study of individual molecules allows one to look beyond the ensemble average and observe the distribution and time trajectory of the structure and motion of molecules. This approach has led to new paradigms in polymer physics. The study of polymer dynamics at the single-molecule level has led to the discovery that identical DNA molecules exposed to the same conditions will follow a multitude of paths to equilibrium as they extend in elongational [1] and shear [2] flows. This "molecular individualism" was not discovered in half a century of experimental work on bulk samples. Nor was it anticipated theoretically. It is possible that biological processes such as protein folding and enzyme activity will also show a rich set of kinetic paths and transient states that can only be fully characterized at the single-molecule level. Thus, it is important to develop techniques that will allow the study of molecular processes at the level of single molecules.

We summarize our current progress in a number of experiments in which fluorescence techniques such as fluorescence resonant energy transfer (FRET), fluorescence quenching, and detection of dye molecules have been used to follow the motion of individual biomolecules. The systems studied include a model RNA system, titin, and the Tetrahymena group I ribozyme.

19.1 Ligand-Induceed Conformational Changes in a Model RNA System

We used fluorescence resonance energy transfer (FRET) [3] to study the motion of an RNA three-way junction containing the binding site for the ribosomal protein S15 (Fig. 19.1). This junction is found in the 30 S ribosomal subunit which is assembled in a cascade of RNA conformational changes induced by a sequence of protein binding events. Since the binding of S15 to this rRNA junction nucleates the assembly of the central domain of the 30 S ribosomal subunit, it has been studied extensively [4].

In FRET, donor and acceptor dyes are attached to two sites of a biological molecule. Donor fluorescence emission is strongly quenched in a distance-dependent manner by the acceptor, while the acceptor emission increases

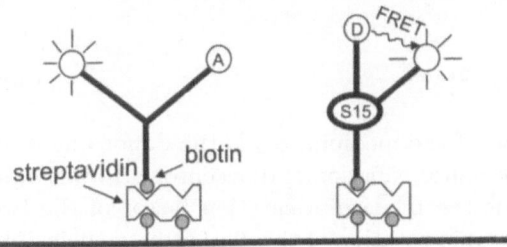

Fig. 19.1. Cartoon of the RNA conformational change induced by protein or Mg^{2+} ion binding

due to the energy transfer. Thus, a measurement of a change in fluorescence from the two dyes can be used as an indicator of a change in the conformation of the host molecule. Since the two fluorophores are on different parts of the molecule, intramolecular motion can be measured in the molecular center of mass frame of reference.

The FRET technique has recently been extended to the single-molecule domain [5], assisted by rapid developments in single-molecule fluorescence spectroscopy [6] and their subsequent application to biological systems [7]. In these earlier experiments, fluctuations in energy transfer could not be interpreted as clearly coming from a known conformational change in the biomolecule. In our model RNA system, we show that the shape and motion of immobilized single RNA molecules, induced by protein binding or ion binding, can be measured using FRET. Furthermore, the observed changes in FRET signal due to the RNA motion are in quantitative agreement with the predicted conformational change.

Previous studies have shown that the free RNA junction is a nearly planar structure with 120° angles between each pair of helices [4]. Binding of S15 protein or Mg^{2+} causes one of the helix arms to rotate by 60°, becoming collinear with another arm as shown in Fig. 19.1. A fluorescein dye [Donor(D)] attached at the end of a helix and a Cy3 dye [Acceptor(A)] attached at the end of another helix were used to detect the resulting distance change. A biotin tag was attached to the third helix to bind the RNA junction to a streptavidin coated surface (Fig. 19.1). A flexible linker arm was used to attach the Cy-3 to the helix so that the relative orientation of the two dye molecules would average out orientation dependent variations in transfer efficiency. Polarization experiments demonstrated that the dye molecules were freely rotating and indirectly immobilized via the RNA junction. Three additional RNA junctions were also constructed for control studies, where one of the three tags was omitted.

The acceptor arm is 15 bp long and the donor arm is 16 bp long. Assuming an angular change of 120° to 60°, the inter-dye distance changes from 8.5 nm to 5 mnm upon ligand binding. Energy transfer distance R_0 was determined to be 5.3 nm using

$$R_0 = (8.79 \times 10^{-5} n^{-4} \phi_D \cdot J \cdot \kappa^2)^{1/6} \tag{19.1}$$

where n is the index of refraction of the medium, ϕ_D is the donor quantum yield, J is the spectral overlap of donor emission and acceptor absorption in [nm^4 M^{-1} cm^{-1}] units, and κ^2 is the relative orientation factor of the two dipoles(3). We assumed free and rapid rotation of the dyes around their linkers, which gives $\kappa_2 = 2/3$ (the polarization anisotropy is 0.17 for fluorescein and 0.22 for Cy3, so this is only an approximation). J and ϕ_D were experimentally determined from the RNA junction. The energy transfer efficiency E, the amount of donor quenching due to the energy transfer, is estimated using $E = 1/(1 + R^6/R_0^6)$, and is 4% for the open RNA and 60% for the folded one. The agreement with the single-molecule experiment (donor signal quenched by 70% upon RNA folding) is good, considering the fact that there are uncertainties due to κ_2 and the length of the dye linkers. Donor average intensity for the RNA molecules was indistinguishable within the experimental uncertainty (10%) from the donor intensity for the RNA molecules with donor only, indicating that RNA spends less than 10% of the time in the folded conformation in the absence of the S15 protein or Mg^{2+}.

Fluorescence from the donor and acceptor on the same RNA junction was measured simultaneously using two detectors in a confocal scanning optical microscope of our own construction [8]. With some sacrifice of time resolution, we also obtained FRET signals from several hundred individual molecules simultaneously by illuminating a wide field of molecules with laser light totally internally reflected at a quartz–water interface.

We examined folding induced by the S15 protein and Mg^{2+} and showed that the average of many single-molecule observations closely resembles the behavior of unlabeled molecules in solution determined by traditional bulk assays. Thus, the dye labeling and the surface immobilization does not seem to affect the functioning of this RNA system. Surprisingly, we observed an anomalously broad distribution of RNA conformations at intermediate ion concentrations, which may be attributed to foldability differences among RNA molecules or a breakdown of the simple two-state model of folding induced by Mg^{2+} ions [8]. We are looking into possible causes for the anomalous distribution.

In addition, we observed the real-time response of single molecules to changing environments by exposing the RNA complex to Mg^{2+}-free buffer and 1-mM Mg^{2+} buffer every 200 milliseconds. Figure 19.2 shows a time trace of the donor signal obtained in our confocal system. The change in energy transfer due to an RNA conformational change is clearly larger than measurement "noise" due to effects such as dye "blinking," spectral diffusion, and rotational dynamics. The transition between the folded and unfolded states is not abrupt and is due to the finite mixing time of our buffer exchange apparatus.

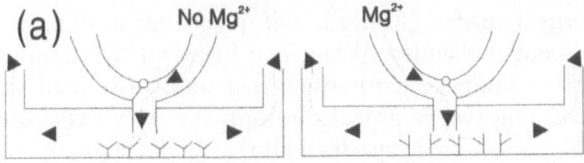

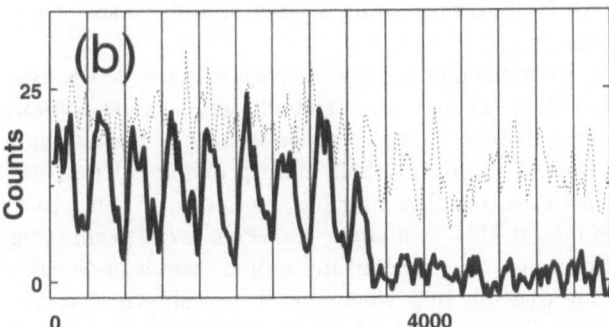

Fig. 19.2. The real-time observation of single RNA molecule conformational changes caused by cycling the buffer solutions every 5 seconds. Solid and dashed lines denote acceptor and donor fluorescence respectively

This work demonstrates that FRET can be used to unambiguously track the movement of a single molecule undergoing induced structural changes. Furthermore, the changes in the fluorescence signal were in good agreement with the predicted energy transfer efficiency based on known structure of the RNA molecule and the parameters of the dye molecules used.

19.2   The Observation of Protein Folding and Unfolding with Fluorescence Quenching

Our second application of fluorescence techniques to the study of single molecules is in the area of protein folding. The dynamics of protein folding are not known. One conjecture is that a protein folds into its final native states through many pathways [9], possibly along a rugged funnel-like energy landscape [10]. The common experimental methods used to study protein folding (e.g. intrinsic tryptophan fluorescence, far-ultraviolet circular dichroism, nuclear magnetic resonance spectroscopy, and x-ray scattering) are bulk studies that average over different folding pathways and intermediate states. Single-molecule experiments may be particularly revealing for protein folding studies, since they can follow different folding pathways taken by different molecules and reveal the existence of a variety of intermediate states.

We have demonstrated the use of fluorescence self-quenching between identical dye molecules to observe the folding of titin proteins [11]. Like fluorescence resonance energy transfer (FRET), self-quenching is due to the interaction of two fluorescent molecules. When two identical dye molecules are in close proximity, their fluorescence is quenched relative to well separated dye molecules. Attaching two identical fluorophores to different parts of a host molecule then allows us to measure its conformational changes. Since self-quenching requires conjugation of only one type of fluorophore instead of two as required by FRET, labeling specific amino acids is easier and the method should be attractive for protein folding studies as well as the study of other biological processes.

In our work, fluorescent dye molecules were covalently attached to the cysteine residues of the protein titin. Titin is an extremely large, multi-domain protein found in muscle tissue. If titin is in the folded state, the fluorescence from the dyes is severely quenched due to their close proximity. Upon unfolding, the fluorescence increases roughly fourfold, as shown in Fig. 19.3. Such a dramatic change is measurable at a single-molecule level, permitting the study of folding and unfolding dynamics of individual protein molecules in real time. By varying the dye labeling efficiency, it was shown that the change in the fluorescence signal between the folded and unfolded states was primarily due to the quenching of dye molecules on the same domain. This test also showed that the fluorescence properties of isolated dye molecules bound to titin were not affected by the two buffers used [11].

Since fluorescence quenching is only an indicator of proximity, it was necessary to show that the conjugation of dye molecules or the immobilization of the protein on a surface does not prevent the tertiary structure of titin from forming. The Ig domains of titin we studied act as an extensible spring that gives an elasticity to muscle tissue. If a molecule consisting of several Ig domains is pulled by an atomic force microscope (afm) or optical tweezers, the domains will unfold one by one. As each domain unravels, the increase of the contour length of the amino acid chain results in a decrease of the entropic elasticity, and therefore a drop in the force is detected by the afm or optical tweezers [12]. We tested for successful refolding of the dye-labeled titin domains by showing that the atomic force microscope signals in the initial folded state and after the dye-labeled titin was denatured (unfolded) and then refolded were the same. In the presence of a denaturing solution, the force vs. extension curves were highly distorted.

We now argue that the quenching is mainly between the dye molecules located on the same module of the titin, as opposed to dye molecules on different modules. The muscle skeletal titin has about 500 cysteines, with more than 90% of them on the 297 Ig or FNIII modules. The ratios between the number of modules that have 0, 1, 2, 3, 4, and 5 cysteines are 1:1.9:1.9:0.9:0.3:0.08 (35). Assuming all the cysteines are equally accessible to dye molecules, we find, for our 85% labeling efficiency, only 25% of dyes are alone on a module (Type-I dyes), while the majority of 75% have at least a partner on the same module (Type-II dyes). Because the dyes on the same

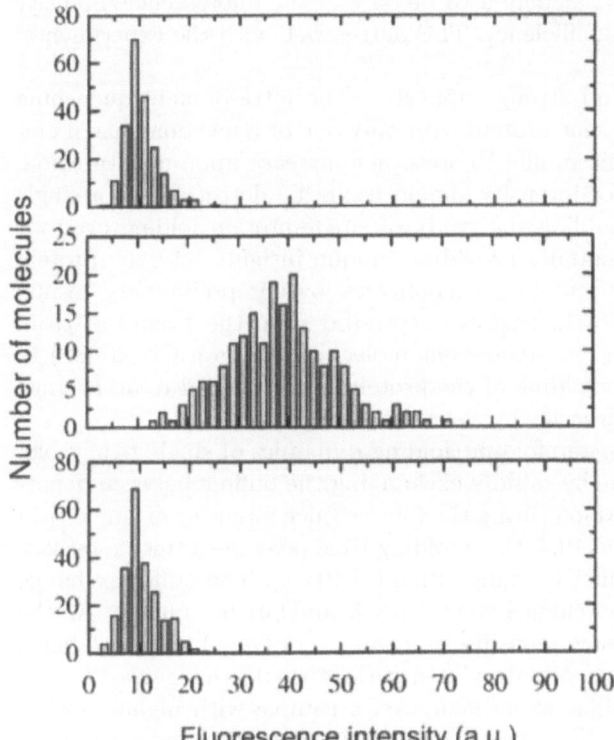

Fig. 19.3. Histogram of fluorescence intensity of single titin molecules labeled with Oregon green and adsorbed onto a gold-coated surface. Top box: fluorescence in buffer solution where titin is in the folded state. Middle box: sample prepared in original buffer, then denatured in urea buffer. Bottom box: same as middle box, and then refolded in initial buffer

module can get significantly closer than those on different modules, we suspect that the intra-domain quenching is responsible for much of the change in fluorescence.

Based on the intra-domain quenching model, we can understand the steep decrease of denaturation-induced fluorescence increase as the labeling efficiency decrease (Fig. 19.3). Assuming that, in the native state of protein, the fluorescence from Type-I dyes is not quenched and that from Type-II dyes is completely quenched, while in the denatured state quenching is totally relieved, we can calculate the denaturation-induced fluorescence increase at different labeling efficiencies. The calculated results are in excellent agreement with the experimental values (data not shown).

Furthermore, the intra-domain quenching model allows us to estimate the absolute fluorescence intensity of native titin molecules as a function of labeling efficiency. The fluorescence intensity of single titin molecules with

8.6% labeling efficiency is calculated to be 35% of the fluorescence intensity of those with 85% labeling efficiency. This agrees well with the experimental value of 31%.

These observations lend strong support to the intra-domain quenching model. They also suggest that protein with only one or a few domains, if efficiently labeled, can exhibit similar fluorescence increase upon denaturation. The change in fluorescence intensity should easily be detectable at a single fluorophore level (36), enabling the study of single-protein folding even for smaller proteins and potentially providing unique insights into the protein folding pathways. This speculation is confirmed by our preliminary results on a single Ig domain protein that we expressed from the I-band of titin. There, we attached up to two fluorescent molecules (Oregon Green 488) to every protein molecule. Unfolding of the protein molecules led to an average increase in fluorescence intensity by severalfold, as expected (34).

We probed the real-time unfolding/folding dynamics of single titin molecules upon buffer exchange by rapidly exchanging the buffers between denaturant and renaturant and monitoring the fluorescence intensity of single titin molecules. As shown in Fig. 19.4, the refolding time (average 32 ms) is reasonably resolved from the buffer exchange time (~ 10 ms). The buffer exchange time was measured by switching between pH 4 and pH 5.5 buffers. As the pH changed, the fluorescence intensity of a sample of Oregon Green-labeled streptavidin immobilized on a biotinylated surface would change.

When we perform similar experiments on a sample with higher surface density of labeled titin, with several labeled titin molecules under the laser beam spot simultaneously (our semi-bulk experiment), the refolding time histogram is a distribution with an average of 31 ms and width significantly broader than that of the buffer exchange time histogram. The broad refolding distribution of the single-molecule data may indicate a distribution of folding pathways. In addition, the same titin molecule occasionally exhibits a change in the refolding time from one buffer exchange period to another, even though its extent is not as large.

There can be other possible reasons for the wide distribution of folding times, and improved experiments are needed to directly observe multiple pathways. We are extending this study to titin molecules that have either one or a few domains. We will also try to attach fluorescent labels to different positions on the protein by changing the location of the cysteine residues with site-directed mutagenesis. Hopefully, this will reveal more information about conformational changes during folding and unfolding.

19.3   The Observation of RNA Enzymatic Activity

Single-molecule experiments may also contribute greatly to an understanding of the RNA folding problem. RNA performs a variety of important functions in the cell, ranging from ligand binding to catalysis and translation. As with

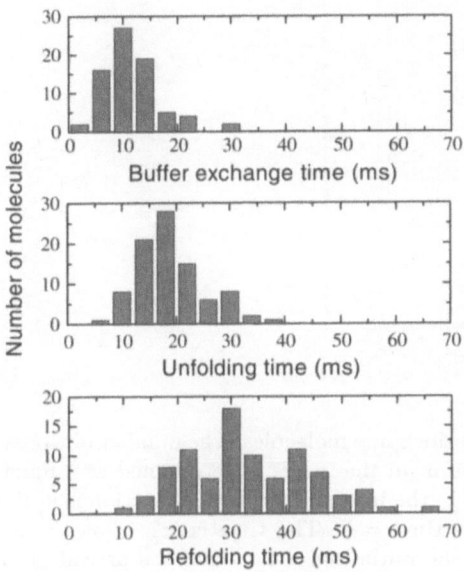

Fig. 19.4. Histograms of the buffer exchange time (top box), titin unfolding time (middle box), and refolding time (lower box). The mixing time of our current flow system is comparable to the unfolding time, whereas the refoding time is clearly time-resolved

proteins, the function that an RNA molecule performs is largely determined by its three-dimensional structure, and it is not well understood how an RNA molecule folds into this structure. Systematic studies of how large RNA molecules such as the Tetrahymena group I ribozymes have begun to emerge only recently [13].

The Tetrahymena ribozyme is derived from a self-splicing group I intron, and is thought to attain its tertiary structure through multiple intermediate folding states [13]. Moreover, data show that the Tetrahymena ribozyme can fold to its native state through at least two distinct pathways. One pathway goes through a kinetically trapped, misfolded state, and another avoids this trap [14,15]. It should be possible to directly observe these pathways and intermediate states by measuring the folding time-trajectories of single molecules. Single-molecule methods may allow examination of unpopulated intermediates and parallel folding pathways with similar folding rates, neither of which can be easily studied by ensemble methods.

As a precursor to the folding studies, we have developed a method to immobilize ribozyme molecules to a glass surface that allowed us to observe the time trajectories of single molecules while rapidly changing buffer conditions or components. The ribozyme was extended at its 3'-end and annealed to a complementary DNA oligonucleotide tether with a 5'-biotin. The ri-

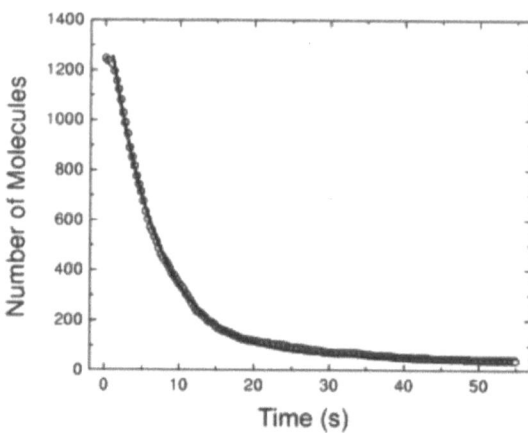

Fig. 19.5. Cleavage reaction of single ribozyme molecules. The number of ribozyme molecules bound to fluorescent S present in the image is determined as a function of time t. The image at t = 0 is given in the inset. The cleavage reaction is initiated by flowing in buffer containing G at time = 0. The time course shows a lag of roughly 1 s due to a delay between the initiation of flow and the arrival of G at the sample. The solid line is a single exponential fit to the data, giving the rate constant $k_{obs} = 0.16$ s^{-1}. A control experiment in the absence of G shows that the number of fluorescent molecules decreases due to photobleaching, but with a rate constant one-tenth that of cleavage

bozyme was then folded to the native state. The folded ribozyme could then be immobilized via the biotin to a surface coated with streptavidin. Single fluorescently tagged ribozyme molecules were detected using two experimental apparatuses, a total internal reflection (TIR) microscope and a scanning confocal microscope.

Dye-labeling and surface immobilization may alter the properties of the molecule of interest. A comparison of the rates of the reaction steps of the modified ribozyme studied here with the activity measured in bulk solution could indicate dye- and surface-related effects. Extensive knowledge exists on the enzymatic properties of the unlabeled and mobile ribozyme [16]. For example, the ribozyme catalyzes the cleavage of the oligonucleotide substrate (S) by guanosine (G). Cleavage reactions were followed using S labeled with Cy3 dye at the 3′ end. Upon cleavage, the Cy3-labeled product is rapidly released from the ribozyme into the solution. The fluorescent molecules remaining on the surface thus represent ribozyme molecules that have not yet reacted (Fig. 19.5).

We measured rate constants for the cleavage reaction at several different G concentrations and compared the single-molecule rates with the results obtained from bulk solution in Fig. 19.6. The identical rate constants with saturating G (k_{max}) indicate that the ternary complex of ribozyme, S, and G is unperturbed by the surface. The concentration of G required for half-

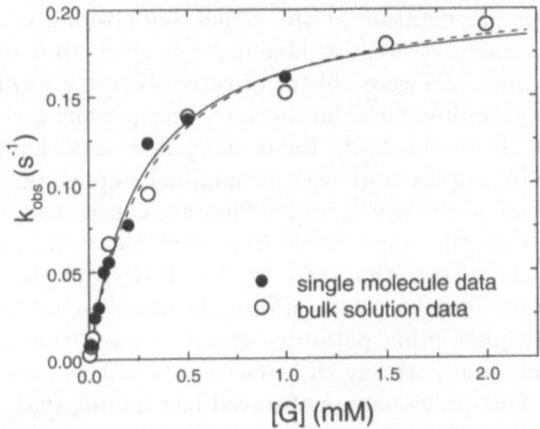

Fig. 19.6. A comparison of reaction kinetics determined from single-molecule measurements (filled circles) to that determined from bulk solution measurements (open circles) at pH 7.0, 22°C. The variability of k_{obs} from experiment to experiment is 20% for bulk measurements and 15% for single-molecule measurements. The solid line is a fit of the single-molecule data to the equation $k_{obs} = k_{max}[G] / (K_{1/2}^G + [G])$, with $k_{max} = 0.21$ s^{-1} and $K_{1/2}^G = 0.26$ mM. The dashed line is a fit of the bulk data to the same equation, with $k_{max} = 0.21$ s^{-1} and $K_{1/2}^G = 0.30$ mM

maximal activity ($K_{1/2}^G$) is also identical, indicating that the G binding site is unperturbed by the surface. In addition, the rate constant for release of S was found to be the same for the immobilized and free ribozyme (data not shown), demonstrating that the ribozyme–S complex is unperturbed by the surface.

The above comparison of the ribozyme with a fluorescently labeled substrate tethered to a surface and in free solution demonstrates that surface immobilization had no measurable effect on the enzymatic reaction rate. The presence of the dye or the extension, however, can still affect the enzymatic reaction rates. In separate experiments, we found that the presence of the 3′-extension does not affect the reaction rate. The effects of dye-labeling were assessed by comparing the ribozyme cleavage reaction of dye-labeled S with that of unlabeled S. It was found that the presence of the dye led to a small decrease (2-fold) in k_{max}. Additionally, we found that the Cy5 dye attached to the 3′ end of the DNA tether has no effect on the cleavage reaction rate of the ribozyme.

We have demonstrated (1) that enzymatic activity of surface-immobilized ribozyme molecules can be followed with fluorescently labeled substrates, and (2) dye labeling and immobilization does not appreciably alter the biological activity. Thus, we now have an assay to measure whether the RNA enzyme has properly folded.

In results obtained after our presentation at the Nobel Symposium, the Stanford Biochemistry/Physics collaboration was able to study the folding of the Tetrahymena group I ribozyme. We were able to directly observe a local folding step in which a duplex reversibly docks into tertiary interactions with the otherwise folded ribozyme. Rate constants for docking and undocking were measured with an all-RNA duplex and with a modified duplex that destabilizes docking by disrupting a tertiary contact. The rate constants for docking of the two duplexes were the same, suggesting that this tertiary contact is not formed in the docking transition state for the all-RNA duplex.

In studies of overall ribozyme folding, we have directly observed intermediate folding states and multiple folding pathways and discovered that a fraction of the molecules fold along a pathway that reaches the native state with a rate constant of 1 s^{-1}. This previously unobserved fast folding pathway is dependent on the formation of the duplex between the ribozyme and its oligonucleotide substrate.

In summary, we have demonstrated that single-molecule FRET, fluorescence quenching, and substrate labeling can be used to unambiguously observe conformational changes associated with ligand attachment to an RNA construct, the unfolding and refolding of titin protein, and the activity of a group I ribozyme. These studies are already beginning to reveal information not seen in bulk studies. We are hopeful that the future application of these methods will give a new window into the behavior of biomolecules.

Financial support for this work was provided by Air Force and National Science Foundation (S.C.), the National Institutes of Health (GM53757, J.R.W.) and (GM49243, D.H.) X.Z. is a Marvin Chodorow Fellow of the Applied Physics Department. J.W.O. is a Fellow of the Jane Coffin Childs Memorial Fund for Medical Research.

References

1. T. T. Perkins, D. E. Smith, and S. Chu, Science **276**, 2016 (1997); D. E. Smith and S. Chu, Science **281**, 1335 (1998); R. G. Larson, H. Hu, D. E. Smith, and S. Chu, J. Rheol. **43**, 267 (1999)
2. D. E. Smith, H. P. Babcock, and S. Chu, Science **283**, 1724 (1999); H. P. Babcock, D. E. Smith, J. S. Hur, E. S. G. Shaqfeh, and S. Chu, submitted to Science (1999)
3. P. R. Selvin, Methods Enzymol. **246**, 300 (1995); L. Stryer, and R. P. Haugland, Proc. Natl. Acad. Sci. USA **58**, 719 (1967)
4. R. T. Batey and J. R. Williamson, J. Mol. Biol. **261**, 536 (1996); op. cit., J. Mol. Biol. **261**, 550 (1996); op. cit., Rna-a Publication of the Rna Society **4**, 984 (1998); J. W. Orr, P. J. Hagerman, and J. R. Williamson, J. Mol. Biol. **275**, 453 (1998)

5. T. Ha, T. Enderle, D. F. Ogletree, D. S. Chemla, P. R. Selvin, and S. Weiss, Proc. Natl. Acad. Sci. USA **93**, 6264 (1996); G. J. Schutz, W. Trabesinger, and T. Schmidt, Biophys. J. **74**, 2223, (1998); T. Ha, A. Y. Ting, J. Liang, W. B. Caldwell, A. A. Deniz, D. S. Chemla, P. G. Schultz, and S. Weiss, Proc. Natl. Acad. Sci. USA **96**, 893 (1999); A. A. Deniz, M. Dahan, J. R. Grunwell, T. Ha, A. E. Faulhaber, D. S. Chemla, S. Weiss, and P. G. Schultz, Proc. Natl. Acad. Sci. USA **96**, 3670 (1999)

6. S. M. Nie and R. N. Zare, Ann. Rev. Biophys. Biomol. Struct. **26**, 567 (1997); X. S. Xie and J. K. Trautman, Ann. Rev. Phys. Chem. **49**, 441 (1998); W. E. Moerner and M. Orrit, Science **283**, 1670 (1999); S. Weiss, Science **283**, 1676 (1999)

7. T. Funatsu, Y. Harada, M. Tokunaga, K. Saito, and T. Yanagida, Nature **374**, 555 (1995); I. Sase, H. Miyata, S. Ishiwata, and K. Kinosita, Proc. Natl. Acad. Sci. USA **94**, 5646 (1997); R. D. Vale, T. Funatsu, D. W. Pierce, L. Romberg, Y. Harada, and T. Yanagida, Nature 380, **451 (1996); H. P. Lu**, L. Y. Xun, and X. S. Xie, Science **282**, 1877 (1998); W. A. Held, B. Ballou, S. Mizushim, and M. Nomura, J. Biol. Chem. **249**, 3103 (1974)

8. T. Ha, X. Zhuang, H. D. Kim, J. W. Orr, J. R. Williamson and S. Chu, Proc. Natl. Acad. Sci. USA **96**, 9077 (1999)

9. R. L. Baldwin, Nature **369**, 183 (1994)

10. J. N. Onuchic, Z. Luthey-Schulten, and P. G. Wolynes, Ann. Rev. Phys. Chem. **48**, 539 (1997)

11. X. Zhuang, T. Ha, H. D. Kim, T. Centner, S. Labeit and S. Chu, to be published (1999)

12. M. Reif, M. Gautel, F. Oesterhelt, J. M. Fernandez, and H. E. Gaub, Science 276, 1109 (1997); M.S.Z. Kellermayer, S.B. Smith, H.L. Granzier, and C. Bustamante, Science **276**, 1112 (1997); L. Tskhoverbova, J. Trinick, J. A. Sleep, and R. M. Simmons, Nature **387**, 308 (1997)

13. D. W. Celander and T. R. Cech, Science **251**, 401 (1991); P. P. Zarrinkar and J. R. Williamson, ibid. **265**, 918 (1994); B. Sclavi, M. Sullivan, M. R. Chance, M. Brenowitz, S. A. Woodson, ibid. **279**, 1940 (1998); D. K. Treiber, M. S. Rook, P. P. Zarrinkar, J. R. Williamson, ibid. **279**, 1943 (1998)

14. J. Pan, D. Thirumalai, and S. A. Woodson, J. Mol. Biol. **273**, 7 (1997)

15. R. Russell and D. Herschlag, J. Mol. Biol. **291**, 1155 (1999)

16. D. Herschlag and T. R. Cech, Biochemistry **29**, 10159 (1990)

20 Studying the Green Fluorescent Protein with Single-Molecule Spectroscopy

A. Zumbusch, G. Jung, and C. Bräuchle

The Green Fluorescent Protein (GFP) of the jellyfish *Aequorea victoria* has recently attracted a lot of attention. This is due to two facts. On the one hand, GFP is hitherto the only genetically encodeable protein that fluoresces without the addition of external cofactors. This means that the chromophore itself is formed from amino acids. On the other hand, GFP can be expressed in many eukaryotic and prokaryotic cells [1]. Both facts together make GFP an ideal marker for highly selective fluorescence labeling of live cells.

As soon as the high potential of GFP as a fluorescence label was recognized, an intensive search for mutants of the protein with specific photophysical properties began. For this purpose specific mutations of either the amino acid sequence composing the chromphore or of that of the surrounding protein matrix were introduced. A recent review by Tsien gives an excellent overview of this field [2]. The two most important factors in this concern were an improvement in brightness and changes in the absorption and emission properties of the mutants.·

It is evident that in order to predict the effect of mutations on the spectral properties of GFP, a detailed knowledge of its structure as well as of its photophysics is indispensable. The primary sequence of GFP was unraveled by Ward and coworkers [3]. They showed that GFP is a 238 amino acid protein with a molecular weight of 27 kDa [3,4]. The chromophore was assumed to be 4-(p-hydroxybenzylidene)imidazolin-5-one. This assumption was finally confirmed by the determination of the crystal structure [4,5], which revealed that the chromophore is part of an α-helix that stretches through a can-like structure formed by 11 β-sheet strands. The chromophore itself is formed in an autocatalytic post-translational reaction of the three amino acids Ser65Tyr66Gly67 [6].

The absorption spectrum of wild-type (wt) GFP in the visible spectral region is dominated by two strong absorptions at 400 nm and 476 nm (Fig. 20.1). Interestingly, excitation at either of these wavelengths leads to fluorescence emission at approximately 510 nm. This reflects the biological function of the protein, namely to shift blue emission from the chemiluminescent protein Aequorin into green emission. The nature of the double absorption peak and the observed unusually large Stokes shift of GFP lately were

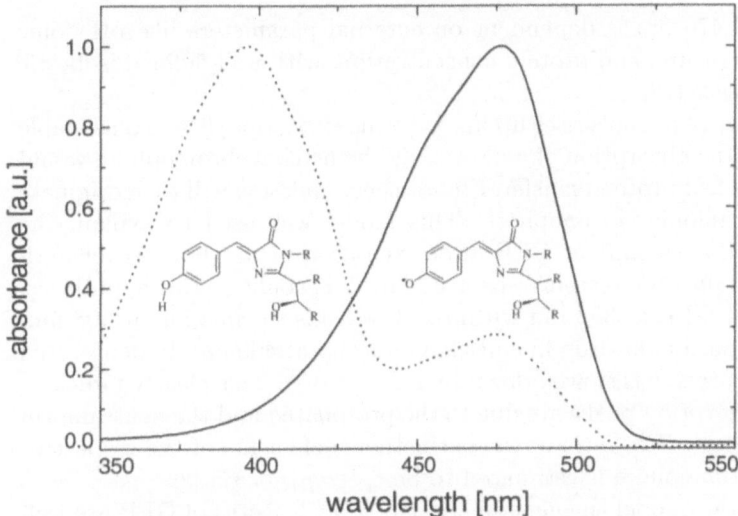

Fig. 20.1. Absorption spectrum of the GFP mutant E222Q (solid line) at pH10 and wt-GFP (dotted line) at pH7. The two maxima of the wt-GFP spectrum correspond to absorption by the neutral chromophore and by the anionic chromophore. The mutant E222Q has its anionic chromophore stabilized and exhibits only one absorption peak. The insets show the different states of the chromophore

subject to many different spectroscopic investigations, encompassing ultrafast spectroscopy [7,8], fluorescence excitation spectroscopy at cryogenic temperatures [9,10], and studies relying on the detection of single molecules such as single-molecule imaging [11–13] and fluorescence correlation spectroscopy [14–16]. The results of these experiments provide detailed knowledge of the photophysics of GFP and show that the chromophore acts as a sensitive probe for its immediate protein neighborhood. In this article we will give a brief review of recent spectroscopic investigations of GFP. After summarizing bulk spectroscopic work at room temperature, we will present work mainly from our own group comparing the specific merits of single-molecule spectroscopic methods with conventional bulk spectroscopy in the case of GFP.

20.1   Bulk Spectroscopy of the GFP

20.1.1   Room Temperature Measurements

With the identification of 4-(p-hydroxybenzylidene)imidazolin-5-one as the light-emitting chromophore of GFP, a protonation/deprotonation equilibrium of the chromophore's phenolic group was proposed to be responsible for the double absorption of wt-GFP [6]. The existence of two different ground states of GFP was implied by the fact that the ratio between the absorption at 400 nm and 476 nm is dependent on external parameters like pH, ionic strength, temperature, and protein concentration, with well-defined isobestic points in each case [18].

As the acidity of phenols rises upon electronic excitation, it was reasonable to assume that the absorption of a photon by the neutral chromophore would be followed by a fast proton transfer. Fluorescence emission will then originate only from the anionic chromophore. This model was used to explain the large observed Stokes shift of GFP after excitation of its short wavelength absorption. Ultrafast fluorescence spectroscopy [7,8] confirms this hypothesis. Excitation of wt-GFP at 398 nm within 2–8 ps leads to emission at 508 nm, accompanied by a similar drop in emission intensity at 460 nm. In deuterated GFP, the same process is slowed down by a factor of 5. This clearly indicates that the two absorption peaks are due to the protonated and the neutral form of the chromophore. The ratio of the extinction coefficients of the respective forms of the chromophore is estimated to be $\epsilon_{R-}/\epsilon_{RH} \approx 1.5$ [8].

While with this model the general spectroscopic features of GFP are well explained, it has to be refined to account for other experimental observations. The rate for the photoconversion of the neutral chromophore into the anionic chromophore is small. Boxer and coworkers [7] therefore point to the fact that the emitting state after short-wavelength excitation cannot be identical to the excited state of the anionic chromophore. Instead they propose an intermediate state I that is photochemically generated in its excited state after excitation of the neutral chromophore. In its ground state, the state I has a low energy barrier for conversion into the neutral chromophore RH (see Fig. 20.3). The nature of this state I can be explained in the following manner: As the photon transfer of the excited neutral chromophore takes place in only a few ps, its protein neighborhood, which originally solvates a neutral chromophore, will not change its structure fast enough to accomodate the anionic chromophore. In this case fluorescence emission will occur from the state I, which electronically resembles a protein with its chromophore in the anionic state while having the structure to solvate a neutral chromophore. Close inspection of the emission spectra indeed shows that the emission after excitation of the neutral form of the chromophore at 400 nm is slightly redshifted compared to the emission obtained after direct excitation of the anionic form in its equilibrated form R_{eq}^-. This effect becomes very pronounced after cooling the sample to 77 K [7,8].

Insight in the rearrangements of the protein structure necessary to solvate an anionic chromophore after photoconversion can be obtained from Stark spectroscopic investigations [17]. These show that excitation of the neutral form of the chromophore indeed affords a large charge displacement with respect to the ground state. In contrast to this, excitation of the anionic chromophore involves only a minor charge redistribution. This difference in the electronic properties of the excited states of the neutral and the anionic chromophore will also be reflected in their respective protein environments. It is therefore reasonable to assume that photoconversion of the chromophore will be accompanied by major structural changes in the protein shell.

20.1.2 Fluorescence Spectroscopy at Cryogenic Temperatures

While room-temperature experiments yield much information about long-lived states of GFP, it is difficult to deduce detailed information about short-lived intermediate states and transition rates between these. Therefore fluorescence spectroscopy at cryogenic temperatures was performed in the group of Völker as well as in our own group. This method is an ideal tool for the characterization of the intermediate states [9,10]. Upon cooling to 2 K, the fluorescence excitation and emission spectra of wt-GFP become significantly more structured. The excitation spectrum at 2 K is dominated by a strong band of the anionic chromophore at 471 nm with vibrational sidebands at 460 nm and 446 nm, while the band due to the neutral chromophore remains broad and unstructured, as shown in Fig. 20.2. The low-temperature emission spectra of freshly prepared samples show two distinct maxima at 482 nm for excitation at 472.7 nm and at 503 nm for excitation at 363.8 nm. These can be interpreted as the emission of the structurally relaxed anionic chromophore and of the unrelaxed anionic chromophore respectively. By contrast to these strong bands, the direct emission of the neutral chromophore is detected only as weak and broad emission around 450 nm.

Apart from these bands, which are already known from room temperature experiments, the illumination of the sample with different wavelenghts leads to the appearance of three new bands attributable to photoproducts that are only stable at low temperatures. Excitation of the sample with wavelengths shorter than 400 nm leads to the production of two new products P502 and P510 with excitation maxima at 502 nm and 510 nm respectively. Another new species P489 with an excitation maximum at 489 nm is produced by longer-wavelength illumination at 473 nm. It is important to note that under no circumstances can one observe direct photoconversion between the protein with the neutral chromophore RH and that with the equilibrated, anionic form R_{eq}^-. This was to be expected from the results of the room-temperature experiments. The emission of all photoproducts exhibits only small Stokes shifts on the order of several nm.

Knowing that the direct emission from the neutral chromophore is observed between 420 nm and 470 nm, it can be assumed that all new photoproducts contain an anionic chromophore. As P489 has only slightly different absorption and emission properties compared to the state with the relaxed anionic chromophore, one might deduce that the two states resemble each other. The two states P502 and P510 however must have different structures. Emission from P502 is almost indistinguishable from emission after excitation of the neutral chromophore. This suggests that it is the structurally nonequilibrated form I of the protein proposed by Chattoraj et al. The product P510

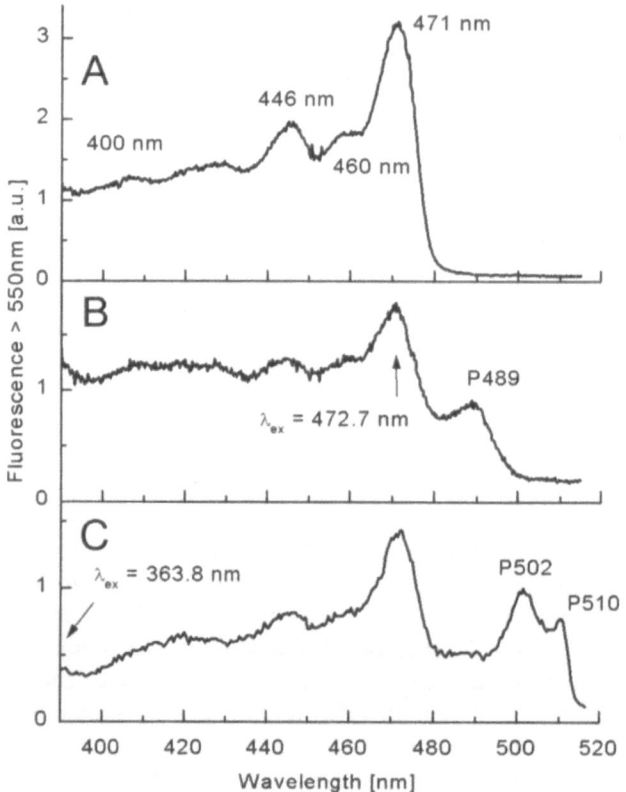

Fig. 20.2. Fluorescence excitation spectra of wt-GFP in 70% glycerol/H$_2$O buffer (pH = 7) at 2 K. Detection at wavelengths >550 nm. **a** Spectrum before illumination, **b** spectrum after illumination with λ_{exc} = 472.7 nm (100 mW/cm^2 for 60 min), **c** spectrum after illumination with λ_{exc} = 363.8 nm (30 mW/cm^2 for 240 min)

is spectroscopically clearly distinguished from P502 but can so far not be characterized in more detail. We have also performed temperature-dependent measurements on the different photoproducts [10]. In these experiments the sample is slowly heated and the fluorescence emission is monitored in parallel. Analysis of the fluorescence emission intensity as a function of the sample temperature then allows for the determination of the barrier heights for conversions between the different substates of GFP. Data are depicted in the potential energy surface diagram in Fig. 20.3.

A promising approach for more detailed low-temperature investigations of GFP would be to perform hole burning spectroscopy or single-molecule spectroscopy with spectral selection of the molecules. While selective excitation of subensembles in the inhomogeneously broadened band was successful and burning of narrow holes in GFP has been reported by us and by the Völker

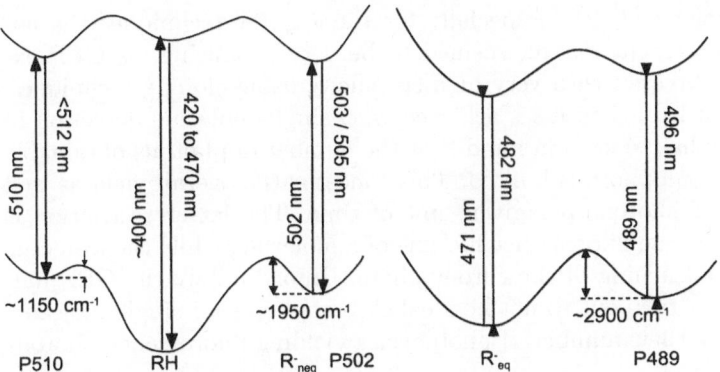

Fig. 20.3. Extended energy potential surface of wt-GFP, including the three new photoproducts and corresponding energy barriers. The state R_{neq}^- corresponds to the state I proposed by Chattoraj et al. [7]

group, efficiencies prove to be low and fluorescence line narrowing was not observed [9,10]. An explanation for the difficulties encountered in hole burning experiments on GFP is given by the observed strong electron–phonon coupling of the chromophore.

20.1.3   Single-Molecule Imaging of GFP [28]

Recent years have witnessed the rapid development of different techniques for the detection of the fluorescence of single molecules. From a spectroscopic point of view these methods are appealing for two principal reasons. On the one hand they make it possible to record data that would not be accessible with bulk spectroscopic experiments. Bulk spectroscopy commonly only yields average values over a wide distribution of different molecules. In contrast, single-molecule spectroscopy removes the necessity to average over ensembles of molecules and allows one to map out distributions of values specific to individual molecules, such as transition energies, lifetimes, and even transition rates. Analysis of these distributions can yield deep insight into the dynamical processes of the fluorophore itself, as well as its interaction with its environment. The other attractive side of single-molecule spectroscopy concerns the possibility of observing new, hitherto unknown phenomena. The most prominent of these examples is an observation termed blinking. This signifies the repeated on and off switching of the fluorescence of a single fluorescence-emitting molecule, which cannot be explained by transitions into known dark states such as triplet states.

Apart from the aforementioned spectroscopic interest in the detection of single molecules, there exists also an analytical interest. This is either motivated by applications affording high sensitivity, as in new schemes for the sequencing of DNA [19], or by high-resolution imaging experiments mainly in molecular biology [20,21]. Especially for these latter techniques the use of GFP as a fluorescent marker seemed to be very promising, as GFP expression can be directed with very high specificity using cloning techniques. Despite the remarkable successes achieved with single-molecule detection in recent years, one has to keep in mind that the number of photons obtainable from a single chromophore is limited. This concerns the overall yield as well as the number of photons per given unit of time. The latter restriction is simply imposed by the fluorescence lifetime of a molecule, while the former is caused by photobleaching of the chromophore. Unfortunately the GFP mutants known so far turn out to not be ideal chromophores for single-molecule detection because their number of photocycles yielding fluorescence photons is too small.

Experiments with single molecules of GFP in the optical far field and in the optical near field were performed in different groups soon after its high potential as a fluorescent marker was recognized. All of these experiments were performed with mutants in which the anionic form of the chromophore was stabilized, such as YFP (yellow fluorescent protein) [11], E222Q [13], and S65T [12,22,23], so that they did not exhibit any significant absorption around 400 nm. It was hoped that the suppression of the stable state of the neutral chromophore in these mutants would result in improved brightness as expressed by the number of photons per unit time. Unfortunately, to date these hopes have not been fulfilled. All the GFP mutants used so far in single-molecule experiments showed very frequent switching between emitting and nonemitting states before a subsequent quick and seemingly permanent photobleaching.

Some observations from these experiments are nevertheless noteworthy. Among the most surprising experimental results was the observation that apparently photobleached GFP molecules indeed emitted fluorescence again after remaining in a dark state for more than two hours [13], a behavior that was also seen in NSOM experiments [23]. A possible explanation for this observation may be found in the optical switching of GFP that was demonstrated by Dickson and coworkers. They showed that a YFP molecule that was first photobleached with excitation light at 488 nm could be brought back into an emitting state by illumination with 405-nm light. The photobleaching step was interpreted as a transition of the molecule into a state with a neutral chromophore that can be depopulated with short-wavelength illumination. A preliminary analysis of the on/off dynamics of GFP by Peterman et al. shows that the time the molecule spends in an emitting state depends on the illumination intensity at the longer wavelength, while the back transition into the fluorescing state is independent of it [24]. A detailed description of these processes based on single-molecule data has however so far been hampered by the short overall lifetime of GFP before permanent bleaching. Because the total number of photons obtained in these experiments is small compared to those using common fluorescent dyes, techniques

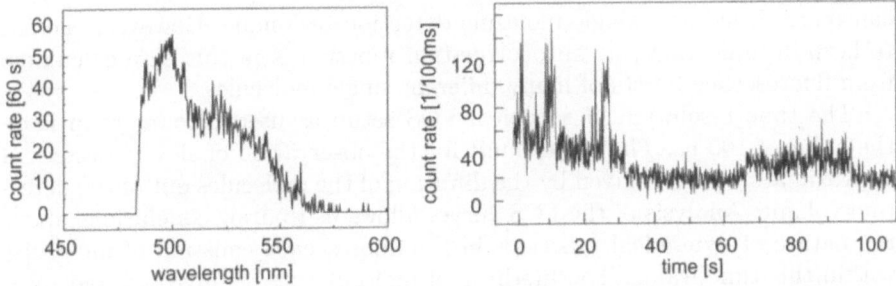

Fig. 20.4. Fluorescence emision spectrum and corresponding trajectory of one single wt-GFP molecule in agarose gel at 30 K (excitation intensity 2.5 kW cm^{-2} at 476 nm)

like the calculation of autocorrelation functions of the fluorescence photons cannot reliably be employed.

The low number of photons obtainable from single GFP molecules proves to be even more detrimental when working at cryogenic temperatures. We performed single-molecule microscopy of GFP at cryogenic temperatures, the results of which are shown in Fig. 20.4 [25]. In contrast to other biological systems for which increased photostability compared to room temperature was observed at low temperatures [26], GFP turns out to be photochemically even more unstable under these conditions. We explain this behavior by means of photochemically induced transition into a dark state, most probably one containing a neutral chromophore, with an energy barrier that is too high for a thermal back-reaction. This model can be reconciled with the room-temperature results described above.

20.1.4 Fluorescence Correlation Spectroscopy of the GFP [28]

If the calculation of autocorrelation functions from the trajectory of immobilized individual molecules is not possible due to an insufficient number of emitted photons, Fluorescence Correlation Spectroscopy (FCS) presents an alternative. FCS yields autocorrelation functions by averaging over fluorescence signals from a small number of molecules diffusing through a well-defined focal region. For low concentrations down to the single-molecule limit, fluctuations in the fluorescence emission intensity can be observed with FCS. These fluctuations can either be caused by diffusion of molecules in and out of the focal volume element or by on and off dynamics of the fluorescence emission itself. The latter effect is observed in the presence of transitions into the triplet state. The contrast of the autocorrelation functions recorded with FCS will decrease and vanish for an increasing number of molecules in the detection volume. In order to maximize the contrast, the experiments described here are therefore commonly performed at concentrations equal to or below one molecule per detection volume element. In this respect, FCS can be regarded as a single-molecule detection technique. However, one has to keep in mind that one autocorrelation function is in this case calculated from fluorescence bursts of many different single molecules.

The time resolution of a typical FCS setup as used in our group is on the order of 100 ns. The upper limit for the observation of slow processes of approximately 1 ms is given by the diffusion of the molecules out of the detection volume. Analysis of the FCS curves allows us to draw conclusions about the nature of dynamical processes in the fluorescence emission of molecules within this time frame. The lifetimes of molecular states involved and rates for transitions into and out of these can be determined using the following theoretical expression [14]:

$$G(t) = \frac{\langle I(t) \times I(t+\tau) \rangle}{\langle I(t) \rangle^2} = 1 + \frac{1}{N} \left(1 + \frac{\tau}{\tau_D} \right)^{-1}$$
$$\times \left(1 + C_1 \times e^{-\lambda_1 \tau} + C_2 \times e^{-\lambda_2 \tau} \right) \tag{20.1}$$

The first part of the product in (20.1) accounts for diffusion within a two-dimensional model [27] that is applicable in our experiment. N is the average number of fluorescing molecules in the sample volume, while τ_D is related to the lateral beam waist ω in the focal plane and the diffusion coefficient D by $\tau_D = \omega^2/4D$ with the factor $\omega = 340$ nm and $D = 8.2(\pm 0.5)10^{-7}\text{cm}^2/\text{s}$ in our case. Intramolecular dynamical processes are represented by the bracket in (20.1). Here the biexponential formulation with contrasts C_1 and C_2 and rates λ_1 and λ_2 accounts for a four-level system, i.e. a system composed of a ground state $|1\rangle$, an excited state $|2\rangle$, and two dark states $|3\rangle$ and $|4\rangle$ (Fig. 20.7).

The fact that FCS autocorrelation curves of the GFP mutants EGFP and S65T cannot be described by a simple three-level system with a triplet state as a dark state was first reported by Haupts et al. [14]. In this work they performed pH-dependent measurements and concluded that internal and external protonation states also contributed to the molecule's dynamics. The observation of internal protonation was however not confirmed in experiments by Widengren and coworkers on the similar BioST mutant [15]. It would be an intriguing result, as both of the mutants in the ground states have stabilized anionic chromophores and do not show any absorption around 400 nm.

A common drawback of FCS experiments is that the nature of the dark states cannot be identified spectroscopically. To find out whether an internal protonation equilibrium is also present in GFP mutants that do not show any ground state absorption of the neutral chromophore, we chose to perform an FCS experiment that offers the possibility of spectroscopically addressing the dark states. In this case, of course, the spectrum of the neutral chromophore as the presumable dark state is approximately known.

As a preliminary experiment we investigated bulk samples using simultaneous two-color excitation at 407 nm and 476 nm, coinciding with the absorption bands of the neutral and anionic chromophore, respectively. We

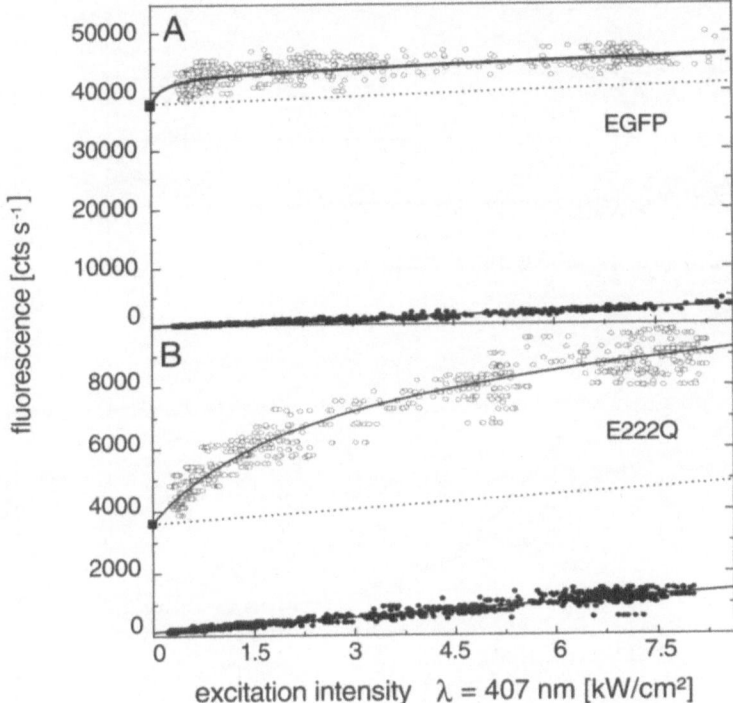

Fig. 20.5. Total fluorescence count rate observed in two-color saturation experiments with **a** EGFP (3×10^{-8}M at pH 10) and **b** E222Q (1×10^{-8} M at pH 10). Solid circles: single-color excitation at 407 nm. Open cirles: two-color excitation at 407 nm with a fixed intensity of 7.6 kW cm^{-2}at 476 nm. Data for solely illuminating at 476 nm are shown as solid squares. The solid lines are fitted to the data using a simple saturation model. The dotted lines represent the expected offsets due to single-color excitation

detected the integrated fluorescence with a wavelength larger than 507 nm. The results of experiments with two mutants EGFP and E222Q, both with stabilized anionic chromophores, are shown in Fig. 20.5. Exciting only at 407 nm, we observe a very small signal that slightly increases with increasing illumination intensity. The origin of this signal can be weak fluorescence from the excitation of the blue shoulder of the anionic chromophore's absorption or scattering. Illuminating only at 476 nm gives the expected relatively strong fluorescence signal due to absorption of the anionic chromophore. Keeping the illumination intensity at 476 nm constant while simultaneously increasing the intensity at 407 nm does however not result in a curve that is a simple addition of the two count rates obtained for separate excitation. Instead we observe an increase in signal intensity of up to 140% in the case of E222Q and up to 13% in the case of EGFP. A similar effect with higher saturation intensities and somewhat smaller fluorescence increases is observed for simultaneous

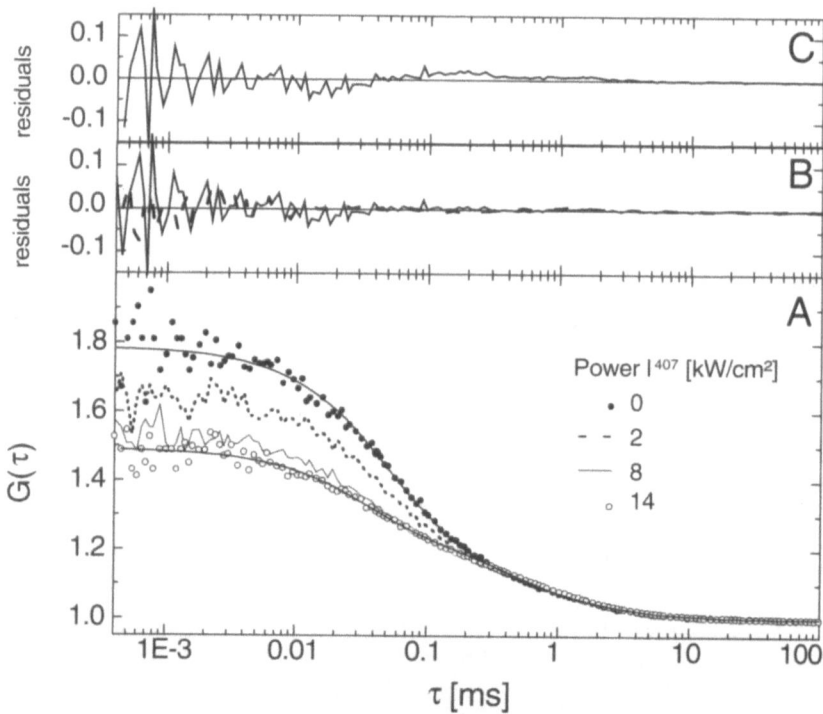

Fig. 20.6. Two-color fluorescence autocorrelation functions of the mutant E222Q at pH 10 with a fixed intensity of 23 kW cm^{-2} at 476 nm. **a** Curves obtained for different intensities of simultaneous excitation at 407 nm. The solid lines represent fits according to equation 1. The lower curve can be fitted using a monoexponential function. **b** Residuals for a biexponential fit to the data depicted as the uppermost curve in **a**. **c** Residuals when attempting to fit these data to a monoexponential function

excitation at 442 nm and 476 nm. In both cases the increase in signal by simultaneous illumination is saturated at low intensities of $I_S^{407} = 2.5$ kW/cm^2 for E222Q and $I_S^{407} < 1$ kW/cm^2 for EGFP with a constant excitation intensity of $I^{476} = 7.6$ kW/cm^2. The estimated population of the neutral state of the chromophore is 60% for E222Q under these experimental conditions. These observations can only be explained by an efficient protonation of the anionic chromophore upon irradiation.

With the knowledge obtained from these bulk experiments we performed two-color FCS experiments on the E222Q mutant to determine the lifetime of the dark states as well as the respective transition rates. Results of experiments with constant illumination intensity at 476 nm and different intensities of simultaneous illumination at 407 nm are depicted in Fig. 20.6. Solely illuminating at 476 nm yields autocorrelation functions that have to be fitted with a biexponential function representative of a four-level system and given by

(20.1). Simultaneous illumination at 407 nm with increasing intensity results in decreasing contrast because of the higher number of fluorescing molecules present under these conditions. For illumination intensities at 407 nm that were saturating the increase in fluorescence signal observed in the bulk experiments, the autocorrelation function simplifies to become monoexponential. This latter effect is of importance for the analysis of the photodynamical behavior of the molecule.

The exact solution of the rate equation system for a four-level system affords data at low and high excitation intensities that are difficult to obtain in our case. Instead we exploit the simplification of the molecule's photodynamics to that of a three-level system when applying two-color excitation. The solution of the rate equations for this system does not pose any problems. From the contrast

$$C_1 = \frac{k_{23}^{\text{eff}}}{k_{31}}$$

and the rate

$$\lambda_1 = k_{31} + k_{23}^{\text{eff}}$$

with

$$k_{23}^{eff} = k_{23} \frac{k_{12}}{k_{12} + k_{21}}$$

the rates k_{23} and k_{31} can be determined. We obtain $k_{23} = 3 \times 10^5$ s^{-1} ($\pm 40\%$) and $k_{31} = 2 \times 10^4$ s^{-1} ($\pm 10\%$). The corresponding lifetime of 50 μs would be typical for a triplet state.

With the rates within the three-level system being determined in this way, we can attempt an approximate solution of the four-level system. The rate constant k_{41} equals λ_2 at low excitation intensities at 476 nm and is determined to be $k_{41} = 2 \times 10^3$ s^{-1} ($\pm 50\%$). From the experimental data available so far, only an estimate for k_{24} can be given. We obtain $k_{42} = 4 \times 10^5$ s^{-1}. This value and the spectral properties of this dark state with absorption at 407 nm indicate that it is identical to the neutral form of the chromophore with a lifetime of 500 μs. The complete description of the four-level system describing the room-temperature dynamics of E222Q is depicted in Fig. 20.7. Our data prove that the neutral chromophore indeed plays an important role in the photodynamics of the E222Q mutant, while it is not detectable in the ground state of the protein. Already at moderate excitation intensities of 8 kW/cm^2 at 476 nm, nearly 70% of the E222Q molecules populate dark states, 65% of which are in the state containing the neutral chromophore and the other 5% occupying the triplet state. Using even higher excitation intensities will further deplete the ground state population. While the effect described here is most pronounced in the E222Q mutant, it is also seen in the widely used EGFP mutant. Two-color FCS investigations on other

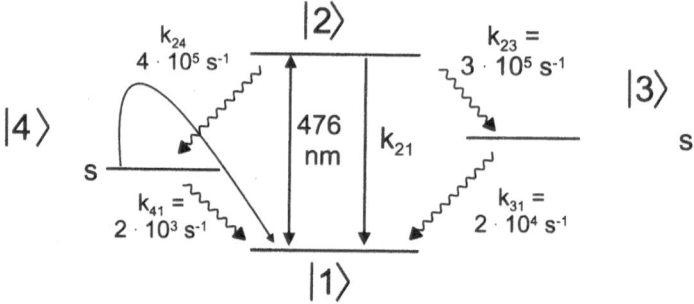

Fig. 20.7. Level scheme for the four-level system describing E222Q and corresponding transition rates and lifetimes. Conversion of state 4 into state 1 after 407-nm excitation takes place via an unidentified excited state

mutants with tunable excitation sources for the second color are currently performed. Apart from gaining insight into the photodynamics of GFP, the results from these experiments might have two practical consequences. On the one hand, awareness of the important influence of the seemingly unpopulated neutral state of the chromophore might contribute to the production of more efficient new GFP mutants. On the other hand, two-color excitation might turn out to be useful even in common microscopic investigation with GFP to obtain an increase in overall brightness [29].

20.1.5   Conclusion

We presented different spectroscopic investigations of the photodynamics of GFP molecules. X-ray analysis of the protein's structure and ultrafast spectroscopic studies provided a basic understanding of the protonation and deprotonation equilibrium of the chromophore in GFP. From these experiments the main features of the absorption and emission spectra of GFP can readily be explained. We performed fluorescence spectroscopy at cryogenic temperatures, which provided more detailed knowledge about the different states present in the molecule and about barrier heights for transitions between them. Unfortunately, the hopes for gaining even more intimate knowledge about GFP's photodynamics from single-molecule imaging experiments at room temperature or at cryogenic temperatures have so far not been fulfilled. This is due to the insufficient number of photons obtainable from a single GFP molecule. These experiments, however, showed some intriguing effects, like the reappearance of fluorescence emission from apparently photobleached molecules after several hours. In order to study the photodynamics of GFP at room temperature, we extended common fluorescence correlation spectroscopy to two-color illumination of one chromophore. This proves to be a very fruitful technique that allows us to gain detailed insight into the importance of the neutral and anionic states of the chromophores for the photodynamics of GFP [30]. In the meantime, this technique has been applied successfully in studies of other fluorophores [31,32]. Most interestingly, we observed that even in mutants with no observable ground state absorption attributable to the presence of a neutral chromophore, this state is nevertheless populated to a large extent after excitation of the anionic chromophore.

References

1. M. Chalfie, Y. Tu, G. Euskirchen, W. W. Ward, and D. C. Prasher, Science **263**, 802 (1994)
2. R. Y. Tsien, Ann. Rev. Biochem. **67**, 509 (1998)
3. D. C. Prasher, V. K. Eckenrode, W. W. Ward, F. G. Prendergast, and M. J. Cormier, Gene **111**, 229 (1992)
4. F. Yang, L. G. Moss, and G. N. Philipps, Jr., Nat. Biotech. **14**, 1246 (1999)
5. M. Ormö, A. B. Cubitt, K. Kallio, L. A. Gross, R. Y. Tsien, and S. J. Remington, Science **273**, 1392 (1996)
6. R. Heim, D. C. Prasher, and R. Y. Tsien, Proc. Nat. Acad. Sci. USA **91**, 12501 (1994)
7. M. Chattoraj, B. A. King, G. U. Bublitz, and S. G. Boxer, Proc. Nat. Acad. Sci. USA **93**, 8362 (1996)
8. H. Lossau et al., Chem. Phys. **213**, 1 (1996)
9. T. M. H. Creemers, A. J. Lock, V. Subramaniam, T. M. Jovin, and S. Völker, Nature Struct. Biol. **6**, 557 (1999)
10. C. Seebacher, F. W. Deeg, C. Bräuchle, J. Wiehler, and B. Steipe, J. Phys. Chem. B **103**, 7728 (1999)
11. R. M. Dickson, A. B. Cubitt, R. Y. Tsien, and W. E. Moerner, Nature **388**, 355 (1997)
12. D. W. Pierce, N. Hom-Booher, and R. D. Vale, Nature **388**, 338 (1997)
13. G. Jung, J. Wiehler, W. Göhde, J. Tittel, T. Basche, B. Steipe, and C. Bräuchle, Bioimaging **6**, 54 (1998)
14. U. Haupts, S. Maiti, P. Schwille, and W. W. Webb, Proc. Nat. Acad. Sci. USA **95**, 1416 (1998)
15. J. Widengren, B. Terry, and R. Rigler, Chem. Phys.**249**, 259 (1999)
16. G. Jung, S. Mais, A. Zumbusch, and C. Bräuchle, J. Phys. Chem. A **104**, 873 (2000)
17. G. Bublitz, B. A. King, and S. G. Boxer, J. Am. Chem. Soc. **120**, 9370 (1998)
18. M. Cutler and W. W. Ward, Photochem. Photobiol. **57**, 63S (1993)
19. W. P. Ambrose et al., Ber. Bunsenges. Phys. Chem. **97**, 1535 (1993)
20. T. Funatsu, Y. Harada, M. Tokunaga, K. Saito, and T. Yanagida **374**, 555 (1995)
21. T. Nishizaka, H. Miyata, H. Yoshikawa, S. Ishiwata, and K. Kinosita Jr., Nature **377**, 251 (1995)
22. A. H. Iwane, T. Funatsu, Y. Harada, M. Tokunaga, O. Ohara, S. Morimoto, and T. Yanagida, FEBS Lett. **407**, 235 (1997)
23. M. F. Garcia-Parajo, J.-A.Veerman, G. M. J. Segers-Nolten, B. G. de Grooth, J. Greve, and N. F. van Hulst, Cytometry **36**, 239 (1999)
24. E. W. Peterman, S. Brasselet, and W. E. Moerner, J. Phys. Chem. A **103**, 10553 (1999)

25. S. Mais, P. Zehetmayer, G. Jung, C. Bräuchle, and A. Zumbusch, unpublished results
26. C. Tietz, O. Chekhlov, A. Dräbenstedt, J. Schuster, and J. Wrachtrup, J. Phys. Chem. B **103**, 6328 (1999)
27. R. Rigler, Ü. Mets, J. Widengren, and P. Kask, Eur. Biophys. J. **22**, 169 (1993)
28. A. Zumbusch and G. Jung, Single Mol. **1**, 261 (2000)
29. G. Jung, J. Wiehler, B. Steipe, C. Bräuchle, and A. Zumbusch, Chem. Phys. Chem. **2**, 392 (2001)
30. G. Jung, C. Bräuchle, and A. Zumbusch, J. Chem. Phys. **114**, 3149 (2001)
31. J. Widengren and C. Seidel, Phys. Chem. Chem. Phys. **2**, 3435 (2000)
32. F. Malrezzi-Canipeggi, M. Jahnz, K. G. Heinze, P. Dittrich, P. Schwille, Biophys. J. **81**, 1776 (2001)

Index

absorption, 38
activation energy, 246, 247
adiabatic following, 100
antenna complex, 62
antibunching, 100, 110, 111
atomic force microscope, 330
autocorrelation, 45
autocorrelation function (ACF), 196
autofluorescence of cells, 204

bacterial photosynthesis, 62
biorelevant reaction kinetics, 130, 134,
    135
blinking, 51
Bloch, 100, 101, 104–106, 108

chemical analysis, 242
cleavage, 334
clusters of receptor, 208
coherent dynamics, 277
concentration reporters, 55
confocal, 167, 169
Confocal Fluorescence Coincidence
    Analysis (CFCA), 212
confocal fluorescence detection, 167
confocal scanning optical microscope,
    328
conformational, 277
conformational dynamics, 237
conformational rearrangements, 89
cross-correlation, 56
cryptography, 99
Cy3, 327
cysteine, 330
cytochrome P-450 dependent monooxy-
    genase, 277

dark states, 343

DBATT, 102, 104, 107
DC Stark effect measurements, 89
DNA, 144
DNA base, 154
delocalization, 76
denaturing, 330
Dextran-SNARF-1, 56
diagonal disorder, 68
dibenzanthanthrene (DBATT), 102
diffusion, 50
dipole–dipole interaction, 63, 75
docking, 336
dual-color cross-correlation, 211
dynamic disorder, 228

electrofusion, 136, 139
electroporation, 131, 136, 139, 142, 143
emission spectra, 48
energy landscape, 329
energy transfer, 63, 74
energy transfer efficiency, 328
ensemble averaging, 33
enzyme kinetics, 227
evolutionary biotechnology, 219
exciton model, 76

Förster transfer, 75
FCS, 345
fluctuations, 56
fluorescein, 327
fluorescence, 100, 102–104, 107–109,
    112
fluorescence correlation spectroscopy
    (FCS), 34, 195, 211, 345
fluorescence decay, 163
fluorescence excitation, 38
fluorescence lifetime, 162, 163
fluorescence quenching, 329

fluorescence resonant energy transfer (FRET), 55, 326
fluorescence spectroscopy, 211
fluorescence-excitation spectroscopy, 64
FM spectroscopy, 36
foldability, 328
four-level system, 348
frequency-modulation, 38

Gaussian, 154
Green Fluorescent Protein (GFP), 51, 53, 338
guanosine, 334

Hanbury-Brown and Twiss, 103
HEK293, 198
hexadecane, 102
high-throughput screening, 217
hole-burning, 44, 82
hole-burning spectroscopy, 342
homogeneous linewidths, 40, 83
hyper Raman, 156

immobilization, 330
inhomogeneous distribution, 83
intercomplex heterogeneity, 68
intra-domain quenching, 331
intracomplex heterogeneity, 68

kinetics, 44

labeling efficiency, 330
LH1, 62
LH2, 62
light-harvesting (LH) complexes, 62
light-induced frequency jumps, 94
lipid domains, 130, 140
lookup table, 165

magnetic resonance, 47
metabolic pathways, 278
metal colloids, 146
micropipette, 131, 136–139
microsphere, 50
misfolded state, 333
molecular heterogeneity, 251
molecular individualism, 326
molecular networks, 278
morphology-dependent resonances, 50

n-hexadecane, 102, 104
nanoenvironment, 33
near-field probe, 50
nile red, 51
nonphotochemical hole-burning, 82

ODMR, 47
on-time distributions, 55
optical switching, 82
optical trap, 134, 136, 142
Oregon Green, 332

p-terphenyl, 36, 83
p-terphenyl crystal, 83
p-terphenyl matrix, 93
paraboloid, 102, 103
pentacene, 36, 93
persistent spectral hole-burning, 82
photobleaching, 207, 344
photon antibunching, 46
photon-counting image acquisition system, 169
photosynthetic purple bacteria, 62
Poisson, 154
polarization, 52
polarization anisotropy, 328
poly(acrylamide) gels, 51
protein folding, 329
protein machines, 277
pulsed excitation, 168

quantum electrodynamic, 50
quantum Monte Carlo, 105, 109
quantum optics, 57

rapid adiabatic passage, 100, 101, 107, 112
Raman effect, 144
rate constants, 336
receptor–ligand, 195
refolding, 330
ribosomal protein S15, 326
RNA conformational change, 326
RNA folding, 332
Rps. acidophila, 62

saturation, 41
SFS, 36
Shpol'skii matrices, 43, 102

signal-to-noise ratio (SNR), 35
single molecule, 99, 227
single photons, 51, 99
single-molecule enzymology, 244
single-molecule fluorescence
    spectroscopy, 327
single-molecule hole-burning, 86
single-molecule imaging, 343
single-molecule spectroscopy (SMS),
    32, 343
single-photon source, 57
small volumes, 277
spectral diffusion, 41, 43, 71
spectral hole-burning, 36
spiking, 283
squeezed light, 99
Stark effect, 102
start–stop, 103, 105, 106, 110, 111
statistical fine structure, 36
stochastic model, 278
stoichiometry of ligand binding, 200
substrate cycles, 278
surface-enhanced Raman scattering
    (SERS), 144

synchronization, 277

terrylene, 57, 83
terrylene in poly(ethylene), 49
tertiary structure, 330
Tetrahymena ribozyme, 333
thermal denaturation, 252–254
30 S ribosomal subunit, 326
three-dimensional orientation, 53
time-correlated single-photon counting
    (TCSPC), 162, 167
titin, 329
total internal reflection (TIR)
    microscopy, 51, 334
transmembrane protein, 195
turnover cycles, 277
two-color excitation, 346
two-state model, 328

undocking, 336
unfolding, 330

vibrational spectrum, 49

Springer Series in Chemical Physics
Editors: Vitalii I. Goldanskii   Fritz P. Schäfer   J. Peter Toennies

1   Atomic Spectra and Radiative Transitions
    By I. I. Sobelman
2   Surface Crystallography by LEED
    Theory, Computation and Structural
    Results. By M. A. Van Hove, S. Y. Tong
3   Advances in Laser Chemistry
    Editor: A. H. Zewail
4   Picosecond Phenomena
    Editors: C. V. Shank, E. P. Ippen,
    S. L. Shapiro
5   Laser Spectroscopy
    Basic Concepts and Instrumentation
    By W. Demtröder   3rd Printing
6   Laser-Induced Processes in
    Molecules   Physics and Chemistry
    Editors: K. L. Kompa, S. D. Smith
7   Excitation of Atoms and Broadening
    of Spectral Lines   By I. I. Sobelman,
    L. A. Vainshtein, E. A. Yukov
8   Spin Exchange
    Principles and Applications in
    Chemistry and Biology
    By Yu. N. Molin, K. M. Salikhov,
    K. I. Zamaraev
9   Secondary Ion Mass Spectrometry
    SIMS II   Editors: A. Benninghoven,
    C. A. Evans, Jr., R. A. Powell,
    R. Shimizu, H. A. Storms
10 Lasers and Chemical Change
    By A. Ben-Shaul, Y. Haas,
    K. L. Kompa, R. D. Levine
11 Liquid Crystals of One- and
    Two-Dimensional Order
    Editors: W. Helfrich, G. Heppke
12 Gasdynamic Laser   By S. A. Losev
13 Atomic Many-Body Theory
    By I. Lindgren, J. Morrison
14 Picosecond Phenomena II
    Editors: R. M. Hochstrasser,
    W. Kaiser, C. V. Shank
15 Vibrational Spectroscopy of
    Adsorbates   Editor: R. F. Willis
16 Spectroscopy of Molecular Excitons
    By V. L. Broude, E. I. Rashba,
    E. F. Sheka
17 Inelastic Particle-Surface Collisions
    Editors: E. Taglauer, W. Heiland
18 Modelling of Chemical Reaction
    Systems   Editors: K. H. Ebert,
    P. Deuflhard, W. Jäger

19 Secondary Ion Mass Spectrometry
    SIMS III
    Editors: A. Benninghoven, J. Giber,
    J. László, M. Riedel, H. W. Werner
20 Chemistry and Physics of Solid
    Surfaces IV   Editors: R. Vanselow, R. Howe
21 Dynamics of Gas-Surface Interaction
    Editors: G. Benedek, U. Valbusa
22 Nonlinear Laser Chemistry
    Multiple-Photon Excitation
    By V. S. Letokhov
23 Picosecond Phenomena III
    Editors: K. B. Eisenthal, R. M. Hochstrasser,
    W. Kaiser, A. Laubereau
24 Desorption Induced by Electronic
    Transitions DIET I   Editors: N. H. Tolk,
    M. M. Traum, J. C. Tully, T. E. Madey
25 Ion Formation from Organic Solids
    Editor: A. Benninghoven
26 Semiclassical Theories of Molecular
    Scattering   By B. C. Eu
27 EXAFS and Near Edge Structures
    Editors: A. Bianconi, L. Incoccia, S. Stipcich
28 Atoms in Strong Light Fields
    By N. B. Delone, V. P. Krainov
29 Gas Flow in Nozzles
    By U. G. Pirumov, G. S. Roslyakov
30 Theory of Slow Atomic Collisions
    By E. E. Nikitin, S. Ya. Umanskii
31 Reference Data on Atoms, Molecules,
    and Ions   By A. A. Radzig, B. M. Smirnov
32 Adsorption Processes on Semiconductor
    and Dielectric Surfaces I
    By V. F. Kiselev, O. V. Krylov
33 Surface Studies with Lasers
    Editors: F. R. Aussenegg, A. Leitner,
    M. E. Lippitsch
34 Inert Gases
    Potentials, Dynamics, and Energy Transfer
    in Doped Crystals.   Editor: M. L. Klein
35 Chemistry and Physics of Solid
    Surfaces V   Editors: R. Vanselow, R. Howe
36 Secondary Ion Mass Spectrometry,
    SIMS IV   Editors: A. Benninghoven,
    J. Okano, R. Shimizu, H. W. Werner
37 X-Ray Spectra and Chemical Binding
    By A. Meisel, G. Leonhardt, R. Szargan
38 Ultrafast Phenomena IV
    By D. H. Auston, K. B. Eisenthal
39 Laser Processing and Diagnostics
    Editor: D. Bäuerle

Springer Series in Chemical Physics

Editors: Vitalii I. Goldanskii Fritz P. Schäfer J. Peter Toennies

Managing Editor: H. K. V. Lotsch

40 **High-Resolution Spectroscopy of Transient Molecules**
By E. Hirota

41 **High Resolution Spectral Atlas of Nitrogen Dioxide 559–597 nm**
By K. Uehara and H. Sasada

42 **Antennas and Reaction Centers of Photosynthetic Bacteria**
Structure, Interactions, and Dynamics
Editor: M. E. Michel-Beyerle

43 **The Atom-Atom Potential Method**
Applications to Organic Molecular Solids
By A. J. Pertsin and A. I. Kitaigorodsky

44 **Secondary Ion Mass Spectrometry SIMS V**
Editors: A. Benninghoven, R. J. Colton, D. S. Simons, and H. W. Werner

45 **Thermotropic Liquid Crystals, Fundamentals**
By G. Vertogen and W. H. de Jeu

46 **Ultrafast Phenomena V**
Editors: G. R. Fleming and A. E. Siegman

47 **Complex Chemical Reaction Systems**
Mathematical Modelling and Simulation
Editors: J. Warnatz and W. Jäger

48 **Ultrafast Phenomena VI**
Editors: T. Yajima, K. Yoshihara, C. B. Harris, and S. Shionoya

49 **Vibronic Interactions in Molecules and Crystals**
By I. B. Bersuker and V. Z. Polinger

50 **Molecular and Laser Spectroscopy**
By Zu-Geng Wang and Hui-Rong Xia

51 **Space-Time Organization in Macromolecular Fluids**
Editors: F. Tanaka, M. Doi, and T. Ohta

52 **Multiple-Photon Laser Chemistry**
By. R. V. Ambartzumian, C. D. Cantrell, and A. Puretzky

53 **Ultrafast Phenomena VII**
Editors: C. B. Harris, E. P. Ippen, G. A. Mourou, and A. H. Zewail

54 **Physics of Ion Impact Phenomena**
Editor: D. Mathur

55 **Ultrafast Phenomena VIII**
Editors: J.-L. Martin, A. Migus, G. A. Mourou, and A. H. Zewail

56 **Clusters of Atoms and Molecules**
Solvation and Chemistry of Free Clusters, and Embedded, Supported and Compressed Clusters
Editor: H. Haberland

57 **Radiationless Transitions in Polyatomic Molecules**
By E. S. Medvedev and V. I. Osherov

58 **Positron Annihilation in Chemistry**
By O. E. Mogensen

59 **Soot Formation in Combustion**
Mechanisms and Models
Editor: H. Bockhorn

60 **Ultrafast Phenomena IX**
Editors: P. Barbara, W. H. Knox, G. A. Mourou, and A. H. Zewail

61 **Gas Phase Chemical Reaction Systems**
Experiments and Models 100 Years After Max Bodenstein
Editors: J. Wolfrum, H.-R. Volpp, R. Rannacher, and J. Warnatz

62 **Ultrafast Phenomena X**
Editors: P. F. Barbara, J. G. Fujimoto, W. H. Knox, and W. Zinth

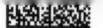